AGRICULTURAL DEVELOPMENT:

PLANNING AND IMPLEMENTATION

W0230396

RAANAN WEITZ/AVSHALOM ROKACH

AGRICULTURAL DEVELOPMENT: PLANNING AND IMPLEMENTATION

(ISRAEL CASE STUDY)

Springer-Science+Business Media, B.V.

ISBN 978-94-017-7076-7   ISBN 978-94-017-7074-3 (eBook)
DOI 10.1007/978-94-017-7074-3

NOTE OF ACKNOWLEDGEMENT

It is our pleasure and duty to express here our thanks to all those who made the publication of this book possible through their experience, advice and assistance.

Thanks are due first and foremost to Mrs. Jean Kosloff, not only for her meticulous editing but for her invaluable comments and advice throughout the preparation of the final version. The first draft has been edited by Mr. Victor Nell, whose contribution is no less appreciated.

Were we to mention everyone, the list of people to whom we are indebted would be very long indeed. At the top of the list are members of the staff of the Land Settlement Department of the Jewish Agency, from whose experience we had drawn incessantly. Among those there are a few who took an active part in our work, providing background material and participating in our discussions: Mr. Y. Abt, Mr. J. Lichtman, Mr. O. Shapiro and Mr. M. Black. The contribution of Mr. A. Hartman, Ambassador of Israel to the United States is also recognized with gratitude. And to all others, even though their names have not been mentioned, their assistance nevertheless has not been forgotten and is greatly appreciated.

The Agriculture and Settlement Planning and Development Centre, The Faculty of Agriculture of the Hebrew University, The Volcani Institute for Agricultural Research, the Production Boards as well as other institutions supplied us with data and information, for which we are grateful.

TABLE OF CONTENTS

PREFACE

Agriculture in Israel has long been subject to careful planning, first on the farm-unit level, and later on also on the regional and national levels. The beginning of agricultural development in Israel dates back to the nineteenth century, when the first waves of immigrant Jews arrived in the country, mostly from Eastern Europe. Their arrival represents a unique case in the history of development where urban educated people settled of their own free will on the land and became farmers. The impact of this settlement is felt in the agriculture of the country up to the present day.

The agricultural sector in Israel is well advanced in terms of both planning and practice. An extensive institutional system is concerned with the planning of agriculture and the rural areas; farmers apply modern technology and make use of an effective organizational structure and modern scientific research. Agriculture contributes 9.5% of the national income, produces a considerable part of exported goods and employs about 13% of the total labour force at a standard of living not far below the national average.

The achievements of Israeli farming may be attributed to a large extent to both the human resources of agriculture and to the careful and flexible system of planning. This system of planning did not spring up 'full blown' but was the result of an historical process responding to the demands of Jewish agricultural settlement in the country from the 1880's onward. By 1948, when the State of Israel came into existence, there was already a Jewish farming population which utilized modern methods and was linked in an incipient organizational infrastructure of co-operative institutions. There was also an established relationship between farming villages and national land and settlement authorities which had some planning functions. In the years which have followed, partly because of general advance in agriculture throughout the world, partly because of having to solve the settlement of large numbers of immigrants, agricultural planning in Israel grew into the skilled art which it is today. In the course of time it expanded to include the development of rural areas in their entirety, involving many other aspects in addition to agriculture. The culmination of this experience in development planning, the concept of the "composite rural structure", which includes services and industry as an integral part of a farming region, has become the base not only of new settlement but a goal in the expansion and development of the older farming communities.

The absorption of over a million and a quarter immigrants, thirteen percent of whom were settled into agriculture in the years from 1948–1960, was indeed an unique phenomenon in the history of rural development. Without the foundations already laid and without planning, such mass settlement on a total population base of less than a million persons in the country could not have come about. Yet though the mass settlement was unique and may not at first seem relevant to those countries whose problem is the modernization of existing traditional agricultural communities, the main idea is relevant – how to draw on and organize resources, especially the human, for the benefit of both the state and the producer.

The foundation on which such a task of mass settlement could take place was in itself unique. There was already long experience in the settlement of new areas in all parts of the country. Co-operative endeavour was already well established; it included smallholders co-operative villages (moshavim) as well as collective villages (kibbutzim), and co-operative networks for credit, for the purchase of farm supplies and for marketing. Agricultural research and extension also contributed to advanced farming. Thus the way was already pointed towards joint effort for large-scale activities which modern agriculture requires, without doing away with the family-farm unit. The retaining of the family-farm was particularly important for absorption of the new immigrants who were mostly of traditional or folk-like origins, mostly unfamiliar with modern agricultural practices and organization. The fact that these people could fit in into a co-operative framework, though adapted to their own needs, is in itself one of the greatest achievements of the new settlement in Israel.

The task of rural planning in Israel was made easier by another feature peculiar to the country, and that is the fact that landownership and water rights of Jewish farming had been in the hands of national authorities for several decades, so that the distribution of land and water resources could be made on a planned basis with no problems of landlord rights or tenancy. Hence it was possible to plan for an equal distribution of means of production to each farmer in a village, avoid land fragmentation and maintain the concept of self or family labour without recourse to hired labour. This in turn allowed a democratic equality among farmers. When immigrants of a traditional or folk background were absorbed into the country, they could be immediately introduced into an egalitarian system without being part of a depressed group or even a tenant class.

The special relationship already developed between the network of farms and the Settlement Authority also formed part of the basis on which planning of immigrant settlements could rely. This was true as well of the pioneer spirit of the old settlers from whose ranks could be drawn the instructors for the new villages and the staff for the much expanded Settlement Authority, including its planning section.

Thus not only was the settlement problem which the planners had to solve unique, but so was the foundation on which it could be worked out. Nonetheless, on many levels, lessons for other countries can be drawn, for the question is not "How did this come about?", but "In what way is it possible to organize people and the means at hand for production in order to realize a developed and scientific system of agriculture?"

The very possibility of absorbing and settling so many people in so short a time is worth consideration. At least two thirds of the immigrant settlers since the beginning of the state were not from western background, at home in modern organization and technology and bearing conventional education, but from traditional or folk-like background with none or varying degrees of familiarity with the 'modern' world. Most had never been farmers, yet they became part of a highly technical and organized farming system.

The old population, the Jewish *Yishuv* into which these immigrants were settled, was primarily of European or western origin and as already mentioned, has established in the course of years an incipient organizational infrastructure for the farming sector. However, this infrastructure had to be modified and expanded to fit both new technologies and the ways of the newcomers. Existing institutional frameworks, including the Settlement Authority had to be drastically changed and expanded. The co-ordination of market demand and of production had to become a function of administration articulated with the general economy based on careful planning. Professional insights of social scientists had to be called in for the handling of crises of the various ethnic groups and eventually for planning how to set such groups in a rural context suitable to them. New extension methods had to be worked out which could teach modern farming methods to people who had never tilled the soil or raised livestock, who were not imbued with ideals of physical labour and whose educational background if any, had not included systematic methods. The classical moshav had to be revised to accommodate kinship solidarity. The diversified farming system on which the older settlements had relied, had to be updated to one of specialized agriculture.

Planning in general became a matter not only of co-ordinating supply with demand, but involved the setting up of villages and later of towns, the organizing of services and eventually the development of the rural network which would include industry connected with agriculture. Hence Israel's whole experience in settlement and agricultural development since 1948 has been one of interaction between various ethnic groups and the institutional framework charged with the development process. Both the groups and the framework have undergone continuous and far-reaching modifications.

The necessity to change and to adapt has been as true of planning itself as of any other aspect of Israel's rural development. The three tiered system – that is, planning on the national, regional and farm levels – and its implementation, was not envisioned in 1948 when the country had just emerged from war and was hardly prepared to receive the large influx of newcomers which would soon be arriving. Yet today the system is able to co-ordinate agricultural resources of land, water, labour and capital with demand, while its five year agricultural plan of 1965–1970 has so far proven up to expectations.

Between 1950–1965, total agricultural production in Israel increased by 500%. Efficiency of production grew steadily, the average rate of increase between 1955 and 1959 being 8.5% and between 1958 and 1963, 4.9%. Labour efficiency in agriculture increased at an average annual rate of 10.8% between 1955 and 1959, and 11.2% between 1958 and 1963. Efficiency of the utilization of capital also increased, 4.2% per annum between 1955 and 1959, and 3.1% per annum between 1958 and 1963.

Along with the great increase in agricultural production and the increase in efficiency, has grown an understanding of the importance of the 'human resource' or the settler, whose ability and willingness to farm is a crucial factor in agricultural production. It was learned that cultural differences could not and should not be wiped out overnight, and that with the right approach new farmers could be so organized that they could keep their traditional identity and yet adapt to a modern way of farming. Those in charge of settlement learned that a strict adherence to western concepts or ideologies was not advisable and would not bring the desired results. At first this approach seemed to lead to cultural isolation, but the general organizational framework gradually brought about a kind of bridging between the various sectors of the population with a mutuality towards solving common economic and community problems.

A ten year plan for Israel is presented in this book as a case study of planning. Behind it is the concept of removing uncertainty from agricultural production as far as possible, reducing waste and increasing efficiency. Agriculture by its very nature can never approach the certainty of industrial production, but the aim must nevertheless be to introduce order as is consistent with free enterprise working in a democracy where the interests of all sections of the population have to be preserved and promoted.

The plan is evolved on three interconnected levels – macro-planning or national planning, micro-planning of the individual farm-unit and the intermediate level of regional planning.

Planning on the national level or macro-planning means a forecast of demand, a survey of the resources available and the selection of the best alternative of resource utilization to meet the demand. Macro-planning has to rely on sophisticated tools of projection which in turn are based on surveys of production means, (land, water, climate and capital) on an evaluation of agricultural know-how and knowledge of the 'human resources'. Macro-planning of the agricultural sector must be tied into and linked with the total economy. Forecast of demand is based on standard of living of the population and its nutritional requirements, as well as on export possibilities, while expenditure on agricultural development must be weighed against other development possibilities. The setting of agricultural targets should not and cannot be an unintegrated piecemeal activity, but must be fitted into the general goals of a country.

Planning on the level of the farm-unit (the farm) or micro-planning is a search for the most favourable scope and structure of farm branches which will combine profitability to the farmer and to the national economy at the same time. The planner is faced with an almost infinite number of resource combinations and must work out the ones which use most advantageously labour and water with land. In Israel this is done by means of farm-type models which are most suited to each region, a technique which allows the planner to point towards the most desirable combination of the available means of production and resources in each region at the prevailing level of know-how, and with existing human resources. The micro-plan is designed to achieve an adequate level of income for the average farmer while at the same time making a contribution towards consolidation of the national economy and closing the gap in the balance of payments.

It is on the level of the region that macro-planning and micro-planning become co-ordinated and implemented in Israel. The inconsistencies between aims defined by planning on the national level and actual production trends on the individual farms cannot be balanced entirely by the free market and price system. There is always a conflict between the dimensions of the various branches of agriculture as determined on the macro level and the expected production of the farm units. It should also be emphasized that the systems of planning on the one hand, and the economic, social and organizational factors on the other, differ basically on the national level and on the individual farm level. On the regional level, therefore, the national plan is adapted to the conditions of the farm units and vice versa. This is where the macro-plan is modified and co-ordinated with real conditions, and where actual implementation is carried out. In a sense, the several regional plans which are implemented become the sum total of implementation on the national level.

Planning and implementation at the regional level not only have the vertical function of linking macro-planning with micro-planning in one sector, but also, and sometimes primarily, the horizontal function of inter-sectoral co-ordination. It is in the region that agricultural development is linked with other sectors of the economy for it becomes only one segment of regional development, which is actually 'multipurpose resource development' with social and political considerations forming part of the purpose.

In the course of agricultural development planning in Israel, it has become clear that planning the agricultural sector alone is insufficient. Modern agriculture, characterized by a trend of specialization, produces goods for industrial processing and demands suitable services for its smooth operation. Modern farmers require adequate social services and amenities in accordance with their rising standards of living. In order to bring about the progress of agriculture it is therefore necessary to tackle all the related aspects at the same time. This may be done for well-defined regions, whose size and nature permit integrated planning of all sectors. In this way agricultural planning in Israel turned from production planning into comprehensive rural planning.

Comprehensive rural planning can be defined as the integrated development of agriculture, industry and services within a rural area, with reference to all aspects of rural life, i.e. economic, social, organizational and environmental. It was first tried out in Israel in 1954 with the planning of the Lakhish region. Today Lakhish comprises several groups of co-operative villages, each one surrounding a rural centre which contains sorting and marketing facilities and other economic and civic services. This constellation of several villages and a rural service centre has been termed a 'composite rural structure'. In the midst of the region, processing industries, credit and commercial facilities, civic and higher educational services were established in a regional town – Kiriat Gat.

This book presents the background to and the many facets of agricultural planning in Israel. It is an attempt to describe, analyse and explain the system of comprehensive rural planning, hoping that its general principles will prove to be useful also for other developing countries, now facing the crucial task of development.

CHAPTER 1

THE DEVELOPMENT OF RURAL PATTERNS
IN ISRAEL

A. LAND SETTLEMENT IN ISRAEL

The following account presents the background to Israel's experiments in forms of settlement and the principles on which its system of agricultural settlement rests. Some of the settlements have been in existence for three generations while others are still in the stage of experiment, though so far successful within the aims which they have set for themselves. Each new settlement scheme has been built on the success and failure of previous experience. Sights have been continually raised to improve agriculture in all its aspects from that of utilization of land, water and other basic factors of production to better organization of the farmers in various rural patterns.

1. First attempts

The forms of agricultural settlement in Israel evolved in response to the needs of the first Jews who returned to the ancient land at the close of the nineteenth century. They strove for a farming way of life, which would satisfy their ideology and provide a rural society of permanence and strength for their own and future generations. The settlement patterns created in the early days and the ideals underlying their evolution were in large measure carried over into the large settlement ventures undertaken by Israel since independence.

Biblical sources, archaeological findings and historical commentaries provide ample evidence that farming was the basis of ancient Israel's economy. Although the Jewish nation was rent from its land and agriculture, to be dispersed throughout the world by the Roman conquest twenty centuries ago, the agricultural mores of the early period were preserved by religious tradition. Throughout the centuries of exile the Jew prayed for rain to fall at the appropriate seasons in the Holy Land, marked dates of agricultural importance in yearly festivals and kept day to day customs, which recalled ancient bonds with the soil.

In exile, however, the Jew became alienated from physical work and the practice of agriculture. Regarded in most countries as a race apart, subject to periodic persecution and even organized expulsion, the Jewish wanderers avoided agriculture, following skills and professions which left them mobile, as their future was always uncertain. As the centuries of dispersion unrolled, legal prohibitions – for example, the forbidding of Jewish land ownership – were added to the practical difficulties in the way of Jewish settlement on the land. The Jewish farming nation, therefore, became transformed into an almost exclusively urban people.

Only in the 19th century did prominent Jewish thinkers agitate for a basic change in outlook. They stressed that the Jew should return to physical work and agriculture and thus 'normalize' his existence; many Jews responded to this call. In 1808, when Czarist Russia planned to settle Jews in remote areas of Southern Russia, thereby reducing the large concentration of Jews in cities, and settling desolate and underdeveloped regions, many took an active part. Settlements were, in fact, established although today only a memory of the scheme remains. Other settlement schemes for Jews were attempted in various countries of the world – in South America, Turkey, Poland, Cyprus, Birobidjan (Russia) – but few were successful to any extent. The absence of any unifying national inspiration or spiritual attachment to the settlement area was largely responsible for the failure of these projects.

The story of Jewish attempts to return to Israel is a long one because of the emotional ties that constantly sought concrete expression. Alex Bein in his study[1] points out the continuity of Jewish settlement throughout the centuries – the Upper Galilee village of Peki'in, for example, maintained a Jewish community without break from ancient times until 1936. At the close of the 15th century many Spanish Jews made their way to Palestine after Spain expelled them; in the 18th century an organized body of 300 Chassidim – members of a mystic movement – settled in Safed; in the late 17th century, a series of 'false messiahs' in Europe offered miraculous redemption stirring long-dormant hopes of a return to Israel. Finally the improved travel facilities and political changes of the 19th century led to settlement on a larger scale, so that there were some 11000 Jews in Palestine in the 1850's and 24000 by 1880.

These settlers formed the 'old community', imbued with religious ideals. They believed their pious way of life on the holy soil would contribute to the speedy return of all their people to the Promised Land in fulfillment of Biblical prophecy. They existed by charitable contributions collected by 'emissaries from the Holy Land' who visited all parts of Jewry abroad or in the 'Diaspora'. From the 1830's, sporadic attempts were made by individuals and groups to free the Jewish community of Palestine from the stigma of living on charity. Purchases of agricultural land were made, while most significant was the establishment, in 1870, of the first agricultural school (Mikve Yisrael) to train young Palestinians in agriculture.

The next important turning point came in 1881 in the wake of severe anti-Jewish rioting in a number of East European countries. A movement called *Chovevei Zion* (Lovers of Zion) was formed, which brought to Palestine about 25000 Jewish settlers whose immigration is generally classified as the First Aliyah (Wave of Immigration).

2. The emergence of modern Jewish settlement

The First Aliyah consisted primarily of students imbued with the ideal of laying the foundations of a Jewish national homeland in Palestine, a goal to be achieved by establishing model farming villages. Among them were a number of moderately wealthy families who shared the same ideals. These early settlers considered it their task to make a beginning, furnishing an example which others would follow. Their lack of experience and agricultural knowledge caused them bitter disappointments, while the constant danger of attack by neighbouring Arabs, coupled with the hostility of the ruling Turkish Government made the existence of their newly-established settlements precarious. Nevertheless, such settlements as Rishon Le Zion, Zichron Ya'akov and Rosh Pina were founded. Shortly afterwards the Turkish Government prohibited the entry of Russian Jews into Palestine; immigration stopped soon after it had begun.

The threatened disintegration of these settlements was averted by the timely intervention of Baron Edmond de Rothschild of Paris. He placed the majority under his care and invested large sums of money, sending out experts from France to teach the newcomers viticulture and 'manage' the settlements. Under the Rothschild administration, the settlers found themselves in the hands of a bureaucracy that was incapable of understanding their outlook and interfered in the smallest details of their lives. The nation-wide movement was replaced by slow small-scale settlement of a philanthropic nature. The settlers lost their initiative, independence, and gradually abandoned the lofty ideals with which they had come to the country.

A decade later, Jewish settlement of Palestine once again assumed national proportions. The incentive given to immigration by the expulsion of Jews from Moscow and the high market price of grapes led to the influx of thousands of Jews within a few months, including the first sizeable influx of Jewish workers. In spite of stringent Turkish laws, large settlements were established. The present town of Rehovot was founded as a village without recourse to charitable grants, while scores of men and women gave their lives to build Hadera in malaria-infested swamps. The village of Be'er Tuvia was founded as a workers' settlement in 1896.

The new settlers undertook ventures of far-reaching importance for the country's agricultural development: by 1900 there were 22 Jewish villages in Palestine; modern methods for growing citrus and vines were introduced; olive cultivation was expanded; local swamp drainage projects were initiated; first steps were taken in afforestation with the introduction of the eucalyptus tree from Australia. Though the untrained settlers tended to introduce European agricultural concepts to a climate and soil not suited to them, they learnt from their mistakes, with a growing awareness that agriculture was not only a science which had to be learnt, but a way of life which had to develop in its own time. The experience gained by the new villages applied to diverse regions. While Petah Tikva, Rishon Le Zion, Hadera and Rehovot were on the coastal plain, Zichron Ya'acov, Rosh Pina, Metulla and Motza were in the hilly country. This diversity of experience gained under varied conditions was to prove later of considerable value.[2]

By the time the first Zionist Congress met in Basle in 1897, modern Jewish Palestine was 15 years old and 18 agricultural settlements were already in existence. The attitudes of the nascent Zionist Movement (see pp. 5ff.) were undoubtedly influenced by the experience that had already been gained in land settlement. The views of Theodor Herzl, leader and founder of the Zionist Movement, were consequently crystallized; "In the place of small-scale unplanned settlement people should be brought into Palestine in large numbers in a well-regulated manner. The country should be built up systematically in accordance with detailed plans based upon a thorough knowledge of its conditions. Philanthropy which deprives its beneficiaries of their most valued possession – freedom of action and economic independence – should be replaced by an economic system under which settlers are assisted not by charity, but by credit.

Unlike the majority of the settlements administered by Baron de Rothschild's officials, the new ones should not be restricted to plantations, but should be based upon mixed farming. They should concentrate on meeting their own requirements, and send only their surplus to market through the medium of marketing co-operatives."[3]

In 1899, Baron de Rothschild decided to entrust the management of his settlements to a Jewish Colonization Association. This organization, on the basis of its experience in agricultural settlement elsewhere, was to introduce a more rational system of farming, so that within a few years the villages might attain economic independence. This step marks the close of the philanthropic phase of Jewish settlement in Palestine.

The substantial influx of Jewish workers into Palestine, which took place in the last decade of the 19th century, continued in the years 1905–1920, when thousands of immigrants reached Palestine in what came to be known as the 'Second Aliyah' (or Wave of Immigration). These newcomers, inspired by the ideals of the Russian revolution, aimed at founding a workers' society.

3. The role of the Zionist Organization

The Second Aliyah has to be considered in the context of the World Zionist Organization and its settlement agencies. Created in 1897, this organization gave concrete form to the desire for the re-creation of a Jewish homeland. While seeking to gain international political recognition for this aspiration, the Zionist Organization also set about creating practical conditions in Palestine for settlement. In 1901, one of its branches, the Jewish National Fund, was established to buy land in Palestine for the organized agricultural development of the country making it possible for Jewish immigrants without means to settle on the land. The Jewish National Fund determined that all land that it bought should be inalienable. The land was to be leased out on favourable terms to prospective settlers, the lease being hereditary and automatically renewable, with the sole condition that the leasee cultivate the land himself.

After the first land purchases by the Jewish National Fund, the Zionist Organization decided in 1908 to establish an office in Palestine (in Jaffa) to direct the work of Jewish settlement in the country. The first director of this office, the late Dr. Arthur Ruppin, devoted himself particularly to agricultural settlement. His aim was to provide a permanent agricultural way of life for the greatest number of Jews. It was he who formulated the policy which is still in force today, without which the agricultural settlement of the country would have been impossible. His policy laid down that agricultural settlement was to be conducted as a partnership between the settler and the settlement agency (the Zionist Organization). The Zionist Organization had to make the land available on lease through the Jewish National Fund, enable the settler to reclaim the land and prepare it for cultivation, and provide him with the necessary investments in housing, livestock, equipment and seed, so that he could farm the land and earn a livelihood from it. The settlement agency had no right to intervene in any manner in the settlers' way of living: the settlers were free to choose the form of settlement they preferred, whether co-operative, collective, or any other. The only criterion of the settlement agency should be that the way of life or social structure should be economically sound and result in the effective use of the land and its development. The settlers, for their part, should undertake to work the land and develop it, accept financial responsibility for all the investments made in the land and also be responsible for the repayment of their investments with a reasonable rate of interest from the moment the investment was completed.

This pattern of relationship between the settlement agency and the settler has remained unchanged. It has contributed greatly to flexibility of social and economic development in the villages, producing an attitude of responsibility on the part of the settlers and has made an essential contribution to the development of the country's agriculture.

4. The impact of the Second Aliyah

The university graduates, office workers and labourers who made up the Second Aliyah were convinced that 'national' interest required the creation of an agricultural Jewish proletariat which would subsist on the basis of personal labour. Integrated with this concept was a pioneering idealism, a determination to reclaim the waste stretches of the country and restore the ravaged soil to its ancient fertility. They were convinced that the reconstitution of Jewish independence in Palestine would be impossible without the creation of a dynamic economy in which Jews would themselves carry out all economic tasks. Because they understood that agriculture was both the essential foundation and the most difficult field to conquer for a people which had been so thoroughly urbanized for so many centuries, it was to agriculture that they devoted their efforts.

At first, the newcomers worked as hired labourers on established Jewish farms. The work was seasonal, wages were low, and competition with cheap Arab labour made it barely possible to live. Until the arrival of the Jewish workers of the Second Aliyah, local Arab labourers – the fellahin – were employed to lay out and till the farms of the Jewish settlers, because of their experience in farm work, their adaptation to the climate, and their willingness to accept minimal wages; whatever they earned through working for Jewish settlers was additional income. The Jewish worker, used to a higher standard of living, had to exist only on his earnings.

The people of the Second Aliyah saw no future in continuing with the system of 'moshavot' – villages in which farmers owned their land, producing cash crops with the help of hired labour. This was a way of life which negated their ideals. They were convinced that it would not lead to the fulfillment of their vision – a Jewish Homeland in which Jewish farmers worked their own land, nor could it ensure the conversion of Jews from the centuries-old urban way of life to farming. The approach of these Second Aliyah pioneers entailed a complete change of outlook in regard to the problems of settlement. The concept of the cash crop had to be abandoned and replaced by the concept of farmers growing their own food, so that grain farming, vegetable production and dairy farming were to be given priority over wine and citrus production. It was only in this way, that new settlers would really become rooted in their land.

This new view of the central purpose of farming – agricultural integration – also involved a radical change of outlook with regard to the social structure of the village. If a man had more land than he could cultivate by himself, he would have to employ hired labour and thus become an overseer instead of a farmer. Thus developed the concept of basing the whole social structure of a village on personal labour, dividing the land so that no man would have need for hired help, but perform all tasks himself with the help of his wife and children.

5. Co-operation and collectivism

A variety of factors thus contributed to the emergence of co-operative and collective patterns. Many of the newcomers had been deeply influenced by the egalitarian ideas of 19th century socialism and dreamed of a society in which all would enjoy the same material level. The inexperience of the workers and the difficulty of the terrain they were trying to develop made necessary the emergence of co-operative and collective forms. Indeed, even the older villages which had developed along individualistic lines had in their early years been marked by a high degree of co-operation in the actual purchase of the land and the initial work of land reclamation. But the Second Aliyah settlers saw great advantage in *permanent co-operation*. They were convinced that, if hired labour were to be avoided, a system of permanent co-operation was essential to take care of seasonal needs. Co-operative workers' societies were founded with the aim of finding work for their members. The 'conquest of labour' became the motto of these newcomers, and it soon became obvious that the 'conquest of labour' could only be achieved by a 'conquest of the soil'.

The first attempt at co-operative farming was made in 1907, when a collective group undertook to carry out on its own responsibility all the work in the fields of a training farm, called Sejera, and shortly afterwards settled nearby, founding 'Degania'.

B. EVOLUTION OF THE SETTLEMENT PATTERNS[4]

1. The kibbutz

The first collective settlement, Degania[5], was founded in 1909, when a group of young men and women received a contract from the Zionist Organization to farm a newly purchased area of land as a collective group assuming financial responsibility for the venture. Unlike the sporadic attempts which had been made previously, this group maintained itself and developed a unique system of collective village living. They called themselves a 'Kvutza' which is Hebrew for 'group' (and is interchangeable with the similar word 'kibbutz'*). They realized that working as individuals on private farms would have no significance in terms of the national revival, and believed that only collectively would it be possible for them to attain the strength needed to overcome the many obstacles and to effect the difficult transformation from urban to agricultural life. The social experiment, although it ran into numerous difficulties, led to the establishment of collective settlements which have since become a familiar but unique characteristic of the social life of modern Israel.

Through the years a generally accepted structure for kibbutzim has developed. Property is collective, goods and services are paid for in kind according to the needs of each member. The collective principle applies to most spheres of life: although married couples have their own rooms, the whole kibbutz eats together in the central dining hall, and during their parents' working hours the children are cared for in children's houses – the pride of the settlement – where as a rule they also sleep. The members thus constitute a single large labour force working the farm and other enterprises, including services and often also industries. The men work in the various farm branches and in the industries recently established in many kibbutzim. The women work usually in the kitchen, laundry, children's houses and other services with some in poultry or other farm enterprises.

Prospective members spend a year's probation period on the kibbutz before acceptance by majority vote – sometimes a two-thirds majority is required – as full members. Members who leave are permitted to take possessions gained during their period as members away with them, usually together with a small sum of money to help them establish themselves elsewhere in the society.

The kibbutz is governed by the members, who voice their opinions and vote their decisions at weekly or bi-weekly general meetings. They decide on

---

* Kibbutz – Plural kibbutzim.

various issues by majority vote. They elect members to the various management posts – secretary, treasurer, production manager, purchasing and marketing secretary and work co-ordinator. Each of these offices carries automatic chairmanship of an elected committee which deals with its particular field on behalf of the whole kibbutz. Members are also elected to cultural and other committees which deal with the many aspects of communal life.

As the kibbutz grows older and larger, the main secretarial functions become full time. Even on the smallest kibbutz the positions of the treasurer and the purchasing agent are usually full-time; often these officials spend the greater part of their working hours in the towns. The production manager also has a full-time job: he must observe the various branches and keep informed on developments. He is the liaison between the kibbutz, the Jewish Agency and the Ministry of Agriculture, and attends regional and national agricultural conferences. He works closely with the kibbutz purchasing and marketing agent and the treasurer. The work co-ordinator has the most unenviable task, since he must be continually available to supply workers for all departments – both for planned and sporadic work.

People elected to such positions must acquaint themselves with the relevant problems, acquire the necessary technical knowledge, and familiarize themselves with financial dealings of the kibbutz with outside organizations. An average monthly turnover may be between $20000 and $40000, an amount which demands skilled management. Originally, members were elected to key positions in rotation, in an attempt to prevent the formation of a 'managerial caste', but as jobs became more specialized and complicated, many of the kibbutzim realized that it was imperative to ensure greater continuity in management positions. In some cases, it may take a year or more to learn the intricacies of a particular post, so that frequent changes lead to wasted money and the slowing-down of development. Today most kibbutzim pick their management personnel a few years in advance, training them for their respective posts in various seminars and courses. Eventually, however, they must be elected.

The implications of this situation in the social sphere are complicated and difficult to assess. To the extent that social status values exist in the kibbutz, they are not necessarily based on managerial position but, to a larger extent, on the type of work undertaken. The experienced and hardworking orange-grower, teacher, dietician, mechanic, or vegetable-gardener is the real backbone of the kibbutz and his influence in policy making is probably as strong as that of those in management.

Participation in planned social activities is not compulsory. Each member's standard of living depends on the financial standing of the kibbutz, while the extent of his personal possessions depends on the length of time he has been a member.

2. The moshav*

The first attempt to set up a planned co-operative village was made at Merhavia in 1910. It failed, but gave a great stimulus to the co-operative idea. In 1921 a group of farmers, some of whom had broken away from the kibbutz of Degania because they did not find collective living to their taste, and others who were trained in agriculture in the United States, established the village of Nahalal in the Valley of Jezre'el. They called it a *'moshav ovdim'* (workers' co-operative smallholding settlement), which later came to be known as a *moshav*. They divided up their land in such a way that each family had an area of land it could cultivate by itself. After the initial clearing of the land which was done collectively, each family lived and farmed on its own account but with organized co-operation in buying and selling, including the purchase and operation of heavy farm equipment and an organized system of mutual help. This moshav system quickly struck root. It met the need for an individual way of life while providing a practical solution to economic problems within a co-operative framework.

Several years later, workers in the larger settlement centres began organizing themselves into co-operatives to which the Zionist Organization, in an attempt to establish a stable community, provided loans for housing and the purchasing of equipment. The idea was to give each settler a house with land to supplement his income as a labourer. In this way auxiliary farming developed, linked to outside sources of employment in the larger centres. In the course of time these auxiliary farms developed into moshavim with family farms based on mixed farming: milk, eggs, vegetables and fruit were produced and co-operatively marketed.

The moshav has become a smallholders' co-operative village of family farms, consisting as a rule of 80–100 farm units[6]. Each farm family is a well-defined economic and social unit, living in its own home, tilling its own fields and making its own decisions. The smallholdings are organized for large-scale activities on co-operative lines to maintain an economically competitive agriculture. A village management is elected by the villagers each year at an annual general meeting. Its task is to maintain municipal and

---

* Plural moshavim.

economic services for the use of the farm families as members of the co-operative, paid for by taxes automatically deducted from the farmers' accounts which are kept by the village. Similarly, all agricultural requisites (water, fertilizers, seeds, feed, etc.) are provided by the co-operative for the farmers on credit, recoverable from incoming market returns. Municipal, medical and cultural services are available to the families as members of the co-operative unit. The village committee has municipal status and is empowered to make and implement decisions, subject to formal confirmation by the district council. The basic resources for farmers in a village are allocated on an equal basis. Each farm unit is provided with the same land and water resources, farm house and outbuildings, implements, and working capital, and all units have the same budget and standard of housing. Farm units are planned to provide as equitable a work schedule as possible throughout the year, but there are peak months when the family has to work longer hours than usual.

The moshav system, unlike the kibbutz, allows different economic status to the members. It offers an equal earning opportunity to all its members on the basis of equal initial living quarters and means of production, but further investment and amount of work are a private matter.

The principles in planning these smallholders' co-operatives will be discussed in greater detail in later chapters.

3. The moshav shitufi and other forms

The *moshav shitufi* (collective village) is a pattern of settlement that has combined some economic aspects of the kibbutz with some of the social aspects of the moshav. The farm operation is conducted by one management as in a kibbutz. Economic decisions are made at the general meetings and the elected farm management committee carries them out. Profits are shared equally and allocated on a monthly basis. The individual family life is conducted as in a moshav; each family has its own house with the children living with their parents. In this way the family has collective economic security with the advantages of individual family life. The first moshav shitufi was founded in 1936 and the form has had limited popularity. The clash of individual and communal interests is the chief reason why this pattern of settlement has not been adopted by more settlers.

There are other forms of settlement. Five per cent of the settlements in the country are of a rural village nature (moshavot), based on private land ownership and individual farming. These rural villages have co-operative features in varying degrees.

Finally, there is a small number of larger administered farms in Israel, run either as commercial enterprises supplying work to a surrounding district or belonging to educational institutes.

Whatever the pattern of settlement adopted, none has prospered by a mere mechanical selection of men brought together by the settlement authority. In the words of Dr. Ruppin[7], "a union of men without a union of souls" cannot make an agricultural settlement successful. It would seem that the principles he formulated for the success of the kibbutz apply to other patterns of settlement: "Firstly, success depends on a natural and organic selection of men. Secondly, on a number of men whose common aims and interests infuse a spirit and soul into the settlement and develop laws and traditions for the benefit of later comers. The success or failure of settlement depends much less on the finding of a suitable system of society for rural settlement than on the finding of a system of agriculture which appeals to the mentality of the immigrant, which is desirable from a national point of view and which will pay its way."

C. WORLD WAR I AND THE MANDATE

On the outbreak of World War I in 1914 there were some 44 Jewish villages in Palestine with a population of about 12000. The majority of the villages were based on mixed farming. Citrus, tobacco and grapes were being produced in growing quantities, Palestine oranges and wine were winning markets abroad, while field crops, vegetables, poultry and dairy production were of continually increasing importance.

World War I was an important testing time for the new villages in a real sense. Cut off from their markets in Western Europe, those villages dependent on a single agricultural branch had a more difficult time than those based on mixed farming. But both types of village proved their worth from a national point of view, for they were able to stand their ground in very difficult circumstances and provide the Jewish community in Palestine with valuable staple foodstuffs, thus reinforcing the growing body of opinion in the Zionist Movement which maintained that the road to Jewish national independence must begin with agricultural settlement.

The Balfour Declaration, which stated the right of Jews to build a national home in Palestine, and the British Mandate in Palestine which resulted from World War I, opened a new chapter in Palestine history. The Zionist idea had won international recognition. The Mandate, as ratified in 1922 by the League of Nations, confirmed that a Jewish National Home was to be established in Palestine and specifically required that Jewish immigration and settlement be facilitated. The Zionist Movement understood the decisive importance of the new opportunities and established a new fund-raising instrument in 1920 known as Keren Hayesod (Foundation Fund) for the purpose of financing the settlement of Jewish immigrants on land to be purchased by the Jewish National Fund. The major effort of the new fund was devoted to agricultural settlement. In 1921 it made a dramatic start by undertaking the agricultural settlement of newly purchased lands in the Valley of Jezre'el. These lands, largely covered by swamp, almost unpopulated when purchased, presented a great challenge. Unlike previous settlement schemes which were based on the establishment of scattered settlements, the Valley of Jezre'el called for the creation of a settlement on a regional scale, clearly a task that required pioneering qualities. The necessary pioneering element was available among the new immigrants coming into Palestine at that time. The Zionist-oriented pioneering youth movement in Eastern and Central Europe had by then achieved an advanced level of organization. When freedom of movement became possible at the end of the war, groups of their members made their way to Palestine. Candidates for agricultural settlement were more numerous than the Keren Hayesod could handle. Indeed, for many years there were to be more applicants than the land or funds available.

The immigration of the early 20's contained also a considerable middle-class element, particularly from Poland, which had means of its own, needing no investment outlay on the part of the settlement bodies. Their new economic initiative found an outlet mainly in urban development and in the creation of secondary industries. However, many of the newcomers turned their attention to agriculture and played an important part in developing the citrus plantations which, by the 1930's, had become Palestine's major export product. The citrus industry also performed a valuable function in providing seasonal agricultural work, thus training new immigrants for their eventual settlement on the land.

During this period agricultural research and experimental activities were for the first time effectively organized. The Zionist Organization set up its first Agricultural Research Station at Rehovot where systematic work was initiated in various spheres. This work was to be of the utmost importance in achieving maximum production with the most effective use of land and water. A system of training courses for new settlers in the various branches of agriculture was developed through the agency of the Research Station creating an atmosphere of modern scientific farming, alert to changes and developments. Another important feature of the period between the two World Wars was the growth of agricultural education at all levels which served not only in training new forces for agricultural settlement but also in raising the level of agricultural production.

The outbreak of World War II once again demonstrated the overwhelming national significance of the Jewish villages. With a larger population to support and with imports severely restricted as a result of shipping difficulties, it was nonetheless possible for the Jewish farmers to sustain the Palestine community throughout the war period at a much higher level of nutrition than had been possible in World War I, while at the same time providing fresh produce for the army camps in the area. The citrus industry, severely hampered by loss of its export markets, recovered to a certain extent by quick expansion of a citrus processing industry. During World War II, processing and canning, at first of citrus and later of other fruits and vegetables, became for the first time an important adjunct to the country's agriculture.

By the end of the Mandatory period in 1948, there were 256 Jewish agricultural settlements in Palestine with a total population of some 105000. The effort to create a Jewish rural society had succeeded. In the older settlements a second generation of settlers had arisen, born and bred on their parents' farms, for whom agriculture was a natural way of life. The Jewish farmer was no longer a curiosity.

D. THE IMPACT OF STATEHOOD

The problems confronting Jewish agriculture were radically transformed by the creation of the State of Israel on May 1948. Jewish purchase of land in Mandatory Palestine had been severely restricted by the Palestine Land Transfer Regulations promulgated by the Mandatory Government in 1940. To all intents and purposes Jews had thenceforth been unable to settle outside a narrow belt of territory not more than half a million acres in extent. After the creation of Israel and the unsuccessful Arab war against it, there was free access to an area of some five million acres, most of it uncultivated.

At the same time, establishment of the State had removed the restrictions on Jewish immigration and the accumulated pressure of Jewish refugees, until then denied entry to the country, could be met. Between May 1948 and the end of 1951 over 685000 Jewish immigrants entered the country.

In this context, rapid and large-scale agricultural development was imperative. The growing population had to be fed, and the import of food constituted a heavy drain on the country's financial resources; secondly, the new settlers had to be employed, and the employment potential of agricultural settlement was great; thirdly, the country had to be defended. Experience, particularly in the War of Liberation, had conclusively shown that the only defensible territory was settled territory. The basic lesson of the war had been that for an invading army a wasteland was not a barrier but a bridge, while for a defending army a chain of villages was not an added responsibility, but a living rampart behind which its forces could group themselves until they were ready for counter-attack.

Accordingly, even while the war was in progress, the new phase of agricultural advance was planned and the first steps were taken to mount what can, in retrospect, be regarded as a great agricultural offensive, to increase both total output and the number of settlers.

E. THE POLICIES UNDERLYING RURAL DEVELOPMENT IN ISRAEL

The following section deals with the policies evolved to solve the socio-economic integration of mass immigration on an unprecedented scale during Israel's first years of statehood. The approach evolved, 'democratic bridging', will be described. The first section outlines the broad principles of settlement policy while those that follow deal more particularly with the human problems.

1. The basic principles of New Settlement

Settlement policy in the immediate post-independence period was guided by Israel's peculiar economic, social and security situation. The young state was surrounded by hostile nations, under-populated, yet faced with an immigration which in relation to the native population was of immense proportions. Its members demanded food, employment, a home and security within the new borders – but at the same time supplied new hands to do the work. A considerable number of the newcomers went to new villages and towns. The settlement authorities were guided by the following *policy aims*:

a) Evolution of a balanced national community by dispersion of the population to new areas and by enlargement of the base of the occupational pyramid. A balance between new large-scale agriculture, and the overall national production effort was to be maintained so that the farmers would earn their living and the urban population be fed.

b) Promotion of general economic development by settling a considerable section of the new immigration in vital employment. The immigrants' need for economic establishment was to be satisfied at the same time as they provided labour for the construction, services and manufacture needed to build new farms and supply their production needs. The demand for agricultural products from the cities, which likewise were fast growing due to the immigration, was to be met by the new farms, saving much expensive food import. The new farmers likewise were to expand the volume of demand for non-agricultural products, increasing employment in the towns.

c) Location of settlement in part determined by defence requirements: A compelling fact in Israel's first years was her perilous security position. The rapid spread of a rooted, determined and organized new settlement to undefended areas was to constitute an extension and consolidation of the country's defences.

d) Establishment of economic and social security for the individual within the new settlement framework: The livelihood of the settler was to be assured by supplying him with sufficient means of production, his social security by firm and continuing laws of possession and inheritance, together with social and educational facilities. The settler could then feel confidence in identifying his and his children's life with the land.

The translation of these aims into practice – the building of new farms and their organization in villages and regions – demanded an executive framework, created in the form of a Land Settlement Authority which had vested in it the authority and resources to mobilize and co-ordinate staff and finances for the planning and construction of new settlement. The authority was and is not governmental, but a subsidiary of the Jewish Agency – which constitutes the executive link between the Jewish people outside Israel and the new immigrants, until their economic position is consolidated. The Jewish Agency raises funds abroad for its operations in Israel in the fields of immigration, absorption, settlement, land reclamation and forestry. In this way world Jewry becomes a partner in the absorption of immigrants and in land settlement.

A further feature of Israel's settlement in practice is that the settler makes his own choice of the type of village in which he wants to live – smallholders' co-operative, collective settlement, etc. Whatever his choice, the Land Settlement Authority undertakes to set him up as a farmer in accordance with certain broad principles:

a) *Nationally owned land:* The land belongs to the nation in perpetuity and cannot be bought, but is leased for forty-nine year periods, automatically renewable by a farmer or his son. The period is based on the seven cycles of seven years each which comprise the Jubilee of ancient times (see Lev. XXV

8–10). New land is thus exclusively the property of the nation, so that the thwarting of settlement plans because of existing land tenure systems is almost non-existent – a matter vitally important for overall planning. The settler cannot own his land, but the terms of his lease are such that he can with confidence build his farm and invest in it. This is the surest way of preventing land fragmentation, as the farmer cannot sublease, nor transfer his tenancy except as a whole. The inheritance goes as a whole to one son and cannot be divided among more than one. Any improvements the farmer makes on the land belong to him and are evaluated on transfer of the lease. The capital goods, but not the land itself, can be mortgaged. The system thus eliminates the danger of fragmentation without stifling initiative.

b) *Family labour:* The Land Settlement Authority grants the farmer resources in the form of a development loan, to an extent that guarantees a living from the family farm for himself and his immediate dependents. The farm's lay-out and production system are designed to make hired labour unnecessary – indeed, many farmers' organizations even prohibit hired labour, thus giving practical expression to the concept of the dignity of personal labour as laid down by the first pioneers.

c) *The income criterion:* The size of a farmer's income is directly dependent on the physical production resources at his disposal. The extent of these resources and their method of allocation are decided by a principle basic to Israel's settlement concepts: the resources allocated to a man settling on a farm will be sufficient to grant him the opportunity of an income at least as large as that of his counterpart in town. The principle of equal income opportunity is applied in all parts of the country and in whatever settlement pattern is chosen. This principle automatically limits the size of new settlements and the number of their settlers to proportions that, having regard for the demand for agricultural commodities, will ensure a certain and equal income opportunity for all.

These principles form the basis of modern settlement and farm planning in Israel. They have their origin in the history of Jewish agriculture in Palestine and the settlement patterns that emerged in the course of its development.

2. Mass immigration: a new human element

The two pre-State generations of Jewish settlers in Palestine established an economic and social structure that was Western in character; the new settlement forms, kibbutz, moshav and moshav shitufi, though they evolved in part because of the settlers' revolt against the social patterns of Europe, nonetheless bore the imprint of European background.

Immediately on the establishment of the State there was a very large immigration from North African, Asian and Middle Eastern countries, much of which was directed to agriculture. Today, settlers from these countries form 57% of Israel's agricultural population; the relative overall impact of the post-State immigration on the country's society can best be grasped by imagining the immigration of 50 million Commonwealth citizens to England within the space of four years, all seeking shelter, work and integration.

The most obvious result of this mass immigration has been the emergence of two elements in Israel society, basically different in their social patterns – one 'Western', the other 'Oriental'[8]. Two worlds were thrown together by the historic events surrounding the birth of Israel; today, the task of the State was to build bridges between them and to evolve common living patterns that would make one people of the 'seventy tribes'.

A new human element, with which the existing institutions had had no experience, was thus introduced into Israel. Certain generalizations can be made with regard to the social and economic patterns these settlers brought with them – though generalizations are unwise. For example the differences between Jews from the Atlas Mountains of Morocco on one hand and from Kurdistan on the other, are as significant as those between the oriental and occidental communities in general, since each Diaspora community was strongly influenced by the way of life of the people among whom it existed. Often the only real link between various Jewish communities was their common religion and the aspiration – often mystic – to return to the Land of Israel and Jerusalem. It is however possible to generalize, with the reservation that a fair range of exceptions be allowed, particularly for people from towns.

The social system which the 'oriental' settlers brought to Israel, was in the main based on the patriarchal family group in which the head of a large family had control and authority, with the individual family subject to group codes and customs. Individual families were large – though usually not as extended as those in the tribes of the countries of origin; marriage customs and inheritance, woman's place in the society and parent-children relationships were those of the previous environment. Such customs were honoured by the tradition of centuries and steadfastly upheld, particularly as other social systems were unknown and therefore regarded with suspicion when first encountered.

The economic system of such societies had been primitive, with little specialization and exchange; capitalist motivations were absent, subsistence was the prevailing way of life, and abundant leisure the goal. The Jews of the oriental countries were not as a rule landowners or agriculturalists: usually, they provided services and performed commercial functions for the surrounding peasants. Otherwise, the communities were usually fairly self-contained. The banking and trading institutions of the West scarcely penetrated their quiet lives, changes were few, while the security of the Jewish group was often precarious. Many immigrants came from crowded urban ghettoes where they had led underprivileged lives in very poor circumstances, were little educated and little affected by Western ways of life.

Palestine, on the other hand, had become transformed by the Jewish pioneers from Western countries, so that on independence, Israel contained the full apparatus of a modern state, based on individual and co-operative enterprise, the full use of money, specialization and the social and economic traditions of the West.

Israel was thus faced with what is becoming an increasingly urgent issue in many countries: the question of bridging the gap between a folk of traditional, non-modern social and economic structure on the one hand, and Western technology and its associated living patterns on the other. In terms specific to Israel, the question was, as we have already seen, that of integrating the large and extremely important new element from the 'Orient' into the essentially 'Western' social and economic life of the country, for the benefit of both groups.[9]

3. Integration: problems and techniques[10]

The new immigrations from the Middle East and North Africa came at a time when the young State was struggling against difficulties to make good the losses of the war, to consolidate its gains and to absorb newcomers from remnants of European Jewry. There was time neither for painstaking, detailed investigation into the complicated problems thrown up by the new folk-like groups nor for the hastily expanded settlement authorities to analyse the causes of difficulties. The immigrants were directed to empty areas in the South and in the hills, where they built villages with only a skeleton staff of veteran Israelis to direct them. Added to the lack of experience, time and manpower, was an acute shortage of funds. It is obvious that mistakes were made, both in the social make-up of the villages and in their design for production. Known to the planners was the success and stability of the settlement patterns created by earlier generations of pioneers: whether they were fitted to the needs of the very different newcomers, only time and bitter experience would show. On the one hand, there was much careful appraisal and soul searching by the planners, and on the other, action by the settlers who, by remaining in the villages or abandoning them, showed their adaptability to the new way of life and their degree of satisfaction with it, or the opposite.

To investigate these questions fully – for it was realized that they were the key to the successful establishment of permanent villages – the Settlement Department of the Jewish Agency and the Sociology Department of the Hebrew University jointly established a research council. The Settlement Department wanted concrete solutions to the social problems it faced in the new settlements it had established, and practical guidance on the lines it should follow in its future work. The question facing the council was: "can the tension between the traditional social patterns of the immigrants and the modern social and economic structures of Israel be overcome while providing the means for a smooth transition from the one to the other?" The council worked through a network of regional rural sociologists and social instructors. Attempts to arrive at general conclusions from a first analysis showed a very complicated picture. It emerged that in each ethnic group, sound agricultural communities had been established with social continuity which had adapted to agriculture and the new way of life. On the other hand, there were villages which had not consolidated – while an objective appraisal revealed that the same facilities and opportunities had been at their disposal as in the successful ones; nonetheless, their economic progress and social development lagged severely. The sociologists, as discussed in Chapter 4, showed quite clearly that the successful establishment of a new agricultural village did not depend only on the provision of adequate economic means, but was tied to many factors including social integration of the community.[11]

By 1954 – when new waves of immigrants from North Africa started to reach Israel – the position was very different. The short period of intense experience that had elapsed had produced men and women with a far better understanding of the personal and the national problems which the process of settlement had to solve and the difficulties raised by it. The impact of the new villages on the country's economy and agriculture had been carefully considered by the planners, who had rethought the question of the traditional settlement and farming patterns and their adequacy for an independent state. Social problems did not go unnoticed, since earlier mistakes had painfully emphasized the inadequacy of simply trying to fit the new human element into patterns moulded by people of Western outlook.

The first lesson[12] was that transit camps – which were hastily set up to provide shelter to immigrants – were a waste both of human resources and of money, leading to idleness, low morale and poor attitudes for future living. They were found to be costly to operate, unproductive and extremely difficult to dismantle, giving rise to a 'refugee' mentality at a time when every incoming person, both for his own good and the life of the nation, had to become a productive citizen in the shortest possible time.

The agricultural settlement authorities therefore decided on the 'ship-to-village' policy: in other words, the whole ethnic group selection for agricultural settlement was to be made on the basis of group factors – demographic make-up, cohesion, state of health and certain other preliminary 'tests'. Though today selection sometimes begins before the journey, the time available is very limited, while the problems of evacuation and travelling add greatly to the difficulties of those organizing the work. The groups resulting from the selection process are sent directly to the new village or its future site. It is realized that this system does not give nearly enough time to either the selected or the selectors, but it has been found far superior to later selection in transit camps.

By and large, the settlement authorities can, in the light of experience, anticipate the problems faced by newcomers in starting their new life. These problems fall under the following main headings:

a) *Social:* Integration into the completely new way of life they find in Israel confronts the immigrants with a new language, different customs, laws and socio-economic outlooks. The position of the head of the family suffers severe strains: the children receive a different and comprehensive education of which their period of army service is an integral part and inter-generation tension is inevitable.

b) *Economic:* The advanced age or impaired health of many settlers, the large number of young children, lack of vocational training and dissatisfaction with the type of work and organization demanded by modern farming techniques.

c) *Organizational:* The introduction to a democratic society with political equality between man and wife and children of voting age, the co-operative framework – organized marketing, payment of taxes and co-operative discipline – and the special attitudes these demand.

These factors make heavy demands on the settlers. Not all of them want a new way of life, while the forces that show up their previous, 'normal' way of life as inadequate in their new environment cause deep frustrations. In the early stages of immigration, the settlers frequently resisted the efforts of the settlement authorities to introduce new practices, either by simple non-acceptance or by violent reaction. This led to the formulation of three very definite conclusions, designed to cushion the crises of change while at the same time pushing ahead vigorously with the policy of integrating the settlers into the life of the country and making each group economically independent within a social framework that would be satisfactory over a long term, – that change should not be rushed, that social patterns must be studied, and that there must be flexibility in dealing with the man and his family in cultural and religious spheres (see Chapter 4).

These goals were achieved primarily through the process of 'democratic bridging' (described in subsequent sections in this chapter) through which the techniques and social organization of the Western world were introduced into the lives of people completely foreign to them.

4. Settlement patterns for the new immigration

In the matter of settlement patterns, the principle laid down by Dr. Ruppin was strictly adhered to: no one could, or would, force the new settler to live within a social framework not of his own choosing. The result of this has been a striking and obvious change in the character of Israel's agriculture: more than 75% of the new settlements established since 1948 are moshavim, because of the marked preference shown by the new immigrants for the moshav framework; the moshav, with its emphasis on individual family living and flexibility in the scope and extent of co-operation, has proved to be by far the most suitable to the new immigrants from Asia and North Africa.

The choice of settlement pattern is shown in Table 1-1.

TABLE 1-1

The relation between settlement patterns in 1947 and 1957[13]

Settlement pattern	Prior to the country's independence		1957	
	No. of settlements	%	No. of settlements	%
Moshava	30	11.5	35	5.0
Kibbutz	145	56.0	253	36.4
Moshav	72	29.0	358	51.3
Moshav shitufi	6	2.3	18	2.6
Other forms	3	1.2	32	4.7
Total	256	100	696	100

The preference for the moshav is clearly illustrated and can be understood by studying the ethnic composition of those who chose the kibbutz and those who preferred the individual farm unit on the moshav.

Table 1-2 illustrates the ethnic composition before and after the country's independence.

TABLE 1-2

The ethnic composition of kibbutzim and moshavim in 1947 and 1957[14]

Geographic Origin	Kibbutzim %		Moshavim %	
	1947 (145 in no.)	1957 (253 in no.)	1947 (72 in no.)	1957 (358 in no.)
Middle East	3.2	15.5	8.5	33.7
Far East	1.9	6.7	2.5	7.6
North Africa	1.5	19.8	3.7	20.7
Balkan States	3.5	5.0	6.0	5.3
Eastern Europe	56.2	25.1	54.6	22.3
Western countries	33.7	27.9	24.7	10.4

The tables show an overall increase in the population of agricultural settlements. The fact that 62% of the moshav population of 358 villages was from the Middle East, the Far East and North Africa as opposed to only 42% of these groups in 253 kibbutzim should be specially noted. The figures indicate that immigrants from the oriental communities went to kibbutzim to a relatively lesser extent than to moshavim. The moshav shitufi – the settlement with individual family living and large-scale farming – rose from 6 settlements in 1947 to 18 settlements in 1957. The new immigrants showed little interest in this form. Most of the moshav shitufi type of settlement was formed by kibbutzim which altered their living pattern but not their work organization. (A fuller discussion of this subject – vital to countries seeking the best form of settlement for their own conditions – can be found in Chapter 6 under 'Organization of the Unit of Production'.)

In the kibbutz or the moshav shitufi it is easier for a new settler to adjust to physical work by gradual introduction through the various farm branches. He does not have any management or sustenance problems, and his responsibilities for the upkeep of his family and – at least in the first years – for the kibbutz are small. Nevertheless, the new immigrants vastly preferred the difficulties, hard work and insecurity inherent in founding their own moshavim to the more sheltered life of the kibbutz or moshav shitufi. The moshav allowed for a gradual process of change in the traditional family patterns as the new life became less strange – a fact of the greatest significance. By the time of the planning of the Lakhish settlement project, the designers had little doubt that the moshav would be the predominant choice of new immigrants opting for agricultural settlement (in fact, it was their only choice). As explained below, security and integration considerations led to the inclusion on the borders and among the moshavim, of a number of kibbutzim and moshavim shitufi'im which were, however, all settled by veteran Israelis or groups from Western countries.

5. The aims of 'democratic bridging' in the Lakhish experiment[15]

'Democratic bridging' as a fully planned and organized project began with the settlement activity undertaken from 1954 with the 'ship-to-village' operation. The Lakhish Region (discussed at length in subsequent chapters) is the area which has served as a testing laboratory for the effectiveness of the approach.

In the first years of independence, attempts to create a homogenous society were made by bringing together groups from various countries and backgrounds in the same agricultural settlement. The attempts failed: the cultural gulfs were too deep to be bridged in so simple a way, and one or another of the groups would abandon the settlement. Social barriers appeared even between town and village people from the same country. The conclusion was that homogeneity had to be sought in the group making up the village community.

At the same time, the creation of backward rural communities had to be avoided at all costs, both because of the effect on the happiness of the individual settler and his children and their high cost to the national economy. The danger was that the social structure of the new villages would be neither that of the 'old country' nor that of the new, creating a 'second Israel' hanging between two worlds. Change, while firm and progressive, had to proceed at a rate in keeping with the group's readiness for it.

Moreover, the ideological principles formulated during the pre-State period of settlement postulated that the incomes and standard of services prevailing in the new communities had to be not less than those achieved in other sectors of the society. This meant that agricultural instruction had to ensure that effective work methods were adopted, while the place of the village in the national economy and questions of national markets and agricultural prices had to be such that both national needs and the income requirements of the settlers were met.

Three objectives therefore confronted the planners:

a) Integration of the newcomers into the full life of the country in the minimum number of years, if possible during the lifetime of the main immigrant generation.

b) Providing them with incomes and services at a level recognized as adequate by the standards of the country.

c) Designing farming systems and settlement patterns suited to the social structure and ability of the farmers which are at the same time in keeping with national production needs.

All three of these aims are interdependent, none capable of achievement without the others. The given factors – particularly preference for the moshav type of settlement and the need for intra-settlement homogeneity – meant that each of these aims was confronted by problems of its own, which are discussed below.

6. Problems

i. *Farming systems*

The land of the Lakhish area was heavy and potentially very fertile, but water was lacking. The completion of the first Yarkon pipeline – from the river just north of Tel Aviv – brought water to the area and made intensive agriculture possible. A decision had to be taken on the crops to be grown and how to organize their production. The planners realized that in a few years' time the market would be saturated with the animal and fruit products of the traditional pre-State agriculture, while the high cost of the piped water in any case militated against fodder production. The soil was in most parts unsuitable for fruit, and the climate not 'early' enough for high quality out-of-season export products, a branch of production which was in any case beyond the ability of the new farmers. In the same period cotton, sugar beet and peanuts were proving themselves on a small scale, so that national policy adopted them as the main basis for industrial crop farming as a boost to Israel's new industry and a widening of the agricultural production programme. These became the main crops for Lakhish. Planning for industrial field crop production started – a new departure in Israeli farming. It was soon realized that the self-contained structure of the old moshav settlements based on individual dairy, poultry and citrus farming with co-operative selling and buying, would not be sufficient to solve the technical problems involved in growing industrial crops.

Another difficulty lay in the adaptation of large-scale industrial farming to the small-scale settlement pattern of the moshav: division of fields into small plots was unthinkable economically. Secondly, heavy machinery had to be provided for working the crops: its use by a group of 70–80 families would mean under-utilization, while for individual farmers it would be impossibly expensive. Vegetables were included in the crop-rotation scheme, both to balance the labour schedule and income over the year and for agro-technical reasons. Cheap marketing of varied vegetable production from 70 small farms had to be organized.

ii. *Social facilities*

Apart from the technical problems involved in the design of the project, a heavy responsibility that rested on the planners was to provide high-grade social services to the new immigrants whose educational and health standards were by and large low, and who had many social and family problems due to complete uprooting from their old environment. Within the villages, building the necessary institutions and guiding the people to independence in the running of village affairs, and training for the women in home economics, hygiene and baby care were tasks demanding patience, skill and dedication. Staff was scarce, adequate provision of schools and medical services could not be given on the basis of a single village because of its limited size, and taxes to pay for the services had to be kept at a minimum.

Another problem was to attract to, and keep in the rural areas skilled managerial and technical personnel, whose background and culture were of another world; ways had to be found of providing the doctors and teachers, instructors and leaders with a community which would be able to keep them close to the site of their work, problems faced by many other countries today.

It was obvious that the high cost of providing individual services to villages, as had been done prior to 1956, could not be continued, since it had resulted in much dissatisfaction both on the part of the immigrant settlers who rightly felt that they were underprovided with services in comparison to their counterparts who went to towns while the cost to the villagers was higher, and on the part of the settlement staff who had no home within the area of their work. Both parties had a right to living conditions equal to those prevailing in the rest of the country.

iii. *Community integration*

Settlement based on homogenous communities was recognized as the only way of ensuring stability of population and intravillage harmony through the difficult years of consolidation preceding the attainment of village independence. However, the creation of isolated villages cut off from the main life of the country was undesirable unless ways of integrating the separate communities could be found. The army alone could not carry out the task for the young generation, while the old settlers also needed strong integrating forces.

7. The solution: composite rural structure[16]

The three sets of problems enumerated above – farming methods adapted to the economic production of industrial crops, the provision of adequate social services, and immigrant integration – led the planners to conclude that the base of the individual moshav-type settlement would have to be widened into a larger framework designed on a subregional pattern. Experience pointed to the fact that 70 or 80 families were the maximum for a new village if internal distances were not to mitigate against community formation. On the other hand, a minimum number of families to support a good school, clinic, recreational and commercial facilities was about 300, or four villages. The idea was thus conceived of settling villages around a centre which would include all the services and have support from the 300 families. Overhead costs could be spread with the services fully utilized. The centre would be the natural meeting place for the different ethnic groups and act as a stimulant towards growing integration over the years, proceeding speedily with the children and youth and more slowly with the adults.

This organization formed the basis of the settlement of the Lakhish area. Rural centres were built to serve four or five villages each. They were provided with attractive dwellings to house the staff who worked in the central and village services. After the school, clinic and local council headquarters had been completed, and the settlers had started to use the centre, coming to look upon it as a necessary and natural part of their lives, a cultural hall, self-service stores and a bank were built. A tractor station was established to supply mechanization on an hourly or per acre basis, followed three years later by a vegetable grading station to help overcome the very difficult problems in vegetable collection and grading from many small farmers. Space was left for small-scale agricultural and other industries to develop – a very important phase of rural centre development (see below).

This composite structure was however insufficient for large-scale agricultural industry. Obviously, four or five villages could not support a cotton gin or sugar factory, nor the vigorous industrial development complementary to agricultural progress and its economic well-being, nor provide a full range of shops and services, nor support county administration and large-scale municipal services. To provide for these needs, the scale was further increased, several subregions merged into a region with an urban concentration as the nerve centre containing the services and industries. This provided an immediate market for the bulky products of the new field crop farms. The town of the Lakhish area is called Kiriat Gat. Its growth, institutions, industry and influence on the surrounding area is described fully in Chapter 7.

Although too short a time has elapsed to reliably evaluate the Lakhish Project, the present stability of the village populations and the yearly increase in production indicate definite prospects for success.

8. Wider applications of composite rural structure

The wide interest aroused in Israel by the Lakhish project's ability to provide cheap organized services and community facilities within the settlement area, has led to discussions of its application in other areas closely settled but with no pattern of regional organization.[17] In such areas, schools, shopping and medical services were poorly distributed while social and recreational life was limited to the individual village (in which these services were expensive) or else farmers had to travel to the nearest town. An experiment was therefore undertaken in a region where settlements of different types and social backgrounds had been in existence for ten years or more, but without a comprehensive regional plan for service provision. Rural centres are now being established there, which concentrate the community and social facilities at central points. The architectural planning has changed with the experience gained from running the rural centres at Lakhish. The eager acceptance of the scheme to revitalize the area by the villages and local councils, and the interest in it by other older areas, is concrete proof of the soundness and necessity of comprehensive rural planning.

9. Conclusions

From the foregoing, it can be grasped that the approach of Israel's planners to rural living standards and to efficiency in agricultural production, has been based on the building of viable communities. The social and cultural background of the people to be turned into farming villagers, has been respected as far as possible in fitting it into modern organization and production methods. The man, his family and traditions has been the foundation on which villages have been planned and services introduced, either in the village itself or nearby service centres. Experience in Israel so far has shown that by maintaining group cohesion, upsetting customs as little as possible, and by providing adequate guidance through extension personnel, it has been possible to gradually introduce the complicated methods of modern agricultural technology to people of a non-modern or traditional background. The physical and economic planning of settlement in Israel has been designed to suit the human and agricultural potential of the region and to introduce to it the professional, technical, and extension personnel, the services and the industry needed for its development.

In considering agricultural development in countries other than Israel, it is suggested that living standards in the rural centre of the agricultural settlement region can act as a stimulant to subsistence farmers to raise the level of their ambitions and therefore the effort they are willing to make for progress. While the main experience of Israel has been in settlement, it is felt that the principles formulated through settlement experience can, with care and thoughtful application to the circumstances of each particular case, be extremely useful tools not only for new settlement but also in revitalizing and building up long-stagnant and problematic areas.

As stated, one of the aims of the Lakhish project was to design "farming systems and settlement patterns suited to the social structure and ability of the farmers which, at the same time, were in keeping with national production needs". In other words, the aim of agricultural settlement is not to produce a show piece with no particular connection with the rest of the country's agriculture but to bring into production empty areas of land, or to reorganize and increase production in existing farming regions. Therefore, *all new settlement must be designed as part of a national plan for agriculture.* Even if the main purpose of the project is to serve as an example, the incomes planned, the size of the farm, the resources allocated and production systems must be capable of application elsewhere. The purpose of settlement is many-sided, but its impact on the national scene both in terms of actual production, social experiment and new farming practices will only be positive if the aims of the settlement are first clearly phrased and understood in a national agricultural context. Settlement does not and cannot stand on its own: its financing, marketing, industrial and service operations all converge at various regional and national levels. A particular scheme can be divorced from these only at great danger to its success. National agricultural planning is in itself essential and most certainly so as an operation preceding settlement planning.

NOTES

1. BEIN, Alex: *The Return to the Soil, A History of Jewish Settlement in Israel* (The Zionist Organisation, Jerusalem 1952).

2. See also:
   [1] BEIN, Alex: Op. cit.
   [2] HARMAN, A.: 'Agricultural Settlement', *Israel Today* No. 2; Third Edition (Published by Israel Digest, 1963).
3. HERZL, Theodor: An opening speech at the Fifth Zionist Congress.
4. See also:
   [1] ORNI, E.: *Forms of Settlement* (Keren Kayemet Leisrael, Jerusalem 1963).
   [2] BEN DAVID, Joseph (Ed.): *Agricultural Planning and Village Community in Israel*, Arid Zone Research XXIII (UNESCO, Paris 1964), pp. 45–57.
   [3] INFIELD, H. F.: *Co-operative Living in Palestine* (London 1946).
   [4] SAMUEL, Edwin: *Handbook of the Jewish Communal Villages in Palestine* (Jerusalem 1945).
5. BARATZ, Joseph: *Village by the Jordan: The Story of Degania* (London 1954).
6. Further details on the planning of the Moshav see:
   [1] YALAN, E.: *The Planning of a Moshav*, International Farmers' Convention in Israel 1959 (The Government Press, Jerusalem 1960), pp. 241–251.
   [2] KADAR, G.: *The Moshav – The Economic Aspect*, International Farmers' Convention in Israel 1959 (The Government Press, Jerusalem 1960), pp. 236–240.
7. RUPPIN, A.: *The Agricultural Colonisation of the Zionist Organisation in Palestine* (Martin Hopkinson and Company Ltd. London 1926), pp. 41–42.
8. EISENSTADT, S. N.: *The Absorption of Immigrants* (Routledge and Kegan Paul, London 1954).
9. WEITZ, R.: 'Sociologists and Policy Makers', *Transactions of the Fifth World Congress of Sociology*, Vol. I, International Sociological Congress of Sociology, Vol. I, Louvain.
10. See also:
    [1] BEN DAVID, Joseph (Ed.): *Agricultural Planning and Village Community in Israel*, Chapter VII, 'Social Integration and Change', by D. WEINTRAUB and M. LISSAK, Arid Zone Research XXIII (UNESCO, Paris 1964), pp. 129–159.
    [2] SHAPIRA, O.: *Moshav Gadish* (Hebrew) (Settlement Department, Jerusalem 1959), (mimographed).
11. WEINGROD, Alex: 'Reciprocal Change: A Case Study of a Moroccan Immigrant Village in Israel, *American Anthropologist* (1963).
12. GOREN, Y.: 'Immigrants into Established Settlers', International Farmers' Convention in Israel 1959 (The Government Press, Jerusalem 1960).
13. GOREN, Y.: 'The Villages of the New Immigrants in Israel, their Organization and Management', Extension Division, Agricultural Publications Division, Publication No. 33 (Tel-Aviv 1960).
14. GOREN, Y.: *The Villages of the New Immigrants in Israel*, op. cit. pp. 6–7.
15. [1] Agricultural Settlement Department: 'The Composite Rural Structure, A Settlement Pattern in Israel' (Jerusalem 1960).
    [2] ROKACH, A.: 'Regional Rural Development', *Israel Today* (The Israel Digest, Jerusalem 1965).
16. [1] Agricultural Settlement Department: *The Composite Rural Structure*, op. cit.
    [2] WEITZ, R.: *A New Concept of Agricultural Settlement Planning*, op. cit.
17. See: The National and University Institute of Agriculture, Settlement Study Centre – '*Regional Cooperation in Israel*' (Rehovot 1965).

CHAPTER 2

STATING THE TARGETS: THE DEMAND FACTORS

The Ten Year Plan for Israel's agriculture which is presented in this and the two subsequent chapters had three aims.

a) to estimate the demand for agricultural products in Israel in 1972–1973;

b) to estimate the resources for producing these products;

c) to calculate the best economic combination of resources to meet the demand, taking probable trade-terms into account.

This present chapter examines factors in the Israel economy which will directly affect the demand for food and fibre products in 1972–1973. It estimates demand in the light of available data, predicts the market situation with regard to specific products in that year, and examines export prospects.

In assessing demand, the factors discussed are population, national income, the influence of capital from abroad, as well as personal income, climate, traditions, produce outlets, and external markets. Although the material deals with Israel, general approaches to planning can be drawn from it.

A. POPULATION PROJECTIONS

Despite the obvious connection between the size of population and its food supply, the situation is more complex than it appears at first glance. Different income levels mean that food provision is the main preoccupation of life for some populations; for others, the cost at farm of its food supply is well within the means of all. Schultz[1] suggests that a classification of populations in relation to their food supply should be in accordance with the drain that food places on their economies: "Populations subject to a high food drain are those where a large proportion of the income is used for food." Following this lead, Schultz goes on to classify populations into three types: High food drain (Type I), Intermediate food drain (Type II) and Low food drain (Type III).

Type I spends 75% or more of income on the acquisition of food. – The situation is sometimes described as 'the pre-industrial demographic class', covered by the Ricardo–Malthus–Mill formulation. Type II, the so-called transitional demographic class, spends between 25% and 75% of income for farm products (leaving aside the services added to food by other producers between the farm-sale and retail-sale stages). Type III populations (the 'advanced industrial' demographic class) spend 25% or less on the farm produced services that go to make up food – so low a figure that population increases or decreases are essentially independent of food supply.

In types I and II, changes in population and changes in food supply are very closely related. A change in food supply results in changes in population until a new equilibrium between the two is reached. A decrease in food supply results in more deaths and in a decrease in population, while any increase in food supply (equivalent to an increase in income) results in an increase of population.[2]

In type III, population increase seems, as mentioned above, to be independent of the food supply. Social factors are decisive in determining the fertility, age of marriage and increase or decrease in the number of children. While the causes of population behaviour are not fully understood for this type, food availability has very little significance in accelerating or retarding population growth.

For types I and II, the planner aims to have the increase in agricultural output keep ahead of the increase in population, so that a rise in nutritional standards and a drop in the proportion of people engaged in agriculture can be effected. For type III, the planner aims to predict the quantity and quality of food the population will demand according to its size and income levels.

Since Israel is a country of immigration and absorption, its population is not homogeneous, but rather a composite of the three types, not only because of the wide range of income but also – and probably mainly – because of different habits. Many ethnic groups, even after earning high wages, do not change their food habits, spending much less on food than the general classification above would indicate. Any projection has to take this basic social factor into account.

Demographic factors have a deeper significance for agriculture than the population-demand correlation, since population characteristics very much influence *general* development and often are the cause of it. Agricultural planning must work within this context. Much of its basic data is drawn from predictions of the future behaviour of the society as a whole. The demographic picture must therefore be studied in its entirety before conclusions specific to agriculture can be drawn.

1. Total population

The rapid development of Israel's economy during the short period of statehood is closely related to mass immigration, which increased her population two-and-a-half times between the years 1948 and 1958. Over 90% of the additional labour force found employment. Table 2-1 shows the population growth – with its 175% increase in the Jewish population – from the establishment of the State in 1948 until 1961.[3]

TABLE 2-1[3]

The growth of population (mid-May 1948–1964)

	% Growth in total over previous year	Total mean population	Jewish population (mean)	Others (mean)
mid-May 1948			650000	
1948		901100	759000	120000
1949		1059000	901000	145000
1950	12.1	1266800	1103000	163000
1951	18.0	1494300	1324000	170300
1952	7.5	1606200	1429800	176400
1953	2.7	1650200	1467700	182600
1954	2.4	1689500	1500600	188800
1955	3.6	1750400	1555300	195100
1956	4.5	1828400	1626300	202000
1957	5.6	1930000	1721200	209300
1958	3.6	2000000	1782700	217400
1959	3.1	2062100	1836200	225900
1960	2.6	2117000	1882600	234400
1961	3.3	2187500	1943800	243700
1962	4.6	2289700	2030500	259200
1963	3.9	2380300	2111300	269000
1964	4.2	2480700	2197100	283600

The population is made up of two groups, 'Jews' and 'Others', the largest proportion of the latter being Arab.

The Jewish group consists largely of immigrants or descendants of persons who immigrated to Israel during the last 75 years. In May 1948 there were 650000 Jews in the country. During the first decade the number increased to about 1780000 (end of 1958).

Because of the importance of immigration, its future extent has to be estimated in projecting the size of population. This estimate is difficult because of the fluctuations in the number of immigrants from year to year; Table 2-2 shows the Jewish immigration for the period 1948–1961.[4]

Since mass immigration of the magnitude of that from 1948–1951 is un-

TABLE 2-2[4]

Jewish immigration to Israel 1948–1961

	Immigrants	Emigrants	Net immigration
1948 (May–Dec.)	101 800	1 000	100 800
1949	239 600	7 200	232 400
1950	170 200	9 500	160 700
1951	175 100	10 100	165 000
1952	24 400	13 000	10 600
1953	11 300	12 500	800
1954	18 400	7 000	11 400
1955	37 500	6 000	31 500
1956	56 200	11 000	45 200
1957	71 200	11 000	60 200
1958	27 100	11 500	15 600
1959	23 900	11 500	12 400
1960	24 510	8 500	16 010
1961	47 638	7 330	40 308
1962	61 328	7 644	53 684
1963	64 364	10 866	53 498
1964	54 716	9 121	45 595

likely, the experience from 1952–1960 can be used as a guide. In no five-year period since 1952 has the average net immigration been less than 18 000 or more than 33 000.

The structure and characteristics of the Jewish group – age and sex, occupational structure, demographic, social and health traits – has been to a large extent shaped by the different waves of immigration by which it was built up. By far the greatest of these waves began with the foundation of the State and continued till the end of 1951; in about three and a half years 687 000 newcomers arrived, nearly half of them the remnants of European Jewry from displaced persons' camps in Central and Eastern Europe and the Balkans (mostly Rumania, Poland and Bulgaria). Approximately the same number entered from Asia and Africa, with a transplanting to Israel of the Jewish communities of Yemen, Iraq and Lybia. Minor contingents came from virtually every country in the world which had a Jewish community.[5] When this massive transfer was ended, and because emigration out of certain countries was not allowed, immigration during 1952–1954 fell off considerably. The years 1955 and 1957, however, brought a renewed movement from North Africa and other countries. The years from 1957 have shown a strong upward tendency in the numbers of immigrants per year, which, it is hoped, will be sustained.

The most obvious stamp of the Jews of Israel is their widespread hetero-geneity of birth-place and of cultural, social and economic background. Only 39% of them today are Israel-born, the rest are natives of Europe and America (32%) and of Asia-Africa (29%).[6]

The age structure has also been formed to a substantial degree by the composition of the several waves of immigration. During the British Manda-tory period (1918–1948), immigration was highly selective in respect to economic and cultural characteristics. A large proportion of immigrants were in the young working ages (15–35). The Jewish community in Palestine consequently had a low percentage of children and a very small one of aged people, with a 'bulge' in the medium age-groups. The bringing to Israel of entire communities (especially Oriental ones) markedly increased the pro-portion of children and created a smoother and more regular age distribution which very much resembles the world average.

An increase of 910000 by immigration in ten years (1948–1958) has been only one of the components of population growth; natural increase amounted to more than 360000. While immigration caused sudden eruptions of growth, changed the structure of the population, increased its heterogeneity and enlarged the proportion of foreign born, natural increase is now working to a greater extent over a larger population, accelerating a trend towards more homogeneity.

TABLE 2-3[7]

Rates of natural increase 1950–1964
(per 1000 average pop.)

	Jews	Others	National average
1950	26.48	33.71	27.42
1951	26.26	37.75	27.23
1952	24.73	34.05	25.68
1953	23.91	38.72	25.44
1954	20.93	35.42	22.45
1955	21.45	37.35	23.12
1956	20.37	37.80	22.18
1957	19.86	37.21	21.63
1958	18.43	40.16	20.66
1959	18.51	40.51	20.78
1960	18.36	42.81	20.91
1961	17.00	42.04	19.61
1962	15.90	43.70	18.60
1963	15.90	42.50	18.90
1964	16.20	45.00	19.40

The non-Jewish group comprises Moslem and Christian Arabs, Bedouin, Druse, and Circassians. Their number grew from 145 000 at the end of 1949 to about 283 000 in 1964.

Table 2-3 [7] gives the natural increase separately for Jews and non-Jews because of the large difference in birth rate between them.

The rate of natural increase among Jews was 19.86 per thousand in 1957 and dropped to 16.2 in 1964. There are great differences in prolificity between Jews of various origins. In 1955, for instance, the gross reproduction rate for natives of Europe was 1.28, for natives of Asia-Africa 2.75 and for Israel-born 1.37. Although far-reaching predictions would be premature, there are indications that the birth rate of oriental Jews is declining as stay in Israel lengthens while that of Western-born Jews tends to increase very slightly. Death rates among the Jewish group are similar to those in most advanced countries: the average life expectancy at birth is about 70. The rate of natural increase for non-Jews is much higher – around 40.00 per thousand.

2. The projected population increase

According to an estimate made as far back as 1958 [8], the natural increase of the Jewish population in Israel – which reached 2.6% in 1950 and fell to 2.1% at the end of 1955 – is likely, during the second decade of statehood to fall to 1.7–1.8%. The natural increase of the non-Jewish population for the corresponding period was expected to be 3.5%. This estimate has so far proved near to correct. It may therefore be assumed that it will be fairly accurate in respect to the ten year planning period.

A forecast of the total increase in population is difficult since it must take immigration, which is an unknown factor and has varied, into account. Predictions of population increase have not proven entirely accurate. For example, a study prepared under the auspices of the Falk Project by Dr. N. Halevy [9] predicted that the population would reach 2.5–2.55 million by 1965 as a result of both natural increase and immigration. In fact, it reached 2.48 million by 1964. Since this forecast, immigration again increased and further Jewish communities in the diaspora were in the process of liquidation by migration to Israel. Population forecasts have of course to be consequently amended. The Planning Authority estimates that the population will reach 2.95 million by the end of 1970 [10], assuming a net immigration of 50 000 per year.

Our estimate, which is based on estimates of the Immigration Department of the Jewish Agency, assumes that after 1966 immigration will become stabilized at 30 000 to 35 000 immigrants per year. With regard to natural increase, it has been assumed that the annual rate of growth of the Jewish population will be about 1.8%.

On this basis, the population of Israel in 1973 will be some 3.2–3.3 million. For the purposes of calculation, the estimate of 3.2 million persons has been taken.

3. Labour force

The employed civilian labour force and its distribution during 1950–1959 are shown in Table 2-4[11].

TABLE 2-4[11]

Employed civilian labour force and its distribution 1950–1964

Year	Total labour force employed	Agriculture %	Industry %	Other %
1950	398 500	16.2	24.0	59.8
1951	469 300	15.9	23.4	60.7
1952	544 100	16.7	22.4	60.9
1953	543 700	16.7	24.5	58.8
1954	563 900	16.7	24.5	58.8
1955	583 800	17.6	23.5	58.9
1956	598 400	17.8	23.8	58.3
1957	642 200	16.3	24.1	59.6
1958	657 000	17.5	24.0	58.5
1959	683 000	16.8	25.2	58.0
1960	701 800	17.3	23.2	59.5
1961	746 500	17.1	23.8	59.1
1962	787 900	16.0	24.8	59.2
1963	813 200	14.3	24.7	61.0
1964	851 200	12.9	25.3	61.8

The distribution of the employed civilian labour force has shown unusual stability (Table 2-5). According to A. Hovneh, the proportion of the population participating in the work force has been around 37% for the last four years.[12] This is high in relation to many underdeveloped countries, and higher than countries like Canada and New Zealand. It is below the figures for Europe which vary from 42 to 49%. The explanation in part is due to the high proportion of children under 15 in the population, most of whom receive schooling to that age or beyond. The proportion in Israel is 36%; in European countries and North America children under 15 years form 22–30% of the population. For the population over 15, the participation rate in the labour force is 60%, almost identical to that for countries of

TABLE 2-5[12]

Population and labour force

Year	Total population (in thousands)	Total labour force (in thousands)	Rate of participation (percent)
1949	1032.6	417.9	40.5
1950	1253.5	500.1	39.9
1951	1480.8	595.0	40.2
1952	1592.7	634.0	39.8
1953	1636.7	648.6	39.6
1954	1676.0	658.0	39.3
1955	1736.9	669.3	38.5
1956	1818.9	696.1	38.4
1957	1916.5	725.3	37.8
1958	1986.1	745.1	37.5
1959	2048.1	761.2	37.2
1960	2113.0	–	–

North, West and Central Europe, and higher than North America (53%). Hovneh predicts that, as the present population gets older and the present 'bulge' in the middle age-group passes into the over-60 group, the participation rate will fall. For the purpose of the population predictions for the 10 year plan discussed in this book, a participation rate of 34.5% has been taken to be realistic.

At this rate, the labour force should reach some 1 110 000 of the 3.2 million population forecast for 1972/3 which is 34.5% of the population.

The next decade will see considerable changes in the occupational structure of the labour force since limitations in internal and external markets and, secondarily, in the quantity of irrigation water, do not allow an expansion of agricultural settlement similar to that of the last decade. However, the market for agricultural products will not grow faster than the expansion of production from the existing settlements. Production in them will undergo considerable intensification in order to reach the income levels of 1972. Newcomers to the labour force will therefore be directed mainly into industry and services with proportionately far fewer finding their employment in agriculture. The percentage employed in agriculture in Israel will henceforth continuously decline, unless important new markets are opened or new sources of irrigation water are found in large quantities. This tendency may be emphasized by increases in the size of some of the uneconomically small farms at the expense of their neighbours who will seek work elsewhere.

B. NATIONAL INCOME AND PRODUCT PROJECTIONS

1. The past course of the economy

In order to project National Income and National Product for the year 1972/3 an examination of the behaviour and growth of these factors in the past is essential. The following table shows the net domestic product of Israel at factor cost for the years 1950–1959 by branch of origin.

The figures show a 263% increase in total Net Domestic Product over the ten years; an increase of 354% in agricultural output; 162% in industrial output and 308% in services. The annual rates of increase of the components, and their share of the total, are of interest in determining whether any definite trends exist. Table 2-7 analyses Table 2-6 to present these figures.

TABLE 2-6[13]

Net domestic product at factor cost, by branch of origin
(millions of 1957 IL.)

Year	Total	Net rent	Total* minus net rent	Agriculture	Industry	Other*
1950	1127	70	1057	124	390	543
1951	1497	92	1405	123	493	789
1952	1540	105	1435	178	398	859
1953	1576	116	1460	179	417	864
1954	1904	133	1771	226	486	1051
1955	2106	154	1952	229	508	1215
1956	2335	164	2171	282	524	1365
1957	2527	186	2341	321	550	1470
1958	2725	196	2529	361	579	1589
1959	2968	219	2749	439	633	1677

* Transportation, contract construction, public utilities, trade, banking, finance and real estate *less* net rent, other services, general government and non-profit institutions.

The table shows that agriculture and services have increased their percentage contribution to the Net Domestic Product, while industry's relative contribution has decreased. The annual rates of increase show no distinct trend for the three components but reflect particular features of the last ten years, with its fluctuation in immigration and the needs which it brought into being. The great 'offensive' has been in agricultural settlement with its increasing entry into production, and in the rise in productivity of the old settlements. Since the population is now larger and its economic position stronger, future immigration and political events will probably not hinder

TABLE 2-7[13]

The percentage annual contribution to the total net domestic product (less net rent) of the three main components, and the annual increase of each

Year	Percentage annual growth rate of total A	Total minus net rent (mill. 1957 IL.) B	Agriculture as % of B	Agriculture % growth rate/ year	Industry as % of B	Industry % growth rate/ year	Other as % of B	% growth rate/ year
1950		1057	12		37		51	
	24.5					26		45
1951		1405	9		35		56	
	2			45		19		9
1952		1435	12		28		60	
	2			–		5		5
1953		1460	12		29		59	
	21			26		16		22
1954		1771	13		28		59	
	10.5			1		5		16
1955		1952	12		26		62	
	6			23		3		12
1956		2171	13		24		63	
	8			14		5		8
1957		2341	14		23		63	
	8			12		5		8
1958		2529	14		23		63	
	9			22		9		6
1959		2749	16		23		61	

a planned increase of national output and its separate components, provided that the increase is kept within the capacity of the economy.

The economic expansion of the last ten years has prepared the way for future expansion. The large increase in services reflects the effort invested in integrating the new immigration into the life of the country and in building the infrastructure of the new State. This work will continue into the next ten years, but its scale, relative to that of industry, will diminish. Since industry builds not only on infrastructure but on 'social capital', it is likely to forge ahead, greatly increasing its contribution to the National Product on the foundations for industrial expansion which have been lain in the first ten years. The comprehensive regional planning approach to industrial location will no doubt prove an important part of the foundations.

The increase in agricultural output shown in Table 2-7 is a direct result of new settlement and of resource allocation to farms which were not yet fully consolidated. The next ten years will see such farms reach full production. Since there is no possibility for much additional settlement, agriculture will not be able to keep up the high rate of expansion of the last decade. The stage is set for a more 'normal' economic life in Israel: while it is expected that the rate of growth of all three sectors will continue to be high, they will come to reflect the country's maturing economic and social trends. In Israel's case, prediction of the future economy on the basis of the raw data of the past is complicated by the fact that the first ten years represent the upheavals of population and the consolidation of independence, so that the performance of the various sectors of the economy on the basis of their past records can be assessed only with difficulty. However, the economy is planned and controlled to a certain extent, so that the future is not left to work out haphazardly. National policy is aimed at settling as much of the Negev as possible through exploitation of its industrial potential and decentralization of population and industry: the longterm plans indicate that a high rate of general development will be maintained.

2. Development forecasts

There have been three forecasts made in recent years of significance: that of the Bank of Israel, that of ministries based on the forecast of the Bank, and that of the Economic Planning Authority.

(a) The Bank of Israel forecast[14] of 1960–1965, was based on an immigration of 40000 people per year, and a decline of capital imports to 200 million in 1960–1965. It recommended an annual output increase of 8%, an export increase of 28%, and an import increase of 10%, in order to finance other developments. It recommended a continuous investment increase and a restriction of per capita consumption growth to 2% per year[15].

(b) On the basis of the Bank's report, the various ministries prepared preliminary development plans for the 1960–1965 period which were similar to the report of the Bank but with important exceptions. They advocated less import-export expansion, more capital imports, and did not stress the need regarded as vital by the Bank of Israel, of limiting consumption increase to 2% per year.

(c) The forecast of the Economic Planning Authority[16] is later than that of the Bank of Israel, covering the period from the devaluation of the Israel pound in 1962 until 1969/70. It predicts a higher rate of increase in GNP per year, – 9% – even though changes in the composition of increased production were anticipated. Whereas in the past a considerable portion of added resources was destined for consumption, it was predicted that this would change in the future when a greater proportion would be set aside for export or import-substitutes. Whereas in the years 1958–1963 about 90% of the added product was directed to private and public consumption, it was assumed that by 1969, private and public consumption would drop to 44% of the added resources.

The distribution of the added resources, as forecast by the Planning Authority, compared with its distribution in the past, is given in Table 2-8.

TABLE 2-8

Distributions of the added resources in the past and as forecast[16]
(Annual mean as percentage of the total added product)

	1958–1963 %	1965–1969 %
Private consumption	42	31
Public consumption	16	13
Gross investments	17	11
Exports	25	45
Total added resources	100	100

According to the forecast of the Economic Planning Authority, the gross national product will increase from IL. 7.720 million in 1963 to IL. 13.150 million in 1969/70. Hence the figure, 9% increase per year. By 1972/73 the gross national product will total about 15.620 million at 1963 prices.

On the basis of the forecast, the gross national product per capita should increase from IL. 3200 in 1963 (at 1963 prices) to IL. 4500 in 1969/70, and to IL. 4880 in 1972. The envisaged increase in gross national product per capita for the 1963–1972 period will accordingly be around 50%.

3. The import surplus

Israel's development has been marked by imports outstripping exports. Without an import surplus there would have been no development. The response of world Jewry with contributions and loans aided the new State to accept and absorb its newcomers. At the same time, Israel took many international loans to finance her economic growth. By borrowing on the world market, or from special funds set up for development, productive capacity could be so expanded that loans could be repaid. The rate of expansion was of far greater importance than the volume of indebtedness.

The importance of foreign capital in aiding development should be stressed. The main need in newly developing countries is for long term capital loans which are not 'self-liquidating', but which go to build the infrastructure of the economy and the framework for its functioning. Some countries are hesitant to take foreign loans because they raise their indebtedness, making their position *vis-à-vis* the rest of the world appear less favourable. Such a policy cannot be supported in the light of need for rapid development. Developing countries must demand aid from those able to give it, taking advantage of any bargaining power to obtain foreign credits for development. Only when their economies begin to expand is there room for concern about debt repayment, for only then is it possible. Small nations have nothing to fear from aid if they are careful not to be tied up politically. Their main concern must be with rising welfare of the population, which, as it rises, increases the resistance of the people to political subversion from outside and increases their overall production out of which they can repay loans.

In Israel, warning voices have been heard against increasing the import surplus. There are strong indications[17] that the marginal activities financed by the import surplus of the last ten years were for consumption as well as investment, and that difficulty will therefore be experienced in meeting the mounting interest and capital repayments. This is a warning not against capital import as such, without which Israel could scarcely have established herself and progressed, but against its use for financing activities which do not lead to a long term increase in the national product. What is important is not the trade gap or the level of indebtedness itself, but the contribution it makes to economic progress and its own final liquidation.

4. Private consumption

As we have seen, the Bank of Israel in its forecast recommended that private consumption be limited to 2% increase per year and the development plans of the ministries concurred. In 1958 Prof. D. Patinkin considered that in order to offset a reduction expected in foreign capital imports, it would become necessary to channel the main part of the added national product to investment intended to increase exports, reduce imports and curtail expenditure on consumption goods.

Agricultural planning carried out by the author in the past was based on this assessment; food consumption trends and the means of production required to secure the desired income for the farmer were fixed accordingly. In 'Agriculture and Settlement'[18] (1958) it was assumed that no increase would take place in the standard of living and that no far-reaching changes would therefore occur in the demand pattern of the internal market: "Any expansion of the production complex and all the added product must be directed exclusively towards increasing exports and eliminating imports by means of local production at competitive prices, rather than to any increase of internal consumption over and above its present magnitude."

The Fourth Report of the Falk Project for Economic Research, published in 1959[17], again assumed that the rise in private consumption would be curbed, since most of the added national product had to be channelled to productive investments. In 'Towards Specialized Farming'[19] the author again based his calculations for the entire agricultural sector on this expectation. Again, in 1960, it was predicted by the author that consumption would not increase to any considerable extent.

All these estimates were proved wrong since real consumption increased by 5% in 1960, and by 6% in 1961, 1962, 1963 and 1964. This meant that the forecasts of added national product and of national consumption required revision. The long term projections of supply and demand for agricultural products in Israel by Dr. N. Halevy, published in the Fifth Report of the Falk Project[20], assume an increase in private consumption at a rate approximately equal to the rate in previous years. The recent study of the Economic Planning Authority[21], which analyses expected economic developments, predicts that in the years ending 1969 there will be "an annual rise of 3.5% in private per capita consumption. Consumption will be held to half its present scale by means of greater private and public saving." Without such increased saving, consumption would rise at a greater rate owing to the expected rise in the national product. Estimates made in 1964/5 put the annual rise in per capita income at an even lower rate – 2.5–3%.[22]

For our calculations we may therefore assume that the rise will be around 3% per annum.

5. Projections for 1972/3 and their implications for agriculture

In view of all the above, and relying mainly on the Economic Planning Authority forecasts, the table on p. 45 is our economic forecast for 1972/3.

In 1963, the average income of a hired labourer was IL. 385 per month[23], i.e. IL. 4600 per annum. According to the same source, the average wages in industry were IL. 365 per month, i.e. IL. 4380 per annum. Since an increase of 50% in the national per capita income over the coming ten years is anticipated, the gross average income (prior to deduction of taxes) of the urban wage earner can be expected to also rise by 50%. As we shall see later, this rate of increase in the income of the average urban wage earner should guide us in calculating the desired level of income of the farmer. In order to ensure the stability of agriculture and of the rural community, the farmer has to be assured a rise in income at a rate parallel to that of the average wage earner in the country as a whole. Unless this principle is strictly adhered to, the foundations of agriculture are liable to be seriously undermined.

	End of Forecast Period 1972/3
1. Average population (in thousands)	3200
2. Number of persons employed	1100
3. Annual rate of increase of the national product	9%
4. Gross National Product, in millions IL. at 1963 prices	15620
5. Net Domestic Product, 1963 prices	12800
6. National per capita income in IL. of 1963	3420
7. Total percentage of increase in national per capita income (from 1963 until the end of the forecast period)	50%
8. Annual increase in per capita consumption	2.5%
9. Cumulative increase in private consumption from 1963 until the end of the forecast period	32%

The developments in the structure of the agricultural sector in general, and in the income from farming in particular, must be predicted in advance, in order to prevent unnecessary upheavals. This should and can be done on a long-term basis. A period of 8–10 years is the most suitable for making such forecasts, in that it is not so long as to prevent the planner from making reasonably accurate estimates, while at the same time it is the shortest space of time in which far-reaching changes in the structure of the different types of farms may be planned.

Accordingly a start must now be made in assessing the expected modifications in the structure of the agricultural sector likely to take place on the assumption of an average income of about IL. 6700 per annum, i.e. about 50% greater than the average gross income of the urban wage earner in 1963.

C. FACTORS INFLUENCING THE DEMAND FOR AGRICULTURAL PRODUCTS

The demand for food, whether grown for family consumption or purchased, is influenced by several factors.

1. Income

The first factor, income, influences traditional food habits. In fact, a traditional diet based on only one or two staple foods is usually a function of income. Such foods are mainly the starchy ones. They are cheapest to produce or buy, for low income gives little choice in consumption. As income – or productivity on the subsistence farm – rises, outlay on food usually increases, new and better foods eaten. Animal products are consumed along with a lower proportion of staples; vegetables and fruits are bought and make their contribution to a more healthful and nutritious diet. As societies move from a 'high food drain' situation to an 'intermediate' and 'low food drain' situation, the total calorie consumption per head and its quality tend to rise.

The percentage of the food budget which is spent on services added between the farm gate and the consumer also rises. The rapid increase in the sale of washed, peeled, prepacked vegetables and frozen ready-cooked foods represents, in the affluent societies, the determined decision of the housewife to emancipate herself from kitchen chores. The impact of her choice of food production on the demand for food at the farm gate is therefore far less direct than that of her counterpart in a less fortunate country who purchases only very simple distribution services with her food. The greater part of her dollar spent on processed food does not get back to the farmer. Similarly, with rise in income, the proportion of restaurant meals in the food budget rises. However, this has little effect on demand at the farm gate, since one who eats in restaurants does not consume more food even though he spends more money.

2. Prices

Within the expenditure allowed for food in the general family budget, the relative prices of different foodstuffs very much influence the demand for them. Substitution is possible between one food and another, and it is undertaken largely on the basis of relative prices. If the price of beef goes up relative to mutton in the long term, demand for mutton will tend to rise at the expense of demand for beef. If new farming practices lower the costs of production of vegetables relative to grains, then a substitution towards vegetables at the expense of grain takes place. The consumer tends, within the family food budget, to maximize satisfaction in the quantity, variety and quality of food bought. Long term changes in price relationship have considerable effect on the demand patterns of agricultural products.

Fluctuations associated with changes in income and in technical food producing and handling processes, can in a very short time alter the basic food habits of a people and hence the demand patterns for its agriculture. A general rise in the price of foods may raise the proportion of outlay on food, and may result in less food being consumed. Such a price rise is, however, equivalent to a drop in real income, with a consequent retrogression to simpler foods in accordance with lower levels of purchasing power.

3. Tradition

Although general food habits of populations change slowly, there is a strong relation between food habits and agricultural knowledge and possibilities. If the climate, soil conditions and knowledge enable only rice to be produced, then the basic diet is rice. When new possibilities arise for introducing other crops which would complement the diet, resistance is met because the population is set in the food habits of generations. On the other hand, need or convenience may overcome tradition. There are many examples of this, the most striking being the way in which foods indigenous to the Americas, such as maize, potatoes, cassava, groundnuts and tomatoes have, since the sixteenth century, become important constituents in the diets of peoples in other parts of the world.[24] These foods early entered into the agricultural patterns of even primitive tribal societies. Today wheat is gradually becoming more important in rice growing areas, while vegetables revolutionized the diet of Europe when they were first grown on a large scale. Thus, as new agricultural practice becomes accepted by a traditional society, its products enter the diet. Knowledge sometimes spreads very rapidly, sometimes very slowly, but there is no doubt that the work of agronomists in areas of traditional diets will result in gradual changes in a relatively short space of time, as habit yields to better nutritional possibilities.

4. Climate

Climate is an important factor in determining the amount and types of food consumed. In hot climates, less food is necessary for body maintenance since there is less wastage of heat from the body. In cold climates, on the other hand, more fats and carbohydrates are needed to increase calorie intake and to keep warm. Climate is also important in determining food consumption habits since the range of crops which can be grown in a country is influenced by its climatic features.

The agricultural planner is interested in the demand for farm products during the period of the plan he draws up. This demand may take the form of a derived demand rather than one stemming directly from the consumer. In more advanced countries the farmer sells an increasingly large part of his products to factories, processing plants and buffer stocks, rather than directly to the consumer. The price structure facing him is thus likely to be very different to that which faces the consumer who buys not only the farmer's product but also that of a host of other marketing and processing services. These must be taken into account when translating the total consumer demand for 'food' or 'textiles' into the actual demand for farm produced commodities.

The planner, therefore, in assessing the demand for agricultural products in order to aim at supply planning and control, must take into account both the market and the commercial and institutional framework within which it functions. An important part of his planning procedure is to seek ways of improving this framework to remove uncertainty from agricultural production as far as possible, reducing waste and increasing efficiency. Agriculture, by its very nature, can never approach the certainty of industrial production (except, perhaps in certain livestock products), but the aim must nevertheless be to introduce as much order as is consistent with free enterprise working in a democracy where the interests of all sections have to be preserved and promoted.

D. THE DEMAND SOURCES FOR AGRICULTURAL COMMODITIES

The outlets (or demand sources) for farm commodities fall under the following headings:

(1) Food for subsistence which is consumed by its producers and does not enter commercial channels. (2) Direct food sales to consumers through marketing channels with a shorter or longer period of storage or processing to retain quality during the marketing process, e.g. vegetables, liquid milk, fresh meat, grains. (3) Sales to the food processing industry, e.g. for milling or canning. (4) Sales to the non-food industry, e.g. cotton, potato starch.

Apart from the first, the other three demand sources may be for internal consumption, for export or for both, and the respective demand functions vary within the classification. All four outlets have different significance for the agricultural planner trying to predict future demand for agricultural products or to introduce new crops or systems. The outlets are examined below to determine the particular features with which they present the planner.

1. Subsistence production

In societies which have not advanced into a commercial stage of specialization and exchange, there is little possibility of obtaining food except by growing it directly. Subsistence farming cannot care for the needs of rising urban populations and has nutritional defects as well. Very often land, labour or other limiting factors do not allow the farmer to produce enough food for his own needs by existing methods. Even if enough 'food calories' are produced, the resultant diet may be badly deficient in the quality of its nutrients and therefore inadequate for good health. Products may be seasonal at harvest and liable to serious deterioration in storage, so that the availability of nutrients varies greatly throughout the year, with periods of starvation and of overabundance. The farmer may have to produce certain crops for home consumption under climatic or other conditions totally unsuitable for them, making their production uncertain. In the event of crop failure, an essential food item will be missing. Other problems of subsistence farming arise from the shortage of money or facilities with which to buy goods necessary for healthy living or for more efficient production.

The long-term solution for problems of subsistence farming is to raise farmers above the subsistence level so that they can produce for the market as well as for home consumption. With the proceeds of their sales they can fill out their needs in nutrients and other goods. In the short-to-medium term, the economic pattern has to set home food needs as the first claim on the farm. In such a situation, the aims of the agricultural planner should be to organize the work of the farm family so as to raise the output of food to that necessary for adequate nutrition, to introduce crops or livestock which ensure a balanced diet, to arrange a cropping system which keeps the family in work and food all the year round, and to orient part of the productive effort towards markets for cash crops. The planner sets 'demand targets' to the levels of nutrition he considers necessary and capable of achievement, and only secondarily to price and marketing movements.

The level of nutrition can be expressed in terms of *a food basket per consumer unit per year*. The size of the basket represents the total number of nutrient calories aimed for; the contents are the products of the farm, grown in quantities designed to provide the total amount and quality of food. By comparing existing output with this ideal food basket, dietary imbalance can be seen and careful thought applied to planning the output to meet the food basket requirements. The necessary changes may be simple: introducing a winter crop suitable to the area; planting fruit trees; utilizing crop residues through sheep or goats, or other activities supplementary to the old farm routine by which the missing nutrients can be supplied.

A group of African visitors to Israel was asked which crops grew easily in their home area. The list offered was long and impressive, including several fruits and several staple crops. To the question, "And what do the farmers grow?" the answer was – "maize and beans". In this particular case, a food basket could be made up to meet nutritional requirements over the year, both in quantity and quality, and yet remain within the agricultural possibilities of the farmers. These requirements could then be carried back into the design of a simple farm system which would produce them.

2. Market sales direct to consumers

The demand for agricultural products sold directly to consumers, depends largely on the ways in which both supply and demand are organized and on the institutions controlling them. If farmers bring in their produce to a central market individually and sell there, price is decided simply by the supply and demand in the market. The price at one place may bear little relation to that at another. With economic progress, regular marketing channels become built up, embodying first commercial and later legal controls. Several clear stages are apparent in this progress.

When marketing firms operating in areas where they have a complete or partial monopoly come between producers and consumers, the price the farmer receives has little relation to that at retail, since wholesale and retail margins show little tendency to fluctuate widely with demand. The farmer who sells to a wholesaler is in effect selling in a market essentially different to that of the farmer who sells directly for retail. Moreover, the price he receives differs more than can be justified from the price his produce fetches in the distant urban market because the wholesaler, who is commercially skilled, deals with a number of farmers bidding against one another, while the farmers often do not know the price received for their produce on the retail markets. This system also tells against consumers, since the relatively limited number of wholesalers can fix prices in the urban centres within certain limits. In such a situation, planning has little application. The farmer does not know what price he will receive at harvest and is therefore uncertain about what to plant or in what amounts. He will tend to over-diversify production in the hope of hitting on something profitable. The consumer, on the other hand, cannot obtain the full range of products he desires over the seasons, because the producer cannot know his wants except through the fluctuating prices with which the middleman faces him.

The number of marketing firms tends to diminish as large firms drive the smaller ones out of business. Their larger scale of operations makes considerable economies possible, which are only very slowly passed on to farmers and consumers. The market, however, becomes more regular since the marketing companies try to regularize their supply by contracting with farmers either legally or by custom.

At this stage, bodies representing other interests tend to form in order to oppose the possible exploitation of monopolies. Producers on the one hand

and consumers on the other, form unions or co-operatives which aim, in both cases, to gain some degree of control over production and distribution and product-prices which directly affect their welfare. Government may enter the picture by imposing controls (sometimes through government-supervised marketing boards which are granted a complete monopoly) by providing incentives, by legislation regularizing buying and marketing operations (an alternative to the granting of monopolies) or even by entering the market itself on a large scale and thus radically effecting it.

Each of the two last stages lends itself to planning, whether by private firms, public or private companies or by government. Each can eliminate waste caused by under- or over-production of farm products. The task of planning – sale to the consumer without major processing* – is to assess the wishes of consumers, passing these back to the producers through a suitable marketing framework which matches supply to demand at a price level satisfactory to both producers and consumers. Demand prediction can therefore be of use to agricultural planners only if they have the means for feeding back demand data to producers. This entails examination of the marketing process to ascertain whether it meets these needs and recommending ways of achieving its more efficient working.**

3. Sales of food and processing industries

Sales to the food and non-food processing industries represent a market very different from that of fresh, non-processed products for direct consumption: the factory buys one product and sells another, whether wheat for bread or cotton for fibre. The market for bread determines the demand for flour which in turn determines wheat demand. The farmer, in this and similar cases, is very much insulated from the sale price of the final product, in which the cost of the farm product is only one of many components. This is even more marked in the case of preserved products such as canned vegetables, the price of which tends to be stable in the shops and unaffected by seasonal variations. The market can be assessed through factory-sponsored research and expanded through advertising.

The long and complicated marketing-chain associated with fresh products tends to disappear for processed food. Because of the need of the factories to determine the quantity and quality of their raw materials in advance, and thus to contract directly with farmers for fixed and quality-inspected quanti-

* Pasteurization of milk is not here regarded as 'major processing'.
** These marketing and organizational problems are fully discussed and illustrated in Chapter 6.

ties of produce, harvest prices can often be stated at planting time. Secondly, production is planned well ahead and the demand exerted by the factories for products for processing is known in advance. In the case of large enterprises such as sugar-beet factories, demand is connected with and underpinned by national policy with regard to imports, national interests, etc. The demand for farm products exerted by processing factories has therefore a large element of steadiness, control and planning. In fact, the factory marketing estimates, often more accurate than those which the planner can produce, should be utilized by him.

The demand for certain fresh products can show a similar certainty with full organization of production and distribution. The four demand sources face the planner attempting to estimate overall demand for agricultural products at a future date. The success of his planning is closely related to the accuracy of his estimation of them.

E. PREDICTING THE LONG-TERM DEMAND

The long-term demand for agricultural products has two main components – internal demand and export demand. Each will be examined separately.

1. Internal markets

i. *The food basket*

Long-term trends in food consumption and habits can be expressed in terms of a food basket per capita per annum (see above, p. 49), i.e. all the food consumed in a year by an 'average' person.[25] A study of present and past food baskets reflects any changes in eating patterns. By taking into account the variations between different income groups, an 'average' food basket for the end of the planning period can be estimated at the income levels predicted for that time.

Such a calculation is simple; if well founded, it reveals the relationships between agricultural commodities at the end of the planning period. These can serve to guide the planner in preparing the agrarian structure to meet the eventual demands upon it, give a base for long-term investment policy and indicate the general lines of needed development. Such general lines are a frame in which current year-to-year planning can gradually fill in a detailed pattern according to the actual market demands as they reveal themselves. Corrections are thus made in the long-term plan without destroying it. The plan must in other words be flexible to allow changes in emphasis if reality differs from expectation. The main frame of the plan does, however, remain basically the same, provided that first calculations were founded on realistic analysis and projection.

ii. *Setting food basket standards in Israel for 1972/3*

In establishing a food basket for the Israeli population for the planning period, it is necessary to consider the following factors:

1. Accepted nutritional customs within the population.

2. Supply values, i.e. the number of calories necessary to ensure satisfactory health and efficiency of the population under prevailing climatic conditions.

3. Nutritional values (proteins, fats, etc.) necessary to ensure the health of the population, especially that of the young generation.

With regard to the first factor, nutritional customs differ greatly within the population and between one ethnic group and another. National eating habits are usually more or less 'standard', but with Israel's variegated population there is much variety and it is difficult to find a common denominator as a true basis for weighted calculations. For instance, immigrants from the Middle East, especially from Yemen, are not used to drinking milk, which they use only as a baby food; on the other hand, they use a large amount of vegetable protein (legumes). The opposite is the case with the people of European origin; their eating habits are more complex. Studies are being carried out to determine how length of stay in the country affects the traditional dietary patterns of the country of origin. It would seem that there is an effect, mainly based on income, that as the population matures and the average length of time per immigrant in the country extends, the dietary pattern becomes more uniform. The total effect of the diversity on the total buying habits of the nation is not therefore likely to be great when weighed against the powerful factors tending to break down the influence of habits brought in from the country of origin.

The second factor endeavours to establish the energy values which a suitable food basket should supply. Research has been done in this field by the United Nations[26], which has drawn up tables for the establishment of calorie requirements.

1. From the age of 25 on, calorie requirements decrease by $7\frac{1}{2}\%$ for every 10 years of life.

2. Every 10 degree increase in average temperature above a standard temperature of 10 degrees Centigrade, decreases calorie requirements by 5%.

3. With increase of body weight, calorie requirements increase according to a certain definite ratio.

4. There is a difference in calorie requirements between men and women and between children of different ages.

As a result of a population survey based on age and sex, the Israel Central Bureau of Statistics established in 1951 that the daily calorie intake per person was 2158.[27] However, calculating that nutritional losses amount to 15% through food preparation and wastage, the gross calorie requirement was estimated at a daily average of 2500 per person. This is one of the highest standards in the world. In a second survey made in 1956/57, this calorie intake had slightly increased.*

With regard to the third main factor, nutritional values, which is important in determining the cost of the food basket, the quality of nutrition depends largely on the amount of proteins, especially of animal protein, which is the most expensive nutritional element all over the world. FAO, in conjunction with the World Health Organization and the Macy Foundation[28], has established that 65 grammes of protein are needed daily for a working man of 25; 85 grammes for a youth aged 10 to 12; 100 grammes for a youth aged 16. The average general protein requirement is, therefore, 75–85 grammes per person per day. Of this amount, 25% should be animal protein. These figures cannot be regarded as modest when it is borne in mind that 80% of the world population exists on a much lower nutritional standard. In this respect, Israel falls within the highest 20% of the world's people.

There is thus little ground for complaint in Israel with regard to either quality or quantity of the food basket. Israel, when compared with Italy, a country with a stable, experienced rural population, is in a very good position and succeeds in filling her food basket at a considerably higher nutritional level. Table 2-9 shows a breakdown of the components of food baskets in Mediterranean countries in 1956.

Not only was Israel's food basket nutritionally sound, in respect to total food, but it compared favourably with that of France in quality except, perhaps, in respect to animal protein intake.

Israel's average food basket is therefore adapted to physiological and dietetic needs – though some improvement is necessary in certain sections of the population. Since, however, our calculations are based on the average food basket (total food consumption divided by population) it can thus be regarded as physiologically adequate. Expected improvements in the food basket such as the inclusion of better and more expensive foods instead of

* For instance, the average daily intake in 1958 was spread between a minimum of 1950 calories per person in India, and 3510 per person in Ireland. In the same year the figure for Israel was 2860 calories per person.

TABLE 2-9

Food basket in selected countries (1956)

Product	Calories per person per day (in 1956)[29]				
	Israel	*France*	*Greece*	*Italy*	*Turkey*
Cereals	1353	1018	1452	1373	1900
Potatoes and starch products	93	249	78	93	62
Sugar and honey	325	284	120	183	138
Legumes and nuts	112	50	144	89	124
Vegetables and fruit	185	134	189	154	165
Meat	100	405	81	105	67
Eggs	56	39	20	33	6
Fish	23	19	25	21	5
Milk and milk products	235	269	170	165	74
Fats and oils	394	423	375	361	126
Total calories per day	2880	2920	2650	2580	2670

cheaper products will thus result not from physiological need but rather from the expected rise in the standard of living, which will in turn lead to changes in the demand patterns of part of the population.

iii. *Method of demand projection*

Demand projection is based on an estimate of internal market capacity and of personal expenditure patterns with regard to food and other commodities. The capacity of the internal market is obtained by multiplying the contents of the food basket projected for the end of the period, by the number of the population. If such a food basket can be reliably estimated, the method is very useful for long term planning. The first stage in food basket determination is estimating the average expenditure on food for the end of the planning period. Table 2-10 illustrates the relationship between total expenditure and the proportion spent on food for a number of countries.

TABLE 2-10

National per capita income and the proportion spent on food

Country	1952–1954	Proportion spent on food and tobacco[30]
	Average per capita income[31] in $	
India	60	84
Greece	220	68.6
Holland	500	34.5
United States	1870	32.4

Oppenheimer[32] has classified some countries according to income levels. He points out that in the first group, which includes the less developed countries of Southern Europe, the yearly per capita consumption expenditure is 220 dollars of which 120 dollars or 55% is for food. In the second group belong countries such as Holland, Norway and Western Germany, where the average expenditure per capita per year is 460 dollars and the outlay on food 200 dollars or 43.5%. The third group includes countries of Western Europe such as France, England, Belgium, Denmark, Sweden and Switzerland. In this group, average yearly expenditure is 700 dollars, of which 250 dollars or 36% are for food. The fourth group includes the United States and Canada. There total average yearly expenditure is much more than a thousand dollars, of which 28.5% is spent on food (not including tobacco).

This matter has been well investigated by many research workers for many years and has been found to hold for many countries and for levels of income over periods of time. It has held true in Israel, where a continual reduction in the proportion of personal expenditure used for food purchase, has been observed since the founding of the State.[33] In 1952, the proportion spent on food (not including tobacco) was 37%; in 1957, 34.5%, and in 1960 it was 33.7%. This reduction is a consequence of continually rising incomes throughout the period. In addition to this reduction in the percentage outlay on food, changes also took place in the composition of the food basket. The amount of inferior foods diminished and the quantity of more expensive foods, such as animal products, increased. The nutritional standard of the food basket at all stages was, however, satisfactory.

General expenditure on food in Israel rose (at constant prices) by 34% from 1952 to 1959, while general consumption per person increased by 47%. This brought about the relative decrease in the amount of expenditure on food. Estimates for the future show that economic development over the coming plan period (1960–72) will increase the private consumption per person by 30%.

While there was a general rise in food expenditure of 34% between 1952–1959, the rise in the expenditure on staple food was only 26.4%. Since the food basket already contains a fair proportion of high value foods, even in comparison with many European countries, it is expected that the rise in food expenditure in the next ten years will be only gradual. In the first years of Statehood the outlay on food increased more, proportionally, than the increase in general consumption. In the last few years only a small proportion of income increases has been used for food consumption because the food basket has improved continually, and other items have priority in expenditure patterns. This is shown in Table 2-11.

TABLE 2-11

The increase in consumption and the proportion spent on food [34]
(At constant prices – 1959)

Year	Total consumption/ capita IL.	Food consumption/ capita IL.	% of food in total consumption
1952	945	356	37
1953	933	340	37.5
1954	1076	359	35
1955	1084	376	35.5
1956	1149	404	36
1957	1213	416	34.5
1958	1299	454	35
1959	1392	476	34

The projected increase in disposable income for the years 1963 to 1972/3 is 32%, 40% from 1960 to 1972/3. If these trends continue, then, at the end of the planning period, outlay on food (not including tobacco) should be 29–30% of total private consumption which will reach IL. 2100 at 1960 prices. In 1960, the outlay on food was IL. 486; in 1972/3 it should rise to IL. 600–630 at 1960 prices. This is the estimate of what the 'average consumer' will be willing to spend on food.

The division of expenditure between various commodity items is considered below. Several methods exist to estimate demand for each product.

In Israel, the method normally employed for estimating changes in the food basket over a period, is based mainly on past experience: it is based in other words, on variations in the consumption of various products in accordance with consumers' choice and purchasing power, on fluctuations in the consumption of products and the mutual effect of the consumption of one product on another and finally, on the selective purchasing power of the average wage earner.

All estimates of demand for food in the development programmes drawn up by the Ministry of Agriculture and the Joint Planning Centre have been made according to the above principles, e.g. the seven-year plan [35], the four-year plan and most recently the five-year plan for 1960–1965 [36]. In 'Agriculture and Settlement' [37] the author included food basket projections and estimated the consumption of different foodstuffs by the end of 1968 – the development period there considered. These estimates were also based on past trends, the basic assumption of the author having been that no radical changes in the food basket, in comparison with the then existing consumption patterns would take place, as no rise was expected in the standard of living and in the level of income. The author also showed that from the physiological dietetic point of view, the food basket was satisfactory. At approximately the same time Dr. L. Oppenheimer published a comprehensive study of the food basket as a basis for agricultural planning.[38] Oppenheimer also assumed that the food basket would remain "within the framework of the present level of disposable income for consumption, taking into consideration its adequate nutritional value".

Since as we have seen, Israel's economic development differed from that anticipated, with incomes and standards of living rising rapidly, there was a consequent increase in food consumption.

A research group at the Falk Project for Economic Research led by Y. Mundlak recently determined the demand curves and demand coefficients for a variety of foodstuffs by use of an analytic method.[39] The demand coefficients were determined according to the price of the product and the size of the consumers' income. Products subject to an elastic demand were those for which the demand fluctuated to a great extent with a change in the level of income of the population. Inelastic products (for which the demand was more stable) were less affected by these two variables. The demand coefficients were determined by collecting data relating to a sample number of families and analysing it in a particular way.

The accuracy and reliability of any demand analysis depends largely on the quality and scale of the data on which it is based. Owing to the general rationing of food in Israel until 1954, no useful statistics could be accumulated in respect to previous years. For certain products, rationing was lifted only after 1954. Statistical treatment over a short period, by the use of monthly observations, is only possible with commodities which are subject to monthly or seasonal variations in prices and quantities. Hence no monthly analysis is feasible for commodities such as wheat and wheat products, oils, cotton, tobacco, etc., the consumption of which is stable throughout the year. Work on these commodities is therefore still in its early stages, or is based on studies carried out in other countries with conditions similar to those prevailing in Israel. For such products as fruit, vegetables and milk, seasonal data may be used.

First results of the analysis of demand for agricultural products appeared in the Fifth Report of the Falk Project for Economic Research. The final report was published in 1964; it covers most of the commodities included in the food basket.[40]

The food basket proposed here for the end of the period under consideration uses all available material on this matter, i.e. past trends in consumption, the estimated food basket for 1965 as given by the Joint Planning Centre (Tables 2-12 and 2-13) and the report of the Falk Project for Economic Research.

Later on, in 1964, a new development plan for the years 1965–1970 was prepared by the Joint Agricultural Planning Centre.[41]

The estimated food basket for 1970 in this new Development Plan took into account the latest trends in food consumption in Israel, and the assumption that per capita consumption would rise at an average rate of 3% per annum. We should bear in mind that our assumptions are based on an increase in consumption of 2.5% per annum only.

Let us now compare the various estimates for the food basket with the past and present consumption of the various foodstuffs.

Projection techniques include careful analysis of the individual components of the food basket. For our purposes it is sufficient to consider two commodities (vegetables and animal proteins), while detailed discussion of the other components can be found in the author's 'Agriculture And Rural Development in Israel: Projection and Planning'[42] which concludes with a projection of the food basket cost.

Vegetables and potatoes: Vegetable consumption in Israel has shown only

TABLE 2-12

Projected food basket for 1965 according to the Joint Planning Centre[41]

Commodity	Kg per person	Remarks
Cereals	140	
Vegetables and potatoes	160	
Sugar	26	not including jams and sweets
Oils	17	
Fresh fruit	130	not including canned fruit and olives
Pulses and groundnuts	7	
Beef and mutton	12.7	equivalent to 15 kg dead-weight or 30 kg live-weight
Poultry	18	
Cow milk and milk products	107	This was subsequently increased to 120 litres
Sheep and goat milk and their products	124	
Total milk	31	
Fish	14	
Eggs	350 eggs	

TABLE 2-13

Actual and estimated food baskets

Commodity	Food basket in 1952[a]	Food basket in 1956[b]	Food basket in 1960[c]	Estimated food basket for 1960 in seven-year plan[d]	Estimated food basket for 1970/71 in new 1965–1970 development plan[e]
Cereals (flour)	146.7	141.0	114.5	150	105.7
Fresh vegetables	111.2	117.2	114.6	120	110.0
Potatoes and starches	46.6	43.8	36.8	60	37.4
Sugar	19.7	25.7	29.1	24	36.7
Oils	16.0	18.0	18.3	14	18.7
Fruit (including melons)	124.8	115.3	131.0	108	122.1
Pulses and nuts	6.8	8.1	7.1	10	5.6
Beef (live weight)	13.2	19.3	24.5 ⎱	9 ⎰ carcass ⎱	40
Poultry	4.1	11.9	19.9 ⎰	⎱ weight ⎰	24
Milk[f]	93.6	106.0	134.7	143	138.8
Fish	15.5	11.4	9.9	15	10.8
Eggs	220	258	343	200	380

[a] Source – *Statistical Abstracts of Israel* No. 5 (1954), pp. 78–80, published by the Central Bureau of Statistics.
[b] Source – *Statistical Abstracts of Israel* No. 8 (1956), pp. 86–89, published by the Central Bureau of Statistics.
[c] Source – *Report on Agriculture submitted to the Knesset by the Minister of Agriculture,* pp. 161–164, published by the Ministry of Agriculture (February, 1961).
[d] From: *Proposed Development Plan for Agriculture and Settlement 1953/4* (1959/1960).
[e] *1965–1970 Agricultural Development Plan* (The Joint Agricultural Planning Centre, 1964) (mimograph in Hebrew).
[f] Quantities for 1952 and 1956 in absolute weight and not in milk equivalents; for 1960 – in milk equivalents.

slight fluctuations, both upwards and downwards, during the past ten years. These fluctuations seem to have occurred in cycles, due to the fact that following surplus years, cultivation has generally been restricted. These fluctuations which are slight do not indicate any radical changes in consumption patterns.

The annual fluctuations in vegetable consumption are given in Table 2-14.

The above statistics tend to confirm the vegetable consumption trends forecast by the Planning Centre in February 1960, with which we concur, – that under present conditions the per capita consumption will remain steady at 115 kg per year.

The supply of vegetables is of course not stable throughout the year. Taking the marketing figures for tomatoes in 1961 as an example, as com-

TABLE 2-14
Yearly fluctuations in vegetable consumption[43]

Year	kg per person
1950	111.1
1951	103.1
1952	112.5
1953	121.0
1954	113.7
1955	114.4
1956	117.2
1958	122.8
1959	114.6
1960	114.6

pared with a record quantity of 9474 tons sold in August (not including surpluses) and of 5800 tons each, in June, July and September, a much lower quantity was sold in winter: 3580 tons in February, 3570 tons in March, 1800 tons in April and 2800 tons in May.[44] The situation in other years was similar. In 1959 the quantities sold in winter (February–May) ranged between 2000 and 2500 tons per month, while in summer they averaged about 6000 tons per month. Although the demand for vegetables in general was greater in summer, in view of the heat, the low consumption of tomatoes in winter was not due to any equivalent decrease in demand, but to the fall in supply.

As will be seen below, our plan envisages a considerable extension of the areas for export vegetables (especially tomatoes) which would be grown during winter in particularly well-suited parts of the country; however, the percentage of exportable vegetables (especially with regard to tomatoes) is not great. This means that if cultivation is extended to particularly suitable new areas where vegetables can ripen in winter, the winter supply on the local market will likewise increase. According to our estimates, the output of special export tomatoes for instance, will be 60 000 tons, of which 15 000 tons will be exported, the remainder marketed locally either to consumers or to the canning industry. This entire harvest will be in winter and early spring, a season when tomatoes are, at present, in very short supply. The per capita consumption will thus be increased by about 5–10 kg per year, the exact amount of the increase depending on the proportion of exportable tomatoes in the crop. All this will, of course, apply only if new areas will be developed.

Consumption of *potatoes and starches* has remained fairly stable during the past few years, except for 1953 when a serious shortage occurred. Annual per capita potato consumption from 1950 to 1959 was as shown in Table 2-15.

TABLE 2-15
Annual per capita potato consumption in Israel[45]

Year	kg per person per year
1950	46.3
1951	41.8
1952	46.6
1953	31.3
1954	38.6
1955	43.5
1956	43.8
1957	46.1
1958	41.3
1959	38.5

Although in its development plan for 1960–1965, the Joint Planning Centre assumed that by 1965 per capita consumption of potatoes would be 45 kg per year, this estimate was high, conflicting with the trend of a drop in consumption of potatoes. The forecasts of the Falk Project, on the other hand, assumed that consumption of potatoes would not rise, while the new development plan of the Joint Planning Centre assumed only a slight rise from 1965–1970. We have therefore assumed that potato consumption will not exceed 40 kg at the end of the planning period.

Our estimate of total annual consumption of vegetables and potatoes per capita is expected to be approximately 165 kg, whereas that of the Joint Agricultural Planning Centre is 147.4 kg and that of the Falk Project 143–146 kg. Since the last two estimates do not take into account the vegetables cultivated in particular seasons in the newly opened Bsor region, they are lower than ours, otherwise they would be closer to it. In any case, the difference between the three estimates is not of great significance since ours deals with an 8–10 year projection, and the other two with estimates referring to periods of five years.

Animal proteins: The level of consumption of any animal protein (other than milk) is partly dependent on the consumption level of other animal proteins. It has been found, for example, that in Israel the consumption of fish increased during periods of meat shortage and declined with an increase in the supply of poultry or beef. We shall therefore first examine the total expenditure on various types of animal proteins (apart from milk) both in the past and as estimated for the future.

TABLE 2-16

Per capita expenditure on proteins at fixed 1959 prices, and compared with the total food budget [46]

Commodity	1952	1954	1956	1958	1959
Beef and mutton	3.8	7.3	18.9	23.0	38.0
Poultry	12.2	14.8	28.1	39.1	42.8
Fresh fish	10.2	11.4	8.9	11.4	11.7
Eggs	19.3	24.0	30.5	35.3	34.8
Total expenditures on animal proteins	45.5	57.5	86.4	108.8	127.3
Total food expenditure	356	359	404	454	476
Percentage of protein	12.7	16	21.1	23.9	27.1

From Table 2-16 it appears that the expenditure on animal proteins (apart from milk) has increased almost threefold; while in 1952 it accounted for only 12.7% of the total expenditure on food, it accounted for 27% of total expenditure in 1959. The quantities are comparable, and in some cases even higher, than the protein consumption of European countries. We are therefore of the opinion that no great change will take place during the period under review in the percentage of total expenditure on food spent on proteins. Mr. Y. Taub in his article 'Changes in Consumption' states: "It appears that in the general group referred to as foodstuffs, two opposing trends may be noted: both a relative and an absolute decline in consumption of the groups of foodstuffs which are generally associated with a low standard of living, such as cereals, and a staggering rise in the consumption of animal proteins, especially of meat. There is no doubt that this process will come to an end within a couple of years, when the decline in expenditure on food will become steeper". In our opinion this saturation point has already been reached for certain animal proteins.

We shall now examine the different kinds of animal protein foods in greater detail, and the changes likely to occur in their consumption.

a) *Eggs:* The rate of consumption of eggs in Israel is one of the highest in the world; since 1950, a steep increase in per capita egg consumption has occurred. The present rate is 380 eggs per year per person as opposed to 220 in 1952. This over 50% rate of increase has now levelled off. During the three years 1961–1964 egg consumption remained practically unchanged, despite reductions in prices intended to raise consumption. As a result, the Planning Centre anticipates little change in the per capita rate in the future, and in its latest Five Year Plan (1965–1970) estimates demand at 380 eggs.

This estimate is somewhat different from that of the Falk Project, which in its Interim Report no. 2 assumes that egg consumption will rise to 390 per person in 1965.

In our view, both the general stability in the rate of consumption during the past few years despite a fall in price, and the relatively high rate of consumption, give reasonable grounds for assuming that there will not be a marked increase in the future, and that even a small decrease may be expected. We have accordingly estimated per capita consumption of eggs at the end of the forecast period at 350.

b) *Fish:* Following on the great increase in meat consumption (i.e. from 1957 onwards), consumption of fish has been fairly steady. Previously, the great shortage of all kinds of meat and the low price of imported fish had brought about a temporary rise in the level of fish consumption. Thus in 1951 – the year of the greatest meat shortage – consumption of fish per capita was 18.6 kg while by 1956 the per capita consumption had dropped to 13.4 kg and has levelled off in recent years at 10 kg per person.

The Falk Project does not anticipate any marked increase in fish consumption in the future, nor does the Planning Centre. Developments in the past few years tend to confirm that there will not be any substantial rise in fish consumption. We have therefore assumed that per capita consumption of fish will be 12 kg per annum. In point of fact, the level of agricultural production will not be affected by local consumption because a considerable part of the fish is imported from abroad.

c) *Meat (poultry, mutton and beef):* As we have stated, a considerable rise in meat consumption has taken place in the past few years. This applies to poultry, mutton and beef.

Table 2-17 illustrates the changes in meat consumption during the last decade (beef and mutton per weight of carcass; poultry, after removal of feathers but prior to cleaning).

The greatest increase occurred in the consumption of poultry, although beef and mutton also showed substantial gains. Poultry consumption rose from 5.1 kg in 1951 to 18.4 kg in 1960, and to 28 kg in 1964, while other meat increased from 9.8 kg in 1951 to 18.9 kg (dead weight) in 1964.

The Joint Planning Centre in its Development Plan for 1960–1965 esti-

TABLE 2-17

Per capita meat consumption in Israel[47]

Year	Kg per person per year
1950	18.6
1951	14.9
1952	10.7
1953	10.5
1954	11.2
1955	18.3
1956	21.6
1957	22.0
1958	25.9
1959	30.5
1960	32.2
1961	34.7
1962	40.8
1963	43.1
1964	46.9

mated the per capita beef and mutton consumption for 1965 at 12.1 kg edible meat, i.e. 15 kg dead weight or 30 kg live weight. In its 1965–1970 plan the estimate for 1970 was 40 kg live weight. As the rate of consumption of poultry is already high, we do not anticipate any further increase by the end of the development period, but on the contrary, a slight decrease may be expected because of the rise in beef consumption. Accordingly, we have estimated that beef and mutton consumption at the end of the forecast period will be 18 kg per person dead weight (36 kg live weight), while poultry consumption will be 24 kg per capita.

d) *Milk and milk products:* A distinction has to be made between two kinds of milk: cow milk and sheep and goat milk, the latter being used mainly for the farmer's own consumption and for cheese production.

No radical changes in consumption of cow milk have taken place during recent years. The quantity of milk for drinking has hardly changed, varying only between 65 and 70 litres per person. The consumption of other forms of milk products has increased somewhat. At present the total consumption of cow milk for drinking and processing amounts to about 120 litres per person. The basic change that has occurred in the past few years is that whereas powdered milk was at one stage widely used, particularly in processed dairy products, fresh locally produced milk is used now exclusively. The Planning Centre originally estimated cow milk consumption at no more than

107 litres. In its subsequent plan for the dairy branch, however, milk consumption was estimated at 120 litres.

In its 1960–1965 Plan the Planning Centre estimated that per capita goat and sheep milk consumption would be 24 litres. This is an artificial figure in that certain farmers who keep sheep and goats drink only their milk and consequently consume much greater quantities than 24 litres. On the other hand, the urban population consumes considerably less. For the purposes of our calculations we have assumed an average for the entire population.

According to the above estimates of the Planning Centre for 1960–1965, the total per capita milk consumption would be 144 litres. For 1965–1970 the estimate was 138.8 kg. On the other hand, the Falk Project assumes a rise in milk consumption to 164 litres per capita. In the light of the stability of milk consumption up to 1964, an increase of such a magnitude in the period under review cannot be assumed. We have therefore assumed that consumption of all types of milk at the end of the development period will be 145 litres per capita. Additional milk products in the form of imported cheeses might be consumed, but these would not affect the local dairy industry.

In conclusion it should be stated that the food basket on which we have based our plan should not be regarded as completely authoritative. However, any deviations from our estimates will not be such as to affect the size of an agricultural branch, either because they will be relatively minor, or because the commodities in which changes may occur, are mainly or partly imported.

We shall now examine the cost of this food basket and the proportion of foodstuffs in total per capita consumption, examining whether this corresponds to our previous estimates of the sum that the average wage and salary earner in Israel will be able and willing to spend on food (see Table 2-18).

From the table we see that total expenditure on food is in the vicinity of IL. 620, constituting 30% of total per capita consumption. These figures correspond to the amount which, according to our estimates, the average family is expected to be able and willing to spend on food. According to the projected food basket, real expenditure on food will increase by 27% between 1960 and the end of the development period, while total per capita consumption will increase by 40% over 1960. These figures show that the rise in food consumption will be more modest than it has been in the past, as predicted by Mr. J. Taub. This moderate increase is due to the fact that the Israeli food basket is gradually reaching its saturation point.

TABLE 2-18

Cost of food basket in 1960 and of the estimated food basket at the end of the
development period
(1960 Prices not including tobacco)

Commodity	1960		1972/3	
	Quantity kg.	Expenditure IL.	Quantity kg.	Expenditure IL.
Cereals and rice (incl. products in the form of flour)	120	64.4	120	64.5
Potatoes and vegetables	151.5	55.7	165	90.0
Sugar and honey	29.5	14.9	35	17.0
Oils and Fats*	18.1	23.2	18	26.0
Fruit (excluding wine grapes)	130.0	60.5	147.5	65.0
Pulses and groundnuts	7.1	4.5	7	4.5
Beef, mutton and other meat (dead weight)	12.0	44.1	18	66.0
Poultry	19.9	48.5	24	58.0
Milk and milk products	134.7	51.3	145	61.0
Fish	9.9	18.4	12	22.5
Eggs	343	32.5	350	33.5
Miscellaneous (tea, coffee, spices, choice preserves, etc.)**		39.0		65.0
Total		457.2		573.0
Light and alcoholic drink		29.0		48.0
Total food and drink consumption		468.2		621.0
Total per capita consumption		1440		2100
Percentage of per capita consumption spent of food		33.7		30

* It is assumed that a rise will take place in the consumption of expensive fats, especially
butter, while generally consumption of oils and fats will decline.
** Miscellaneous includes mainly luxury items.

2. External markets

i. *The future of Israel's agricultural exports*

To date, Israel's agricultural exports have consisted mainly of citrus fruit.
Citrus exports account for 85% of Israel's agricultural exports and 42% of
its total exports, being the main earner of foreign currency. The production
costs of citrus include a relatively small foreign currency investment.

Although exports of other agricultural products have been small, this by
no means indicates lack of opportunities. Israel's natural conditions of soil
and climate, especially in the Jordan Valley and the Western Negev, have
definite advantages for many agricultural export products. Although there

are seasons of the year during which it is impossible to compete with countries such as Spain, Sicily, North Africa, the Canary Islands and others, Israel in other seasons is virtually the only country able to offer certain fresh agricultural produce to the world market.

In West European countries, fresh vegetables are produced in summer, the climatic conditions making it virtually impossible to produce them in the open air during the winter months. The result has been that a substantial proportion of the West European population is accustomed to a much lower consumption of fresh vegetables in the winter than in summer. The last two decades have seen far-reaching changes in the marketing and transport of fresh products. The expansion and improvement of international trade and transport facilities are rapidly breaking down the concept of a 'closed economy', while the tendency to build economies based on free trade among countries is on the increase. Consequently every country is trying to produce agricultural products in which they enjoy a comparative advantage and increasing the import of those products that cannot be profitably produced locally. West European countries, for example, have greatly increased the import of vegetables for winter consumption from Southern Europe and the Canary Islands. As the economies of African and Asian states develop, trading will increase and foodstuffs that cannot be profitably produced locally will be imported from abroad. Such policies are bound to have an effect on the eating habits of their peoples.

In the early years of this century Jewish agriculture in Palestine aimed at self-sufficiency in all essential foods. Farms were planned to produce for home consumption and only later was the emphasis shifted to the production of citrus for an export market. With the increase in the number of farms and their rising production, specialized farming is on the increase, while it is essential to aim for export if Israeli agriculture is to be economical. Agricultural exports can undoubtedly be increased considerably during the coming decade – but not without changes effecting the entire agricultural economy. The fact that Israel produce is of high quality is not in itself enough to ensure large export. The basic principles in developing export trade in agricultural produce, are specialization and planning for export production, with a vigorous and reliable sales organization.

Production must be encouraged in those branches which are Israel 'specialities', and for which there is specific demand in European or other markets; advantage must be taken of season, production and high quality produce. This needs a special programme for exports, co-ordinated with general agricultural planning, which will lead to specialization for export production on the basis of the regions which are best suited for the different products. The planning must carefully work towards this end, and be made to realize its aims through a strong export policy. If the nation hopes to obtain a foothold in foreign markets, other activities must be developed around the planning of agriculture: adequate mechanization in grading, suitable shipping, cold storage and packing facilities and the requisite sales organization.

The shipping of seasonal produce requires an adequately equipped fleet with facilities for refrigeration and ventilation according to the commodity. Accurate co-ordination with local and international shipping companies is essential. The success of any export trade depends largely upon export auxiliaries; unless these are developed, no amount of sound agricultural planning will avail.

The two main factors in developing Israel's export trade in fresh agricultural produce are therefore *specialization* and *planning for export*. Export planning depends upon the following elements:

1) Market research to determine the demand in terms of quantity and quality in the importing countries. Two examples are given later to illustrate how this factor has been used to determine policy.

2) Correct timing of marketing and shipping. The export of table grapes can serve to illustrate the importance of this factor. Table grapes have a ready market in Western Europe, especially Britain and Germany. During most of the summer Israel faces strong competition from Italy and Spain. During the short one-and-a-half month period of June–July, Israel is virtually the only country in a position to offer table grapes on these markets. Any delay in delivery means that Israel's grapes arrive later than August and profitability is greatly reduced.

3) Export of varieties acceptable to overseas markets: The British market, for example, requires a potato of the 'Scraper' type, i.e. early white potatoes with a thin skin; on the other hand Northern European countries require a yellow variety of potato.

The supply of suitable varieties for specific overseas market demands, ensures higher prices and better profits. Information based on regular market surveys is essential in order to clarify exactly what the demands are.

4) The export of stable qualities. A successful export trade cannot be based on periodic surpluses. It is vital for success to acquaint the consumer with Israel's products. The export trade must to a large extent be based on firm contracts.

5) High quality production: The stiff competition on the world market

can partially be overcome by guaranteeing top quality produce. Quality production must be the foremost aim of the extension service in providing instruction in agro-technical and marketing methods.

6) The reduction of export expenses: By increasing the scale of export production, grading, packing and shipping expenses can be reduced. This is especially the case if the production of an export product is concentrated in a specific region, where it is possible to erect specially constructed grading and inspection centres which assure high quality consignments.

Two examples of policy making on the basis of market research in the importing country, that of citrus and tomato export, are presented below:

ii. *Citrus exports*

Most of Israel's citrus is exported to Western Europe and especially Britain; exports to East European countries are limited. Exports to Western Europe reached a peak in 1938–1939, when 15 million cases* were shipped. With the outbreak of World War II, exports of citrus fruit ceased almost completely, and were resumed only after the war. Since the establishment of the State, export has increased from 4 million cases to 9 million cases in 1958–1959, and to more than 10 million cases in 1959–1960. In spite of the ravages of war, the annual per capita demand for citrus fruit in Western Europe has increased by 60%, although the average annual per capita consumption is only one third of a case, which amounts to 11.5 kg (9.5 kg net oranges, 1.5 kg of other citrus products, and 0.5 kg of various citrus juices). By way of contrast, the consumption of citrus products in the United States has reached an average of 35–40 kg per capita annually.

In Northern Europe, demand varies greatly. Sweden for example requires 13.5 kg per capita annually, while Finland only 5.2 kg. Generally speaking, demand for citrus in Scandinavian countries has increased over the past year. Citrus fruit consumption in West Germany is similar to that of other countries in Western Europe.

To date, Britain has purchased 40% of Israel's citrus produce and products with other West European countries taking up 15%. Exports to Eastern Europe and especially Russia, important as they may be in the future, depend on political factors which are largely beyond our control.

Canada is another important potential market. In 1938, 91 000 tons of oranges were exported to Canada, and in 1954, 197 000 tons. Grapefruit

---

* 25 cases per metric ton.

exports increased from 20000 tons in 1938 to 63000 tons today. In 1957 and 1958, Canada took up considerable quantities of Israel citrus.

There are two opposed schools of thought on the question of the future of the Israel citrus industry. One sees little justification for expanding plantings for the estimated potential of the European markets, and predicts a fall in prices of 25–30%. The other holds that the markets in Eastern Europe and Asia have been very much underestimated: these, it feels, will become important citrus buyers and thus reduce the downward pressure on prices, leaving the traditional markets for Israeli fruit well able to absorb all production of existing and newly planted groves. Forecasts have not yet been made as far ahead as 1972, but there are projections for 1965 which show future trends in the world's main citrus markets.

FAO[48] has published a paper on the prospects for production and export of citrus fruit from Mediterranean countries which quotes figures of great relevance to Israel: all available information – according to the report – shows an increase in plantings in the Mediterranean region and South Africa which will lead to a 55% increase in output in 1965 over the 1955–1957 average. Continued plantings, the rise in average yields and the fruiting of many young orchards will lead to further steep rises in output after 1965. A conservative estimate puts total 1965 output at 5.5 million tons, which will yield 3.1 million tons for export as against 1.9 in 1955/57 – an increase of over 60%. Countries outside this region, particularly the U.S.A., have also greatly expanded plantings and the amount for export will certainly reach 3.7 million tons in total in the world as a whole. These countries will constitute strong competition, particularly in the processing field for juices and other products.

Examining consumption in the main Mediterranean citrus markets, the report points out that world consumption of citrus products has increased 200% over the last thirty years, and world trade by 100% or more. Canada, Scandinavia and Western Europe were in the forefront of the rise. In Western Europe, the average consumption per head is 7–14 kg, much less than in the U.S.A. with 40 kg per head of fresh fruit equivalent. However, the figure is considerably more than before the war: the large increase was a result of the rise in population and in incomes. The income elasticity in Western Europe is 0.8, which means a rise in expenditure on citrus products of 8% for every 10% rise in total income. This is higher than for most other foods. Consumption is very sensitive to price, since citrus is still regarded as a luxury product. Price elasticity is a little under 1, but the absolute price is important in deciding the response to price changes. Demand for citrus is also greatly influenced by the supply of other fruits on the market and their relative prices. Marketing practice and advertising also influence sales.

The report estimates a total increase in the citrus market in Western Europe of 20–30% over the average of 1955/57 – if prices do not change. Estimating income rises at a minimum, demand should rise from 2.3 million ton average of 1955/57 by 20% to 2.8 million tons in 1965. With a more optimistic rise in incomes, the demand may rise to 2.9 million tons, an increase of nearly 30% over past years.

If the demand and supply estimates are close to the truth, those countries exporting citrus to Western Europe face a serious situation. Supply estimates from the Mediterranean area and South Africa forecast a rise of 50–60%, while the demand rising from income and population increases in Western Europe will be of the order of 20–30%. In addition to this supply, other countries of the Western Hemisphere are likely to enter the European market in one way or another. This difference between supply and demand is likely to make prices drop 25–30% below those of 1955/57. This fall in prices, will lead to severe economic problems in the citrus industry of those countries dependent on the West European market, which are generally not able to pay subsidies to their growers and may find their balance of payments position badly affected.

One suggested way of expanding the market is the manufacture of citrus juices and other products; another is expanding the market in Eastern Europe and possibly Asia. Advertising and health education are thought to be other means of increasing overall demands for citrus products. The report concludes by stressing the importance of co-operation between the producing countries in supplying information and working out common policy and advertising in the face of common problems. The importance of the citrus policy of the producing countries vis-à-vis the importing countries is also stressed.

Dr. Levi of the Economic Planning Department of the Ministry of Agriculture made a forecast of the prospects of Israeli citrus for 1965 and came to similar conclusions[49] to those of the FAO report. He shows that the output of pre-1957 plantings for export in 1965 will be more than twice that of 1957 (600000–675000 tons against 302400). A comparism of these figures to those for the whole Mediterranean region and South Africa given in the FAO report, shows that Israel's share of total citrus exports to Europe will rise from 15.9% to 21.8%. Levi, on the basis of research carried out by him on the British and other markets, says that income elasticity is tending to drop for fresh fruit and to rise for fruit juices. He forecasts that prices will have to be reduced by 25-30% of those of 1955/57, if the 100% increase in the amount of Israeli fruit on the market in 1965 is to be sold.

He contrasts this conclusion with the FAO report which forecasts this reduction for a 50% increase of the total regional crop. He therefore sees Israeli fruit as being able to hold its own to a considerable extent in spite of the severe drop in prices by sustaining the same price drop as other countries but with double the relative increase in output.

He maintains, however, that any increase over the projected 600000–675000 tons will bring a more severe drop in prices which will render all the marginal groves unprofitable and reduce the country's total earnings of foreign currency. Plantings over and above those needed to renew the old orchards should be very restricted, as the cost in terms of land, water, labour and capital will not be economical. Better uses could be found for these scarce resources.

The optimism inherent in Levi's view is supported by the experience of the most recent export seasons. The quantity exported in the winter of 1959/60 reached 400000 tons which fetched 33 shillings a case in England – a price similar to that achieved in previous seasons. The average price of fruit exported to Europe was $ 130 per ton f.o.b. In the 1962/63 season, exports rose to nearly 500000 tons. Far from dropping, prices in fact rose markedly, reaching an average $ 140 per ton f.o.b. One of the reasons for the rise was frost damage to the Spanish crop, but market surveys in Europe nonetheless showed that demand for the Israel fruit exceeded supply and from the point of view of European markets there was therefore a possibility of exporting larger quantities without depressing prices. In 1963/4 and 1964/5 seasons prices dropped somewhat, being on the average around $ 120 per ton f.o.b.

In 'Citrus in Israel' (1963), J. Horin takes an optimistic view, pointing out that the present citrus area in Israel, some 360000 dunams, can be considerably increased without danger of a fall in income. Though he also mentions the possibility of a 20% drop in prices, this applies only to quantities which greatly exceed those mentioned by Dr. L. Levi. In Horin's view, the potential of the European citrus market is large and only partially exploited.

In a report published in 1963 by Y. Pat, Director of the Ministry of Agriculture Citrus Department, on the citrus juice industry in the U.S.A., he recommends that Israel begin preparations for the manufacture of citrus juice for export (to date, Israel's citrus export has been almost entirely confined to fresh fruit). Processed citrus may thus become a significant export item.

On the basis of the various export researches and recent experience, we believe that export can be expanded considerably and even doubled in 10 years, reaching 1–1.2 million tons of fruit (fresh and processed) by 1972/3, although prices are expected to fall by some 20%. Over a longer period, we feel that exports can be even further expanded.

iii. *Tomato exports*

The export of tomatoes from Israel has potentialities which can be examined through market surveys and the determination of production costs.

Though the main market for fresh tomatoes is Great Britain, other West European countries import large quantities of the fruit, with Britain, Western and Northern Europe together absorbing some 600000 tons of imported tomatoes annually; one third of this total goes to Britain. Britain's tomato imports have averaged $ 70 million annually over the last few years, while citrus imports have been around $ 65 million, which gives a measure of the former's importance. Two thirds of the total tomato consumption is imported as shown by Table 2-19.

TABLE 2-19

British local production and import of tomatoes

Year	Local tomato production (tons)	Import of tomatoes (tons)	% local production of the total consumption
1954–1955	104000	195000	34.8
1955–1956	112000	183000	38
1956–1957	96000	202000	32
1957–1958	99000	202000	32.9

Britain imports tomatoes throughout the year, while other European countries tend to do so only when prices are low, mainly in the summer months: for example, 80% of Germany's imports are in June, July and August. These facts hint at the great possibilities for Israel tomato exports to Europe and especially Britain. As a result, a research team headed by Dr. N. Kedar was established at the National and University Institute of Agriculture to study the problem of tomato export in all its aspects.[50]

The team examined export prospects, both in regard to the foreign market potential and likely returns to the farmer. It considered the problem of the low percentage of fruit fit for export (which has until now been the principal obstacle to export), dealing with ways and means of increasing this percentage, such as controlling diseases and other pests which cause it. An economic investigation, carried out by S. Ben David of the Department of

Agricultural Economics of the Faculty of Agriculture, has shown that the export season may be divided into the months of October–November, when the f.o.b. price is lower, and into December–May, when the price is higher.

With regard to costs of sorting and packing, the actual expenditure for 1960 was assessed by the Agricultural Export Company at IL. 465 per ton. However, the calculation showed that a proportion of exportable tomatoes of no more than 15% (as against 5% at present) would reduce these costs to IL. 282 per ton. It was pointed out that with the expected rise in the pro-

TABLE 2-20

Expected prices of tomatoes in England
(According to prices of tomatoes imported from the Canary Islands)

	Wholesale price in dollars [a]	c.i.f. price in dollars [b]	f.o.b. price in dollars [c]	f.o.b. price in IL. [d]
October	354.2	258.4	208.4	624.2
November	285.6	221.3	181.3	543.9
December	371.3	287.8	257.8	773.4
January	403.2	312.5	282.5	847.5
February	425.3	329.6	299.6	898.8
March	491.7	381.1	331.1	993.3
April	528.9	409.9	359.9	1079.7
May	523.0	354.6	304.6	913.8
June	411.6	258.1	208.1	624.8

[a] The wholesale price constitutes an average of 1958–60 prices. The monthly price for each year is a weighted average calculated on the basis of weekly quantities.
[b] The c.i.f. price was obtained after deduction of customs, commissions and transportation costs from port to warehouses.
[c] The f.o.b. price in dollars is after deduction of shipping expenses.
[d] The f.o.b. price in IL. is according to a uniform rate of exchange of IL. 3 = $1 as assumed for dollars exchanged and of IL. 3 for added dollars. Hence the IL. price here is somewhat higher than in Mr. Ben David's calculations.

portion of exportable produce, sorting and packing expenses, depreciation, and interest charges on capital investment in equipment and packing installations, could be reduced to IL. 350–400 per ton. The farmer would thus be left with a return of IL. 350–400. Since this estimate was made before the devaluation, the present expected returns should be raised to IL. 400–450 per ton.

The economic study concluded that in view of foreign market requirements and the prices obtainable, exports should be carried out between January and May rather than in autumn. However, the main prerequisite for penetrating and gaining foreign markets would be an increase in the exportable percentage of tomatoes. The low exportable percentage was due to hollowness and irregularity of the fruit, lack of uniformity in colour and overripeness.

Investigation has shown that by adequate organization, overripeness may be avoided. In order to eliminate irregularity and hollowness of the fruit, a suitable strain producing the desired kind of tomato had to be found. Lack of uniformity in colour seemed to be due to virus diseases which could be overcome. It was further shown that transportation and sorting were incorrectly carried out, so that the tomatoes suffered frequent 'knocks'. The general conclusions of the investigation were that as far as foreign markets are concerned, tomatoes could well be exported during the winter months, when a ripe crop could be obtained, particularly from the Jordan Valley and the Western Valley or B'sor Region.

The main direction of most of the experiments carried out so far followed the conclusions of the investigation. Many strains of tomatoes were examined, apart from the one hitherto regarded as particularly suited for export purposes, the 'Moneymaker'. Experiments, which were begun in 1961, were extended in scope, the results serving as a general indication rather than as final conclusions. Of the strains tested[51], 'Potentate' was found to yield tomatoes of the desired form, i.e. round tomatoes, similar to the Moneymaker. Potentate tomatoes were sufficiently uniform in shape and, above all, were less hollow – the defect which had prevented the export of the Moneymaker. Both in total yield and in percentage of exportable fruit, this strain improved on the results obtained with the Moneymaker strain.

TABLE 2-21

Comparison between 'Potentate' and 'Moneymaker'
Cultivated in Mivtahim Settlement in Winter 1961/62

Property	Potentate	Moneymaker
Relative yield	9.3 tons	6.7 tons
Proportion of hollow fruit in February	8%	8%
Proportion of hollow fruit in March	6%	38%
Proportion of exportable fruit in February	63%	31%
Proportion of exportable fruit in March	52%	28%

Not only the percentage of hollow fruit of the Potentate was low, but the *degree* of hollowness was so much less that it may not affect exports.

Estimates of the Agricultural Attaché in London show that provided

Israel exports arrive at the right time (winter and early spring) and are of the desired quality, thousands of tons of tomatoes may be exported without difficulty.[52] Our estimate that 15000 tons of tomatoes will be exported by 1972 is dependent upon their being grown in suitable areas of Israel where tomatoes mature at the right season, and also of course on the cultivation of a suitable strain which yields a high proportion of exportable fruit. Exports on a much larger scale would have been considered if the areas fit for such cultivation were not limited.

The work carried out in connection with the export of citrus and tomatoes has been described by way of illustration. Our estimates regarding other exports are similarly based on the various studies and surveys which have been carried out.[53] Despite the high degree of uncertainty inherent in export forecasts, a reasonable estimate can nonetheless be drawn up without exceeding the normal range of error; details of figures at the end of the development period will be given in Chapter 5.

In order to ensure the projected increase of exports there is definite need for a specially organized body to deal with agricultural exports, more or less along the lines of the Israel Citrus Board in order to co-ordinate planning implementation. There is already a company for the export of fresh agricultural produce, excepting citrus, peanuts and cotton, which was formed by the Jewish Agency, the Ministry of Agriculture and the local marketing agencies. It handles orders, contacts growers, and deals with the handling and shipping of produce. Another institution is the export department of the Ministry of Agriculture, which issues export licences and to a lesser extent deals with professional and business aspects of exporting.

To attain maximum efficiency, an overall body must be formed to co-ordinate the promotion of agricultural exports and maintain contact with professional, organizational, business and economic bodies. Such a body would determine and implement policy, organize research teams to determine which crops are suitable for export, investigate which varieties are best for export, and solve the many packaging and transport problems involved. A professional advisory service and a Bureau of Standards would be an integral part of the proposed organization; credit and subsidies for export products would be controlled by it. This body would deal with the business aspects of the trade, excluding citrus exports, set up contact offices in the importing countries to check on standards and product presentation, and to promote advertising.

There is undoubtedly a future for the export of a wide variety of dried, canned and processed agricultural products from Israel which is only now being explored through market surveys. Estimates have been included in the commodity targets which form the basis of the ten-year plan in Chapter 5, but apart from certain specific products which have been carefully investigated, there is little information on which they can be reliably based. In the near future, projects will be initiated to find out more about market requirements in order to provide suitable factory capacity and to make allowance in agricultural planning for the requisite production.

3. Demand variables in long-term production planning

There is a large difference between the techniques of assessing demand for short-term planning from year to year, and for long-term planning from over five to ten years. In the short term, experience of the previous two or three years, indicates whether supply was below or above demand as reflected in prices and their fluctuations. Short-term changes in demand are not likely to be large unless a new factor appears (new industry, for example, or more factories). The requirements of such factors are usually known, and short-term planning is therefore an effort to bring supply better into line with demand, which is expected to be similar in overall proportions to the previous year or two. Market statistics of prices, amounts of commodities produced and sold, the geographical pattern of demand and supply for them, and seasonal shortages or abundance, aid in planning for a better supply pattern for the next year or season.

Long-term planning must on the other hand, be based on overall trends which can be determined only by careful analysis. Apart from major factors such as size of population and income which greatly influence future demand, changes in consumer preferences must be expected, and as far as possible predicted. Such changes are greatly influenced by income, but the wider choice of products offered by improving agricultural techniques may also lead to changes in eating habits as price relationships between products alter and enable substitution of preferred products for some staples.

Moreover, long-term planning must work in terms both of overall increase or decrease in 'all food commodities' consumed and try to predict future demand relationships between particular products. On this point, evidence from consumer surveys conducted in various parts of the world indicates probable changes in food habits as income rises or new products become available. Thus an increase in real income, together with a more developed agriculture and marketing system, tend to lead to higher consumption of livestock products, fruit and vegetables per head, at the expense of starchy staple foods. Similarly, increasing urbanization leads to a shift in demand from raw agricultural products to factory processed foods and higher consumption of luxury products from specialized farms, as well as imports of food commodities from abroad for confectionery, special fruits, etc. At the same time the consumer is less content with what the market offers. He demands new products and a constant supply all the year round, a factor which has a strong effect on specialization of supply areas within a country and competition from imported products.

These trends can be predicted with a fair degree of certainty; they do not have to be of a very high degree of accuracy since agriculture is flexible regarding substitution between products. Farmers can shift production out of one commodity into another in a season, and often do so in response to price changes. Where this shift is not controlled, but is due to each individual farmer's assessment of his market in response to the previous season's price structure, wild fluctuations in output with the well known price cycles result. But if planning is carefully used to inform farmers of future market demands, this flexibility can be of great advantage in correcting mistakes in long-term planning.

The data obtained through forecasting must, however, be sufficiently accurate in its predictions of long-term trends in the demand pattern for agricultural products, to be able to guide planning in evolving a basic structure of production which will meet the demands. A mistake in assessing the demand of wheat against barley, or one vegetable against another is insignificant; but if the relationships between the future demand for livestock products as a whole are misinterpreted, and an arable farm economy built up to meet a market which demands a high proportion of livestock products, the results will be extremely serious.

Long term planning must therefore adapt the production income structure of farming and its regional specialization patterns to the demand picture forecast for the end of the projection period; it is this overall demand pattern which must be determined as accurately as possible for the main branches of farm products.

NOTES

1. SCHULTZ, T. W.: *The Economic Organisation of Agriculture* (McGraw Hill Book Co. Inc., New York 1953), pp. 32–33.
2. SCHULTZ, T. W.: Op. cit., p. 36.
3. *Statistical Abstracts of Israel* (1959/60), p. 7; (1962), p. 31; (1965), p. 20.
4. *Statistical Abstracts of Israel* (1959/60), pp. 69, 86; (1962), pp. 97, 111; (1965), pp. 95, 107.
5. See also: SICRON, M.: *Immigration to Israel 1948–1953*, Statistical Supplement (Jerusalem 1957).

6. *Statistical Abstracts of Israel* No. 16 (1965), p. 46.

7. *Statistical Abstracts of Israel* (1959/60), p. 32; (1962), p. 62; (1965), pp. 55–57.

8. WEITZ, R.: *Agriculture and Settlement* (Am-O ved Ltd., Tel-Aviv 1958), p. 39 (in Hebrew).

9. The Falk Project for Economic Research in Israel: Fifth Report 1959 and 1960 (Jerusalem 1962), p. 52.

10. Prime Minister's Office, Economic Planning Authority: 'Proposed Directives for a Five Year Development Plan for the National Economy 1965/6–1969/70' (Jerusalem, November 1963).

11. [1] GAATHON, A. L.: 'Capital Stock Employment and Output in Israel: 1950–1959' *Bank of Israel, Special studies*, No. 1 (1961), Appendix Table C-1.
    [2] *Statistical Abstracts of Israel* (1962), (1963), (1965).

12. HOVNEH, A.: The Falk Project for Economic Research in Israel: Fifth Report 1959 and 1960 (Jerusalem 1962), p. 110.

13. GAATHON, A. L.: Op. cit., Appendix Table B-1.

14. Bank of Israel, Research Department: 'Forecast of Developments in the Israel Economy, 1960–1965', Working Paper No. 1, (Jan. 1959), (mimograph).

15. Bank of Israel Research Department: Working Paper No. 2 (September 1959), (mimograph).

16. Economic Planning Authority: 'Proposed Directives for a Five Year Development Plan 1965/6–1966/9' (Jerusalem, October 1964), p. A 10 (in Hebrew).

17. PATINKIN, D.: 'The Israel Economy: The First Decade', The Falk Project for Economic Research in Israel, Fourth Report 1957 and 1958 (Jerusalem 1959), Chapter 5.

18. WEITZ, R.: *Agriculture and Settlement* (Am-Oved Ltd., Tel-Aviv 1958), p. 60.

19. WEITZ, R.: *Towards Specialized Farming*, Agricultural Publications Section; The Joint Agricultural Extension Centre (Tel-Aviv, January 1960), p. 10.

20. HALEVY, N.: The Falk Project for Economic Research in Israel: Fifth Report 1959 and 1960 (Jerusalem 1962), p. 152.

21. Proposed Directives for a Five Year Development Plan', op. cit., p. 10.

22. Estimates of the Economic Planning Authority and the Ministry of Finance (unpublished).

23. Bank of Israel – Annual Report 1963 (Jerusalem, May 1964), p. 158 (in Hebrew).

24. FAO: *'The State of Food and Agriculture 1957'* (Rome 1957), p. 71.

25. FAO: 'Dietary Surveys, Their Technique and Interpretation', *FAO Nutritional Studies*, No. 4 (Rome 1953).

26. Food and Agriculture Organization of the United Nations: 'Caloric Requirements', Report of the Second Committee on Caloric Requirements (Rome 1957), p. 68.

27. LEVAVI, Y.: *Food Balance Sheet of Israel*, Central Bureau of Statistics (1956), (mimographed in Hebrew).

28. 'Human Protein Requirements and Their Fulfillment in Practice', Proceedings of a Conference in Princeton, USA, 1955. Sponsored by The Food and Agriculture Organization (FAO), World Health Organization (WHO) and Macy Foundation (New York 1957).

29. Sources: FAO Yearbooks of Agricultural Statistics, Production: 1957, 1958, 1959, Rome, Italy.

30. NITZAN, S.: 'Living Standards in Israel', *Economic Quarterly* No. 13–14 (Tel-Aviv 1960), p. 71 (in Hebrew).

31. 'Per Capita National Product of Fifty-Five Countries 1952–1954', *Statistical Papers* No. 4, Series E.

32. OPPENHEIMER, L.: 'Food Basket Estimates, as a Basis for Agricultural Planning in Israel', Ministry of Agriculture, Tel-Aviv (January, 1958), p. 51 (mimograph in Hebrew).

33. TAUB, Y.: 'Changes in Consumption', *Economic Quarterly* No. 28 (Tel-Aviv 1960), p. 386.
34. TAUB, J.: Op. cit., p. 386.
35. 'Proposed Plan for the Development of Agriculture and Settlement for 1953/4–1960', Joint Planning Centre, Tel-Aviv (October 1953), Table 10 (in Hebrew).
36. 'Agricultural Development Plan for 1960–1965', Joint Planning Centre (Tel-Aviv, February 1960), (mimograph in Hebrew).
37. WEITZ, R.: *Agriculture and Settlement*, op. cit.
38. OPPENHEIMER, L.: Op. cit.
39. The Falk Project for Economic Research: Fifth Report 1959 and 1960, pp. 189–196.
40. The Falk Project for Economic Research.
41. The Joint Agricultural Planning Centre: 'Agricultural Development Plan 1960–1965', (February 1960), p. 20 (mimograph in Hebrew).
42. WEITZ, R. with the assistance of ROKACH, A.: 'Agriculture and Rural Development in Israel: Projection and Planning', The National and University Institute of Agriculture, Division of Publications, Bulletin No. 68, Rehovot, (February 1963).
43. Source: For 1950–1958 – *Statistical Abstracts of Israel* No. 10 (1959), p. 128; for 1959 – *Statistical Abstracts of Israel* No. 11 (1960), p. 132; for 1960 – 'Report on Agriculture submitted to the Knesset by the Minister of Agriculture' (February 1962), (in Hebrew).
44. 'Report on Agriculture submitted to the Knesset by the Minister of Agriculture' (February 1962), p. 105 (mimographed in Hebrew).
45. *Statistical Abstracts of Israel* No. 10, p. 128, p. 132; Report on Agriculture submitted to the Knesset by the Minister of Agriculture (February 1961).
46. TAUB, Y.: 'Changes in Consumption', pp. 386–387.
47. *Statistical Abstract of Israel* No. 10, p. 128; No. 11, p. 132.
48. FAO: *Monthly Bulletin of Agricultural Economics and Statistics*, 9, No. 4 (April 1959).
49. LEVI, E. L.: *Forecast for Citrus for 1965*, Ministry of Agriculture, Department of Economic Planning (Tel-Aviv 1959), (mimograph in Hebrew).
50. 'Investigation on Export Tomatoes 1960/61', Preliminary Reports, National and University Institute of Agriculture, No. 357, 360, 361, 362, 363, 364 and 368 (Rehovot, January 1962).
51. KEDAR, N., FALBETZ, D., RATIG, N., and FELDMAN, S.: 'Preliminary Conclusions of Experiments on Export Tomatoes', Winter 1961/62. National and University Institute of Agriculture, Internal Publication (July 1962).
52. [1] PARAN, M.: 'Export of Fresh Agricultural Produce from Israel to West European Countries', Israel Embassy, London (January 1957).
    [2] PARAN, M.: 'Import of Tomatoes to England', Israel Embassy, London (July 1956).
    [3] PARAN, M.: 'Export Prospects for Fresh Agricultural Produce from Israel', Israel Embassy, London (1959).
    [4] PARAN, M.: Reports and Internal Reviews, transmitted to the Ministry of Agriculture (not printed).
53. [1] WEITZ, R.: *Agriculture and Settlement*, pp. 69–98 (in Hebrew).
    [2] WEITZ, R.: *Towards Specialized Farming*, op. cit., p. 21.

CHAPTER 3

MEETING THE TARGETS: THE SUPPLY FACTORS

A. THE IMPORTANCE OF KNOWING THE RESOURCES

Resources – physical, technological-organizational, and human – are the raw material of production. This chapter examines all three in the context of agricultural planning on a national or 'macro' level. As in the previous chapter, Israel's ten-year agricultural plan is the basis for the examination.

The importance of planning lies in its ability to indicate a rational course of action for the attainment of a desired end; it chooses the input combination most likely to yield the desired output at the lowest cost. Planning is therefore based on knowledge of a) the desired end; b) the resources available; and c) the potential of different resource combinations, i.e. input-output relationships. The more perfect the knowledge of these three factors, the closer will be the actual result to that planned.

The end is formulated in relation to resources and various input-output potentials. Essentially, the resources available in their separate categories and in possible useful combinations determine how high the planning aims can be set. The quality of resources is no less important than their quantity since their quality determines their production potential. Any country wanting to get the most from its national effort must know what is at its disposal by building up an inventory of potential production resources. Priorities for development are based on such an inventory so as to achieve the highest utilization of both current and long-term productive energy.

1. Survey

The first step in setting up an 'inventory' of resources is a survey which sets down on paper and in maps, in general terms, the overall physical structure of the country and an evaluation of population in terms of size, geographical distribution, health standards and education. Such crude information points up areas and resources which would repay more detailed investigation and also limits the field of operation to certain projects which seem most worthwhile. If the surveys show abundant natural resources, while the farming population is undernourished and unproductive and capital resources almost non-existent, the first priority will be proper feeding of the population through agricultural development. The rest of the surveying effort should then be applied to investigating agricultural land and its potential with a view to finding the areas best suited for development and the techniques that will be necessary to make use of them. Once such areas have been decided upon, detailed development can begin: the target setting, the resource combinations and the implementation of the plan. Such an approach restricts national effort to immediately practical aims that are capable of realization.

However, it is dangerous to decide priorities without an overall survey: effort not applied at the most responsive points, and unintegrated piece-meal activity without regard for overall possibilities, are similarly wasteful. For instance, charting river courses may show places which could immediately and cheaply benefit from cheap irrigation, while spot decisions may lead to the settlement of sections of land inferior to others from the point of view of climate, irrigation, fishing or access. Development must first have a base upon which to build, planned in respect to geographical location and institutional framework. Particular development schemes must from their very beginning be seen as part of a larger scheme to be mapped out in overall terms. Future generations, building on the foundations laid by their predecessors will alter and finally achieve the goal of the general scheme. The national investment needed for survey work is minute in comparison with the advantages to be gained and the effort and money saved through the comprehensive planning which it makes possible.

2. The general physical features of Israel

Surveys of Israel had been conducted at various times and for various purposes in the past, but to supply data for the planning needs of the new State, the earlier surveys had to be supplemented by additional survey work in the first years of statehood. These surveys indicated Israel's general features: The country has a total area of 20700 sq. km. More than half of the population of two and a half million is concentrated in the densely settled coastal region. The southern region, or Negev, which includes more than half of the total area, is as yet sparsely populated, and is land which requires irrigation for production.

Half of Israel's western boundary is the Mediterranean, the remainder the Sinai Desert, with a small outlet to the Red Sea at the southernmost point of Eilat. Its eastern boundary roughly follows the Jordan Rift Valley running along the eastern banks of the now drained Lake Hule, the Sea of Galilee and the Dead Sea, through the center of the Arava Valley to Eilat. The Jordan Valley is part of a major rift extending from northern Syria, through the Red Sea, to the lakes of Central Africa. Part of the Jordan Valley has the unique distinction of containing bodies of water below sea level. In the north Israel is bounded by the mountains of Lebanon.

Longitudinally, Israel is divided into *three distinct strips*:

a) the area of low land extending along the coast in varying widths from the Lebanese to the Egyptian border, including sand dunes and several streams and rivers, both permanent and seasonal, and extending inland to the Jezre'el Valley and the northern Negev.

b) the mountain ranges which run from north to south, broken between Lower Galilee and the mountains of Samaria by the fertile Jezre'el Valley. The mountains of Samaria and Judea form one continuous mass, prolonged by the Carmel spur to Haifa. The range on which Jerusalem is built, falls gradually towards Ber-sheba, where it begins to rise again in a south-westerly direction to join the mountains beyond the Egyptian border.

An unusual feature of the Negev are the makhteshim* – deep anticlinal depressions surrounded by almost vertical rock walls.

c) the Jordan Rift Valley described above which lies between the mountain ranges of Israel and the Jordan Plateau.

This topography makes for well defined *natural regions:*

a) the northern region reaches from the Lebanese and Syrian borders to Mount Carmel and the Jordan border. It covers 17% of the country, and includes one third of the land suitable for irrigation and 47% of its potential utilizable water resources. The Jordan River contributes well over half of these water resources. The sewage treatment plant at Haifa and planned reservoirs for storing storm run-off are other irrigation sources of the near future.

b) the central region, small but highly developed, extends from Mount Carmel to the Yarkon River, and includes a number of important centres. It covers 7% of Israel's territory, contains 16% of land suited for irrigation and 19% of water resources. The water supply is obtained from streams and the percolation of rainwater into underground reservoirs, plus certain return flows from irrigation and percolation pits. In the future, storm run-off will be dammed.

c) the southern region extends from the Yarkon River to Eilat. It is bordered by Egypt and Jordan, covers 76% of the country's area, possesses

---

* Hebrew 'makhtesh' – literally, a crucible.

50% of land suited for irrigation but only 34% of the currently available water resources.

The Sh'fela (Coastal Plain) is the most densely populated part of the State. Its northern section contains the main orange-growing area and the bulk of urban and industrial development, including many large towns and the biggest city, Tel Aviv-Jaffa. In the Jerusalem corridor, vine and fruit growing and afforestation are prominent and successful. The agricultural area from Tel Aviv to Ber-sheba is developing rapidly, along with the towns of Ashkelon and Kiriat Gat, the new port of Ashdod, the Heletz and Kochav oil fields, and the new Lakhish region. In the northern Negev the city of Ber-sheba and part of the Ber-sheba plateau are undergoing rapid development, as is a strip of the arid Dead Sea area.

The southern Negev is a sparsely inhabited desert containing only the port of Eilat on the Red Sea and the mineral works at S'dom, Oron, and Timna. It is largely arid, difficult to irrigate, and peopled by nomadic Bedouin tribes. Nonetheless, there are a few agricultural settlements including Eilot, Yotvata, Ein-Yahav and others.

Israel's climate is the outcome of its geography and topography. It is typified by a clear separation of the year into the rainy season of winter and the dry summer season. Proximity to deserts involves a rapid diminution of precipitation from north to south and from west to east, with a relatively heavy dew, particularly near the coast. Within the geographical confines, topography produces an extremely varied pattern of climatic conditions of great importance to agriculture. Generally speaking, rainfall decreases with the distance from the sea, increases with height, is higher on the windward or westward side of the hills, and decreases from north to south. There are four *principal climatic zones*, though differences exist within each zone.

a) *The main coastal zone*, including the Jezre'el Valley, which has an annual rainfall varying from 600 mm to 150 mm (decreasing from north to south).

b) *The mountain zone*, which has up to a maximum of 1000 mm of rainfall annually in the north, lessening rapidly towards the Negev.

c) *The rift valley zone*, which has extreme differences in temperature and a rainfall varying from 400 mm in the north to 50 mm in the Dead Sea area.

d) *The Negev*, with rainfall decreasing from north to south and from west to east (from 250 mm in the north-west to some 30–40 mm at Eilat).

Dew is an additional form of precipitation, but though it is considered beneficial to certain summer crops it has little effect on the country's water balance.

Temperature generally increases from north to south and from high to low altitudes; humidity increases with higher temperature near the coast and declines with increased temperature further inland.

Israel has certain regions in which climatic features give agriculture a comparative advantage. For instance, in the Jordan Valley (Tiberias district) the moderate winter climate and the ample cheap water resources, make this area the most suitable in the country for banana production; 50% of the local crop is grown there. A limited but profitable table grape export market in Europe in May and June is partly based on the Jordan Valley, while it can export grapes even earlier than May, as conditions are especially favourable. In the western Negev, the high day temperatures and the absence of frost at night with favourable soil conditions, make possible the early export of vegetables. Similar conditions exist in the Arava. Thus, even in a small country such as Israel, climatic differences are of great importance for agriculture.

In most of the country, *water* is in short supply and rainfall not distributed evenly. Agriculture must depend on utilization of rivers and underground reservoirs for irrigation, which are insufficient to permit irrigation of all suitable land.

3. Conclusions from the survey

On the basis of the survey work carried out in the new State, certain conclusions were drawn:

a) Water is a limiting factor in agricultural development and should be carefully conserved.

b) The available water, largely concentrated in the north of the country, should be brought south to irrigate as much as possible of the potentially fertile land.

c) To enable a decision about the most suitable areas to be irrigated, an intensive soil survey should be conducted, but only as far south as the 60th co-ordinate below Ber-sheba.

d) The Negev south of Ber-sheba would have to develop as an industrial area on the basis of its mineral wealth, fed mainly by the agricultural production of areas further north. Only in certain limited areas of the Arava, where local water-sources exist would it be possible in the future to develop an agriculture based on off-season export products and partial supply of agricultural produce to nearby urban settlements such as Eilat.

A scheme for a general plan was next worked out, using comprehensive regional planning, with an interplay of agriculture, industry and services to determine the location of new agricultural areas, towns and the development of the ports of Ashdod and Eilat. Development was planned in stages so that the human and capital resources of the country could be used to best account; population (particularly new immigrants) was transferred to new development areas, and financial policy was designed to take agricultural and industrial settlement progressively further south.

As the time came to carry out a particular objective of the main plan, the necessary detailed surveying, planning and implementation were done. The results can be seen in Israel today. A continually advancing development front is penetrating well into the underpopulated south, with surveying preceding detailed planning. Construction of the new agricultural or urban settlement areas follows the overall lines of the planning. As surveying reveals additional sources of natural gas, oil or minerals, the planning aims are raised, or the time for their execution brought forward. Within the main planning framework, there is sufficient flexibility to allow unforeseen difficulties to be overcome and unexpected resources to be integrated.

4. Land survey

The techniques of carrying out land survey, coding the information and transferring it to maps, is outside the scope of this book. A survey of Israel was carried out over a period of four years, and the results published in 1955. Since the purpose of the survey was primarily to aid agricultural planners, the area covered was restricted to that which, at the time, could conceivably be brought under irrigation. Mapping scales to show greater or lesser detail according to the stages of the work and the purpose in hand, were 1:20000, 1:40000, 1:50000. Aerial photographs were extensively used to aid ground observations. The following summary is quoted from the report.[1]

Summary of soil survey of Israel

The agricultural mapping of the soils of Israel has been completed by the Soil Conservation Service of the Ministry of Agriculture. The total area mapped is 9 514 380 dunams*, covering the country from Metulla in the North to the 60th co-ordinate south of Ber-sheba.

Purpose of the mapping:

---

* 1 dunam = 1000 m² = $\frac{1}{10}$ hectare.

1) To provide an inventory of the soils of Israel.

2) To classify the special characteristics of each area.

3) To determine the proper use of the land according to its capability.

The country was divided into 21 Problem Areas according to the conditions of soil, topography and climate.

Seven Land Use Capability Classes were used:

Classes I–III – Land suitable for cultivation with proper soil conservation measures.

Class IV – Land unsuitable for annual tilled crops, but suitable for plantations, pasture and perennial crops.

Class V – Land suitable for pasture but not for cultivation.

Class VI – Land suitable for afforestation only.

Class VII – Land not suitable for any agricultural purpose (includes the badlands and sand dunes of the Negev on the southern boundary of the survey).

The Area mapped includes:

	Dunams
Area classified in L.U.C. Classes	8 763 170
Existing Fish Ponds	34 180
Built-up areas	447 590
Rivers, *Wadis*, Lakes, Reservoirs	227 390
Ruins, Mounds*, Antiquities	26 560
Not surveyed	15 490
Total	9 514 380

In defining the Land Use Capability Classes, we found it necessary to treat separately the Land Use Classification under dry conditions and that under irrigation. Arid conditions limit the use of land which may have all other favourable conditions for cultivation. When such areas are put under irrigation, they fall into classes which permit full cultivation, as low rainfall is no longer a limiting factor. On the other hand there are areas suitable for irrigation which, because of their topography, elevation, size, location or distance from sources of water, cannot be put under irrigation. In the light of these limitations four priority classes have been determined for the development of land suited for irrigation:

The following is a summary of Land Use Capabilities (in dunams):

* Archaeological sites.

Under dryland (non-irrigated) conditions:

		Dunams
Classes I–III – Suitable for cultivation of all crops		3395550
Class IV – Suitable for plantations and perennial crops		697540
Class V – Suitable for pasture: included in this area are:		
a) Land which after reclamation can be put into classes II–IV (area: 241720 dunams)		
b) Land in the Northern Negev which, if put under irrigation, will fall into classes II–IV (area: 1122930 dunams)		3317800
Class VI – Suitable for afforestation		882500
Class VII – Badlands and Negev sand dunes		204680
Coastal sand dunes		265100
Fish ponds		34180
	Total	8797350

Under irrigated conditions*:

		Dunams
Classes I–III – Suitable for cultivation of all crops		3940990
Class IV – Suitable for plantations and perennial crops		1339010
Class IV D – Suitable for agricultural use after reclamation		265100
	Total	5545100

The following priorities were established for land irrigation:

	Dunams	
Priority A	1885700	
Priority B	1556660	3442360
Priority C	1238220	
Priority D	864520	2102740
	Total	5545100

Lands under priorities C and D should not be considered for irrigation until those under A and B have been fully developed. It is doubtful whether the lands in priorities C and D can economically be put under irrigation so long as there is not sufficient water supply for that purpose.

Concluding notes

a) The numbers given refer to the gross physical area.

b) To obtain the area net, allowance must be made for roads, waterways and other losses which amount to about 10% of the gross area in each class.

c) Built-up areas and ruins totalling 474150 dunams were not included in the agricultural area.

* Note: the 5.5 million dunams in this classification constitute that part of the 8.8 million dunams of arable dryland *which can be irrigated*.

d) The Land Use Capability Class will not restrict, for example, the use of Class I–III land for plantations or afforestation but will bar annual crops from Class IV land, or the cultivation and planting of fruit trees from Class V land. Class VI land is restricted to afforestation only.

e) The allocation of each area to its most suitable Land Use category should form the basis of planning for agricultural development. The map can help in the future planning of sites for agricultural settlements and towns. This material will be of special importance to the Government and the National Institutions in their national planning and development programme.

This soil survey does indeed form the basis of land use policy in Israel and was of great help in zoning for different types of agricultural development. A scale of 1:20000 is not, however, sufficient for detailed settlement planning. It is sufficient to decide the location of a village or settlement region, after which more detailed surveying on a map of scale 1:5000 must be completed to determine the particular land use patterns of each village – houses, irrigated land, pasture, roads, waste, etc.

5. Land reclamation

In the Land Use Capability classification of the soil survey, the figure of 241720 dunams is given in Class V as capable, after reclamation, of being worked according to Classes II–IV. Since the establishment of the State, the Jewish National Fund has extended its land reclamation work particularly in the hill regions. This was necessary both to increase the area of available agricultural land and to enable settlement in vitally important strategic areas. Before the State, the J.N.F. had reclaimed only a few hundred dunams each year, a rate which increased to several thousand dunams with independence, and enabled a rapid increase in the rate of settlement.

Land reclamation works in the hills are shown in Table 3-1.

The programme of land reclamation will continue, perhaps on an even greater scale. Because of the experience gained during the years and the improvements in mechanization, the J.N.F. has succeeded in reclaiming large areas which previously could only be used for pasture and afforestation and were classified in the Survey as classes V and VI. Now they are capable of intensive farming and have eased the shortage of agricultural land for the hill farms built largely for security purposes. The scale of the work was possible because it was undertaken by a national body which financed and executed the work, while the expense involved represents a net addition to the capital of the country which will bear fruit over many years. The economic

TABLE 3-1

Hill land reclamation

Date	Area reclaimed (dunams)
1953	20000
1954	18000
1955	23000
1956	23000
1957	35000
1958	35000
1959	35000
1960	35000
1961	35000
1962	25000
1963	25000

test of whether or not to reclaim additional areas in the hills depends on whether a system of farming can be found for these problematic regions.

B. AGRICULTURAL LANDS AND POPULATION

1. Population-land ratios

The amount of agricultural land worked in Israel is 0.19 hectare per head. This figure is one of the lowest in the world today, as shown in Table 3-2.

TABLE 3-2

Area of agricultural land per head for a number of countries[2]
(mid-1950's)

Country	Total land area		Total agricultural area		% of land worked of total
	Total in hectares	Per head (hectares)	Total in hectares	Per head (hectares)	
Holland	3316	0.31	1044*	0.1	31
England	24100	0.47	7099	0.13	29
Israel	2024	1.00	390	0.19	19
Italy	29402	0.6	15756	0.3	54
Greece	13156	1.3	3515	0.35	19
Cyprus	924	1.7	434	0.8	46
Turkey	76749	3.1	22543	0.9	29
U.S.A.	770440	4.7	188309	1.1	24

* Not including the intensively worked natural pasture.

If the agricultural population alone is compared with the agricultural area, the ratio for Israel is again low.

TABLE 3-3

Land-population ratio

Country	Agricultural land per head of agricultural population (hectares)
Egypt	0.2
India	0.5
Israel	0.9
Turkey	1.0
England	2.4
U.S.A.	6.7

The above figures do not compare quality of agricultural land or the intensity with which it is worked (in terms of capital investment or technique), so that care is necessary in their interpretation. However, the order of size does indicate the relative shortage of agricultural land in Israel and points to the necessity of carefully conserving it in the future – not essentially because land is in short supply, but because water is only available to irrigate part of it and arable land is therefore limited in amount.

If agricultural incomes are to keep up with the general rise in incomes, the amount of land per head of the agricultural population cannot be allowed to decrease appreciably. It may have to increase, since the present system of farming over most of the land is already very intensive in terms of capital investment. Thus the absolute number of people working in agriculture cannot increase by any great number, and a drive must take place for efficiency and nationalization of farm size where farms are too small for adequate family income.

As the total population increases, the amount of agricultural land per head will correspondingly decrease with consequent changes in the structure of agriculture and its production.

The present export-import pattern for agricultural commodities will probably also change in the long term, with Israel agriculture specializing even more than at present, on high value products for the home and export market, with staple foods and perhaps fibre supplies imported. These matters of specialization and trends in Israel agriculture are further dealt with in the coming chapters.

Population density as an index to the importance of land must be qualified by many factors, both by those connected with the land itself such as labour and capital, or by those which are an alternative to it as a productive source, such as trade, or perhaps fishing.

2. The changing economic importance of agricultural land

Schultz, in a chapter entitled 'The Declining Economic Importance of Agricultural Land'[3], shows that in the United States and the United Kingdom, the relative importance of land as an input factor in agricultural production is declining, and that the proportion of the food dollar paid as a return for the services of land, is very small. Capital, techniques and organization have so improved the output of agriculture per unit of land, while the consumption of non-agricultural services has so increased, that land is not an economic variable of any decisive importance for these countries. In contrast to this situation is the position in high food drain communities. Schultz estimates that for them, if 75% of income is used for food, and $33\frac{1}{3}$% of the cost of food is for the net rent for land used in farming, one fourth of income, at factor cost, is spent for the services of agricultural land. For the U.S.A. at its present stage of economic development, approximately 12% of disposable income is expended for farm products that enter into food, while 20% of the cost of producing farm products, is net rent. Under these circumstances only 2.4% of the income of the community is spent for the food producing services of land. Hence, while for the high food drain communities land is a most important factor in economic life, the low food drain communities have emancipated themselves from great dependence on it.

In Israel it is certain that shortage of arable land will not be a factor preventing the increase of population. Trade patterns will emerge with other countries which will ensure the supply of food in return for industrial products or services. At the same time, agricultural inputs other than land will be used to raise agricultural efficiency and output, to the extent that the role of land will become relatively less important in the cost of supplying the nation with its foodstuffs. Land remains the physical base of farming, but it is the economic use to which it is put, the inputs which are used in conjunction with it, and the type of product produced from it, rather than the land itself, which determine the economic viability of agriculture. Together with trade patterns, these factors determine the dependence of the whole community on agricultural land.

With regard to other countries in the pre-industrial phase (high food drain communities), the degree of dependence on agricultural land in the future, depends on the extent to which they can apply the tools of agricultural science to raising the productivity of land and total output. These tools are extremely effective if they find the right organization and application.

C. CLIMATE: ITS IMPORTANCE, ESTIMATION, AND CONTROL

The two main determinants of climate are rainfall and temperature. Their joint effect which is responsible for the pattern of flora and fauna over the world has great influence on agriculture. Their importance is modified by the reliability of their performance over the year and from year to year. Agriculture is tied to the growing-cycles and the needs of plants and animals, which are usually known for each particular species. The farmer, in obeying the seasonal laws which their requirements impose on him, may be completely frustrated by the failure of climate to follow any similar pattern. The uncertainty of rainfall – both in quantity and time – and of temperature, make farming a gamble with the elements. The farmer has only the beginnings of meteorological science to help him guess correctly; unprotected against the vagaries of climatic behaviour, he may find his fortunes doubled one year by favourable weather, and decimated the next by drought, flood, frost or wind. Similarly, he can never know what diseases and pests even a favourable climate may bring to rob him of unusually successful plantings. Soil quality may be important in deciding fertility but, fertile or not, a farmer knows what to expect from his soil. With climate, the range of expectation is so wide that complete uncertainty can prevail, ranging from expectation of near starvation to abundance. In one county in the U.S.A., a certain wheat area averaged an annual net income of $ 1008 per 640-acre farm between 1912 and 1934. The total net income for the period was $ 21167, of which $ 20472 was for the single year 1920, a year of high yields and exceptionally high prices. The average annual net income not including 1920 was less than $ 35![4]

The evidence overwhelmingly shows climate to be an unreliable accessory to other factors of production. For most areas of the world where agriculture is practised, climate shows certain main rhythms of the seasons, but the actual timing and duration of specific behaviour of the seasons is largely unpredictable in those aspects which most affect the farmer. Because of its variability, climate makes a profound difference between agricultural and industrial production. It tends to play down the importance of the initiative of each farmer as against industrial management, where known labour operations can be split and allocated.

1. The systematic prediction of climate

Despite climatic unreliability, systematic study enables the main characteristics over the years to be described in terms of the amount and distribution of rainfall, temperature fluctuations and extremes over the seasons. Once these factors are measured and known, an 'average' climatic pattern is discernible within which agriculture can be planned with more certainty than that possible for the traditional peasant farmer whose planting and harvest seasons are largely guided by religious and customary laws, based on ancient experience of climatic behaviour. The introduction of new crops is only successful if the species and varieties introduced are suited to the prevailing climate. In order to choose new crops and integrate them into the farm, the climatic conditions must be studied and allowed for.

The study of climate is based on empirical observations and recordings of rainfall, temperature, wind velocity and direction, day length, etc., throughout the year at a number of points over a country. This involves setting up stations to collect and record the information and a centre for documenting and analysing it. In order to have any real significance, such studies must be conducted over a number of years because climate, apart from yearly variations, may show variations within longer periods. The most characteristic climatic pattern emerges out of these records both as to long term and yearly variations – the time distribution of climatic features and their geographic distribution. The range of variation for temperature, rainfall, etc., is plotted from year to year, and the climatic basis for agricultural ventures is gradually made clearer.

Unfortunately, climatic measurements are not available for all countries or for all the regions within countries which want to improve their agriculture. Sometimes climatic data from other regions of similar situation and topography give a basis, but can be only a very unsatisfactory substitute for actual local data, the collection of which should always be started at the earliest possible moment. Over the years, the statistical coverage will become more adequate to the task of describing general climatic conditions and particular area variations within the overall picture.

2. Climate control

With the advance of the science of meteorology, interesting and important work is being done on the macro-control of climate, such as seeding clouds to cause rainfall, or altering the course of typhoons to avoid the immense damage they do in built-up areas and directing their path and the huge amounts of fresh water they contain, to areas where the water can be conserved and used. Such work is in its infancy, but it may have important results in the future. Meanwhile, climate control is confined to micro-control through the use of irrigation, drainage, water storage, wind breaks, glass houses and heating against frost in orchards, livestock housing, etc. These measures are efforts to shelter the productive plant and animal life from the hazards and extremes of climate in order that they can develop under conditions which may basically be unfavourable. Careful siting of farmsteads, cultivation and crop practice against wind, sleet, and gully erosion, can also be seen as efforts by man to temper and control the unfavourable effects of climate.

The data needed by the agricultural planner include the amount and yearly distribution of rainfall and the seasonal variations in temperature. With this information, he can see which crops, livestock and systems of farming are applicable to particular areas and what measures – irrigation, wind breaks, etc., – have to be taken to safeguard them from the difficulties of the climate.

3. The exploitation of climatic variations

National and regional land use patterns can only be decided scientifically with reference to local climatic behaviour. Climate is the factor which, together with, or irrespective of soil type, decides which areas are suitable only for forestry or pasture, and those in which arable farming, fruit growing and intensive livestock husbandry can be practised. Such decisions can only be based on assessment of accurate data collected to show the required magnitudes of the agricultural branches in question. This should be done for long periods and *for specific localities* as far as possible. However, if the data are not available, development should not be held up until the information is assembled.

The general climatic features of Israel were mentioned earlier, and reference made to the way in which the climatic advantages of certain areas are utilized in planning the production of particular high-value commodities. Climate recordings are made at closely spaced points all over the country. They are used both for the study of long-term climatic effects and for current forecasting as a service to help farmers arrange their planting, harvesting, spraying and other operations. The meteorological service is continually expanding both its research and services for agriculture and for fishing. It is expected that such work will lead to greater understanding of climatic behaviour, suggesting ways by which it can be used to the advantage of the populations of Israel and of other countries in the Mediterranean region.

D. WATER RESOURCES AND AGRICULTURE

At the present time, all the world's sources of water stem either directly or indirectly from rainfall. Though efforts are being made to distill sea water cheaply, mankind is still dependent on rain for its agriculture and its drinking and industrial water. Farmers have always recognized this dependency which, probably more than anything else, is the controller of their fortunes in most parts of the world. But urban dwellers are becoming increasingly aware of the importance of water, with water availability proving a limiting factor in the growth of many towns. Planning water supply involves inquiring into the layout of local and national schemes, the smallest cost for the largest amount of water delivered, the reliability of supply and possibilities for future extension. It is important that the source of water be pure and free from pollution.

Most agriculture remains tied to the caprices of rainfall and the seasons, with the farmer planting after the first rains, and hoping the rest will come in season and without floods. Past experience leads him to expect a particular rainfall pattern, around which he plans his crops, the timing of their sowing and harvest. The range of his crop choice is limited to that which fits in with the seasonal rainfall pattern. If rain falls only in the winter, he cannot cultivate in the summer; if it falls throughout the year, grain crops are risky and grass farming predominates. Flooding also limits the choice of crops, and farming systems must adapt themselves to its possibility.

1. Desert agriculture, ancient and modern

The farmer, however, has not always accepted the water supply immediately available to him as a given factor, unchangeable by his actions. From ancient times, farmers in semi-arid lands have sought out water sources, dug wells, changed the course of streams and built channels and terraces to conserve flood waters.

An ancient example of ingenuity in constructing irrigation works has been studied in the Northern Negev in Israel by Professor Even-Ari.[5] There, a pagan Semitic people, the Nabateans who dwelt in the area around the beginning of the Christian Era, practised a flourishing agriculture where today there is arid desert supporting only a few flocks of camels, sheep and goats, and a flora of a sparse desert shrub association. The ancient farmers worked out catchment areas for rainfall that were 20–30 times larger than their field unit, so that the normal sparse rainfall of 70–120 mm (3–5 inches) which would fall in typical desert flood proportions, was increased and retained for the growing of crops. The country is a highland 400–1000 metres above sea level, with rich loess soil in the valley bottoms and wadi bottoms of several metres thickness, but with gravelly immature desert soils on the slopes under a thin loess layer. Since loess forms an almost impermeable crust when moistened, there is quick run-off from the slopes which the ancient farmers channelized to their fields. In deep loess, such a crust usually causes gulleying and erosion, so that in the fertile depressions containing their fields the farmers built terraced walls to prevent it. These walls forced the run-off water to remain on the fields, penetrating slowly into the loess soil. Their height was regulated so as to retain an amount of run-off water sufficient to penetrate 2–4 metres into the loess. Water thus stored was enough for the annual water requirements of cultivated orchards or general field crops. The overflow water was directed from each field to the one below through spillways constructed so as to avoid erosion.

Even-Ari's study of Nabatean irrigation shows the control man can achieve over water supplies if he applies himself with vigour and intelligence to his task. Through seeking out natural water resources and tapping them for agriculture, building artificial reservoirs for water storage, controlling the flow of streams, rivers, *wadis*, and run-off, the farmer and his community can make themselves increasingly independent of the short-term caprices of rainfall.

Rainfall measurement and assessment of the water capacity of natural reservoirs from year to year are very necessary both to ensure that irrigation works will not deplete springs, wells or rivers, avoiding overexploitation and salination, and to obtain the best use possible at lowest costs. By mapping rainfall distribution in amount and season, and relating it to measurement of water resources, deficiencies and overabundance can be assessed. Rational schemes for water transfer and irrigation can then be planned on a regional and national basis at a higher or lower level of investment, scope and complexity. Flood concentrations can be controlled and beneficially used, while dry climates can be utilized to grow crops of high economic value at particular seasons through irrigation waters fetched from afar. The growing cycle of plants can be immensely aided by water control, through irrigation or drainage, greatly minimizing the biological uncertainty of farming. It is stressed again that intimate, detailed recorded knowledge is the only basis for ambitious planning of large water schemes: this does not mean that nothing can be done on the local level unless vast sums are spent on research, but the principle of survey before planning still applies.

2. The irrigation projects in Israel[6]

The development of Israel's water resources in recent years has led from quick-effect, low-cost schemes to more expensive long-term projects.

In the first stage, it was necessary to provide water cheaply and at short notice to hundreds of agricultural settlements spread over the whole country. The solution was groundwater exploitation by drilled wells, which enabled the irrigation of newly developed land, especially on the coastal plain and in the northern Negev. Such wells have rapidly grown in number since the establishment of the State, so that exploitation of the coastal aquifer is nearing the point of maximum safe yield and has, in some locations, even exceeded this point. Remedial works are now under construction to correct overdrafts where they have occurred.

i. *The Yarkon-Negev project*

The need for further water directed attention to the category of sources next on the scale of difficulty, namely, the summer flow of rivers. This led to the implementation of the Yarkon-Negev project, aimed at diverting the water of the Yarkon from its source at the Rosh Ha'ayin springs east of Tel Aviv to the northern Negev. Part of this project has been in operation since 1955, carrying 100 million cubic metres per annum to the northern Negev, through a series of pumping stations and a 106 km-long pre-stressed concrete pipeline of 66-inch diameter. A second line of 70-inch diameter is now being constructed; the completed section already supplies the municipal requirements of Greater Tel Aviv.

ii. *The Western Galilee-Kishon project*

Another undertaking of similar magnitude is the Western Galilee-Kishon project which aims at collecting the yield of wells and springs in Western Galilee for conveyance to the fertile, but water-deficient Emek Jezre'el and to the town of Haifa. This project, of which most sections are already in operation, has as its principal feature a main conduit built of pre-stressed concrete pipe, mostly of 48-inch and partly of 36-inch diameter. The project includes an open reservoir with a capacity of some 7 million m^3, formed by damming up the course of the Kishon river. The project is to supply over 180 million m^3 per year.

An essential feature in the operation of both the above projects is the regulative effect obtained from incorporating in each an important municipal supply. The relatively steady municipal consumption throughout the entire year helps to overcome, with a minimum of surface storage capacity, the difficulties resulting from the concentration of agricultural demand in the dry summer. Thus, in the rainy season, mainly spring water can be utilized for supplying the towns, while wells are given a much needed period of recovery in anticipation of the heavy summer agricultural pumping.

The exploitation of saline springs has been the subject of special hydrological study in the Western Galilee-Kishon project and elsewhere. In some cases (e.g., at the Na'aman springs), the study has led to the interception of the flow by wells before it becomes saline and has resulted in a useful addition of fresh water.

Today, eighteen years after the establishment of the State, it is possible to point to marked achievements in the development of the country's water resources. The area under irrigation has grown from 300000 to 1500000 dunams and the capacity of the country's water installations from 350 to 1300 million m^3 p.a. About 80% of the country's conventional water resources are already under exploitation. The remaining water will be brought under control mainly by reclaiming seasonal flood flows and treated sewage effluent, and by implementing the final stage of the Jordan project, which was first put into operation in 1965.

iii. *The Jordan project*

The Jordan project, the largest of all, is intended to carry at its full capacity 325 million m^3 of water per annum from the Jordan for a distance of 180–220 km to the Negev, and at the same time to provide a means of coordinating and integrating the country's regional projects through the construction of its main conduit. This conduit, which has been constructed partly as a canal, partly as a tunnel and partly as a large diameter (108-inch) pressure pipeline, serves to interconnect and supplement other schemes which it intersects on its course from north to south.

To regulate seasonal and annual variations in the Jordan River's flow, it is necessary to store considerable quantities of water from the rainy to the dry season and from years of ample rainfall to years of shortage. Other alternatives having proved unfeasible, Lake Kinneret* (the natural lake formed by the Jordan River) has become the country's principle storage facility (net usable storage capacity 1000 million m^3) despite the drawback of its low elevation (212 m below sea level).

The most costly item of the Jordan-Negev project has been the 108-inch

* The Sea of Galilee.

pipeline. After a rigorous series of tests, production of the pipes – locally manufactured – began in 1959.

The first stage of the project which began operation in 1965, brings Jordan water to the headworks of the Yarkon project. In its southward course, the main conduit supplements local supplies in the north and enhances the supply available from the Yarkon river for the Yarkon-Negev pipeline. The yield, at this stage, is about 180 million m^3. The complete project will take several more years to construct.

3. The use of sprinkler irrigation[7]

About 90% of the 1.5 million dunams irrigated in Israel today are irrigated by sprinklers – this despite disadvantages of the method: (a) relatively high water-loss due to evaporation compared to surface irrigation; (b) the necessity for water more purified than needed for irrigation purposes; (c) and the cost of additional pumping to provide the pressure necessary for sprinkling. These three factors are offset by advantages summarized below, and by the fact that even in localities where surface irrigation could be employed to good advantage, the experience necessary for its operation is lacking.

The advantages of sprinkling are:

a) The topography of most of the arable lands is rolling, so that extensive earthwork would be required to prepare the land for surface irrigation in general, and for border irrigation in particular. In many cases the layer of fertile soil is so shallow that earthworks would make the soil barren.

b) The increase in the population of Israel due to mass immigration in recent years has called for a corresponding increase in the irrigated areas. As the population increased very rapidly (tripling itself in the past twelve years) and the newcomers had as a rule no experience in farming, it was essential to select an irrigation method having the following features:

i) requiring relatively little experience on the part of the irrigator.

ii) readily adaptable to most topographical, soil and crop requirements.

iii) requiring a relatively short time for the design and construction of irrigation systems.

Under these circumstances, and considering the fact that since the end of World War II good quality portable aluminium pipes have become available in unrestricted quantities, it was decided to install overhead irrigation systems in the newcomers' settlements.

c) The smallholders' settlements – the moshavim – constitute the major part of the rural population in Israel. There the characteristically small tracts adjacent to one another warrant the use of an irrigation method that will ensure effective distribution and metering of small quantities of water, delivered simultaneously to many users. These demands could be satisfied best by sprinkler irrigation.

d) Since both land and water resources in Israel are limited, water and land losses have to be reduced to a minimum. Sprinkler irrigation does not bring about loss of land from canals and ditches. Furthermore, it is easier – as a rule – to arrive at higher irrigation efficiencies when sprinkler irrigation is used than with surface methods.

e) Since most of the irrigation water in Israel is pumped from deep wells, the cost of incremental pumpage required for sprinkling is a relatively small item in the overall price of water.

f) Urban and rural settlements are scattered all over the country and water under pressure is supplied to all urban and farming communities for domestic use. Hence, if sprinkler irrigation is used, one water supply system can provide water for irrigation as well as for domestic and industrial purposes. On the other hand, water for surface irrigation, conveyed in canals and not fit for domestic use, would necessitate an additional supply system.

4. Water norms for agriculture

The most certain and economic way of adding water to the country's water resources is through thrift in the use of the available supplies. There is no doubt that considerable amounts of water can be saved by abolishing wasteful uses and economizing on legitimate ones.

Fish ponds in Israel use at present large amounts of fresh water, and will have to be done away with when resources, in the final stage, become scarce. Only saline water, unfit for agriculture, should in the future be allocated to fish ponds.

Considerable saving in the agricultural use of water can be achieved by employing cultivation methods designed to reduce water consumption, studying better methods and the correct frequency of irrigation, and by suitable use of fertilizers. Present application of water in Israel's agriculture has not yet reached the desired level of efficiency, but nonetheless the amount of water reckoned as necessary for agricultural crops has been on the decrease in the last few years. Table 3-4 shows the accepted norms for several crops in different periods and regions.

The table shows that the main saving has been in respect to crops recently introduced to Israel. For instance, there has been no real reduction in the amount of water required for various vegetables and certain of the fodder

TABLE 3-4

Water norms for selected crops in Israel

Crop	Region	Norms in 1955[8] m³ per dunam	Norms in 1959/60[9] m³ per dunam	Norms in 1961[10] published
Groundnuts	Coast	650	600	550
Clover	Coast	500	400	300
Maize for fodder	Coast	300	300	400
Citrus	Coast	700	720	720
Sugar beet	Lakhish	450	500	500
Citrus	Lakhish	800	750	750
Cotton	Lakhish	800	700	–

crops (such as maize and green fodder) while there has been a considerable decrease in requirements for new crops such as groundnuts and cotton.

Studies made during the past few years clearly show that there are possibilities for further reducing water requirements. For example, experiments carried out by the Irrigation Department of the National and University Institute of Agriculture have led to the conclusion that the amount of water needed for sugar beet can be considerably reduced – to 3.5 m³ in the Negev and 3.3 m³ or even less in the coastal belt. The Irrigation Department believes that it will be possible, after properly directed research, to reach an overall saving of at least 10% of what are at present considered minimum requirements. This saving could be even greater with certain crops such as cotton.

Experience gained by the Settlement Department of the Jewish Agency confirms the possibility of reducing water requirements. Many settlers, for instance, grow cotton in the Lakhish area, reaching yields of 300 kg per dunam, using not more than 450 m³ of water per dunam. A survey carried out in the Negev has shown that plots which in 1961 received only 450 m³ of water gave excellent yields.[11] At the kibbutz of Or Haner, a field of cotton received a total quantity of only 455 m³ of water per dunam and yielded 358 kg raw cotton. At Nir Am, only 400 m³ of water was used to obtain a yield of 310 kg raw cotton, per dunam.

A similar situation exists in regard to sugar beet.[11] At Re'im the sugar beet fields received only 410 m³ of water, while yields amounted to 5.7 tons per dunam; at Bror-Hail, fields receiving 370 m³ of water yielded 5.9 tons per dunam. These examples are not exceptions, but typical of well-run settlements in the Negev and show what can be achieved when efficiency is maximized, correct methods applied.

In the citrus industry, in which Israel farmers have specialized over several decades, it has become clear that previous concepts of the amount of water required for irrigation were considerably exaggerated. While in the past 800 m³ of water per dunam were used for irrigation in the coastal region, and although the Joint Planning Centre in its recent plan for moshav-type farms assumed a requirement of 720 m³ per dunam, recent research has shown that actually only 600 m³ per year are required.[12]

Table 3-5 is a summary of the assessed irrigation requirements of various crops (based on recent research and experience).

TABLE 3-5

Irrigation requirements of various crops

Crop	Zone	m³ per dunam
Citrus	Coastal region	600–650
Vegetables and potatoes	Coastal region	500
Vegetables and potatoes	Lakhish area	550
Groundnuts	Coastal region	500
Groundnuts	Mivtahim Shore	600
Cotton	Southern region	500
Sugar beet	Lakhish area	400
Clover	Southern and Coastal region	500
Green fodder	Coastal and Sharon region	350

It should be pointed out that it is becoming increasingly evident that the optimum water quota is not necessarily that which promises the highest yield. Dan Yaron and his co-workers[13], who recently carried out research in this field at the National and University Institute of Agriculture, have stated that while water requirements are generally determined according to the quantity which assures the maximal yield, research has shown that the optimum amount is in fact that which gives the most desirable economic results, i.e. the highest returns. Sometimes a certain additional quantity of water may raise the yields, but the resulting added input reduces the profitability of the crops. This research should be extended to all main crops in order to determine the optimum water requirements from an economic point of view rather than a purely agrotechnical one. Undoubtedly such research will lead to a further reduction in water requirements.

5. The control and administration of water use in Israel

The land area suitable for irrigation is greater than that which the total water resources – even after their fullest development – will be able to supply in Israel. Since water is one of the factors which limits agricultural development, all possible economic measures have to be taken to reach efficient and correct use of water supplies. Waste has to be prevented and underground reserves guarded against overexploitation and spoilage. Firm control must be maintained through

1) an adequate legal framework;
2) water meters for individual users;
3) care of water installations;
4) suitable pricing for different uses of water.

i. *The legal framework*

The government of the British Mandate in Palestine put forward the first proposals regulating use of water. They were received with distrust by the Jewish population which saw in them political implications which would limit the spread of Jewish settlement. With independence and the rapid spread of irrigated farming, it became clear that the efficient use of water could be guaranteed only by special law. A committee set up by the Minister of Agriculture presented detailed recommendations in January 1955. It was given its first Parliamentary reading, however, only at the end of 1957. The law lays down the principle that the ownership of water is in the hands of the State: "Land rights or ownership do not apply to water resources found on the land or running through it". Indiscriminate use may not be made of water resources. For agriculture, water is allocated according to particular farm purposes and in the particular quantities required. The water user is also protected by this provision, since if allocations are not fulfilled or are reduced he can claim compensation.

The allocation and utilization of water resources are under government inspection. According to the proposal, there is a water committee authorized to determine maximum consumption needs in different areas, according to seasonal and daily requirements. The committee also has power to prohibit all water use for particular purposes, except household use. This supervision of water distribution is designed to ensure full and efficient exploitation for different uses without extravagance and waste. To ensure full inspection of water resources, the law states that national and regional water works be managed by government or by licensed local bodies, and that all who use water for themselves or others from a private source, must be licensed. The amounts of water they can draw and the uses to which it can be put must be specified.

ii. *Water meters*

The law can only be of use if the amounts consumed by individual purchasers are accurately measured. This means not only that water meters must be fitted to each consumer outlet, but that their efficiency also be constantly tested.

iii. *Upkeep of water installations*

Burst pipes and poorly serviced equipment are sources of much waste of water both in agricultural regions and towns. Renewal of installations and constant checking on pumps and other equipment is costly, but must be brought to the stage where water loss is minimal. The choice of materials for irrigation and water supply works is important in lengthening the life of the installations by combating corrosion.

iv. *The pricing of water*

One of the main instruments for water control is pricing. For agriculture, since water is an indispensable raw material for production, its price is a major factor in determining the profitability of crop production. For industry and the household, the water price is a factor in the cost of living. The effectiveness of water meters in restricting urban consumption has been proved in certain trials, hence meters, together with a rising price-scale, are expected to drastically reduce wastage in towns.

With regard to agriculture, policy makers have debated the pros and cons of establishing a uniform unit water price for the whole country or the supply of water at cost. Since the water is nationally owned, a public company supplying most of it to consumers, either policy is possible. However, artificial water rates and subsidies would distort the true profitability of water use in the various regions, leading to areas of high real water cost taking advantage of subsidized water to grow crops which can be more cheaply produced in other areas. Water pricing policy must be simple and in accord with the main costing factors, thus encouraging a thrifty and efficient attitude to water use on the part of consumers. It must be drawn up with the aim of achieving an economic specialization of agriculture in the several regions of the country which differ in their irrigation potentialities, costs and needs.

6. A future water resource: desalinated sea water

The problem of finding a cheap method of desalinating sea water has occupied the minds of many prominent scientists. Over the past decade a number of countries, among them Holland, the U.S.A., Germany, France and Israel, have sought a method of producing fresh water from sea water at a cost approximating that of water derived from conventional sources. At present there are three basic methods for the desalination of salt water.

Electrodialysis: This method is based on the property of chlorine and sodium ions to separate in water in an electric field, with the aid of special membranes. Electric energy must therefore be used for the separation of the salt from the water. The electricity consumption depends on the quantity of salt which has to be removed from the water.

Freezing: This method is based on the fact that when salt water is frozen, the subsequent ice crystals do not contain all the salts found in water. The 'Zarchin' desalination process, evolved by an Israeli engineer, I. Zarchin, and operated in Eilat (in southern Israel on the Red Sea coast) is based on this principle.

Distillation: This method is based on evaporating the water, turning it into steam, and condensing the salt-free steam. The process can be combined with the activities of a power station. Some of the heat may be utilized which remains in the steam after leaving the turbines, once it has completed the mechanical energy-producing process.

The electrodialysis method is suitable for the desalination of water with a salt content of up to 5000 ppm. It is therefore unsuitable for the desalination of seawater, which usually contains 35000 ppm, being applicable only for the desalination of salty wells and springs. The Negev Research Institute in Ber-sheba is dealing with the elaboration of this method, with a view of solving the problem of brackish water found in various parts of the country, especially in the northern and western Negev.[14]

The advantage of the freezing method[15] is that it utilizes only small amounts of energy in comparison with the distillation method. The freezing method employs low temperatures, so that the resulting degree of corrosion to the plant is slight. Wear and tear is thus lower than with the distillation method, where the problem of corrosion is serious, along with a considerable accumulation of precipitates in the plant. One of the difficulties of the freezing method is that the frozen water generally is in the form of small ice crystals in an unfrozen water-salt solution. In order to separate the ice crystals from the solution, the crystals have to be washed with salt-free water. When a method is found to produce larger crystals, this washing procedure will be simpler.

According to the estimates of the Fairbanks Whitney Corporation, the total energy required by the freezing method for the desalination of 1000 gallons of water (4 m³) is approximately 40 kW, with the possibility of reducing it to 30 kW. The plants are planned for different capacities: the largest are capable of desalinating 10 million gallons, i.e., 40000 m³, per day. Their annual capacity is thus about 12–13 million m³. According to present estimates[16], the price of desalinated water will be between 50 and 80 cents per 1000 gallons, i.e., between 37.5 and 60 agorot/m³. The disadvantage of the method is that, after desalination, the water still has a salt content of 500 ppm, which is a high content for agricultural use. A plant which has begun partial operation according to this method is the Zarchin plant in Eilat. Its four desalination units have a capacity of 250 m³/24 hours each, so that the entire plant can have an output of approximately 1000 m³/24 hr. One of these units is functioning, the other three to be completed by the end of 1966. The price of water from the Zarchin plant will be higher than the price estimate above.

Various calculations made in different parts of the world show that desalinated water without any salt can be obtained at a lower cost by the distillation method than by the other two, provided it is combined with the turbines of a power station. The quantity of fuel required for the operation of a plant in conjunction with the turbines is $4\frac{1}{2}$ times less than if it were run separately without turbines. Such a plant operates in Eilat; it has a capacity of 7600 m³ per day, but operates now at half of its capacity, producing 3800 m³/24 hr. The price of the water is now IL. 1.88. The price is high because the plant operates only four months a year. At full time operation the price is estimated at IL. 0.99.[17]

Desalination plants operating according to the distillation process are also functioning in several other places in the world. The Richardson Westgarth Company is presently erecting a distillation unit in Curaçao, with a capacity of 6500 m³ per day. The Wier Company in England, together with the Richardson Westgarth Company, is planning a plant capable of distilling and desalinating 5 million British gallons (22500 m³) per day, for providing water to the Island of Guernsey, one of the British Channel islands.[18]

Recently, the physicist R. P. Hammond[19] carried out comprehensive research on the possibility of combining distillation plants with an atomic reactor operating at low temperatures. His calculations showed that a nuclear plant based on three reactors with a total output of 25 million kW would be capable of desalinating 1300 million m³ of water at a cost of 2.6–5.5 cents/m³. The nuclear power stations now producing electricity have a capacity of no more than 1 million kW each, because larger stations have not been required for power. In view of the potential low cost of desalinated water from nuclear power stations, it has been considered worthwhile to examine the possibilities of both establishing larger power stations or desalination plants in conjunction with smaller ones.

Following a joint communique issued in June 1964 by the President of the United States and the Prime Minister of Israel, calling for co-operation in desalination, a preliminary study was undertaken by a team of United States and Israeli engineers which examined the feasibility of a large dual-purpose electric power and water desalting plant based on nuclear power, to be constructed in Israel. The team indicated that a plant of such a kind could be technically and economically feasible. It specified the following ground rules:

1) A plant of 175–200 megawatts salable power, credited at 5.3 million kW/hr.; 2) 100–150 million m³ yearly output of water; 3) Fixed charge rates of 5, 7, and 10%; 4) The plant to be ready for commercial operation by mid-1975.

An engineering feasibility and economic study was next prepared by Kaiser Engineers and its main subcontractor, Catalytic Construction Company. They confined themselves to the process known as multi-stage flash evaporation process, which is the most developed and most used in commercial operation. The selected nuclear dual-purpose plant was compared with a fossil fuel dual-purpose plant.

The study showed that the cost of water produced in a nuclear plant was lower than that produced in a fossil fuel plant for all fixed charges through 10% when the water plant capacity was 100 million gallon per day (about 120 million m³ a year).

The estimated capital cost for the selected nuclear dual-purpose plant was 187 million dollars. Together with the electrical transmission and water conveyance facilities the cost would reach 217 million dollars.

The price of water was estimated at 22.8 agorot (7.6 cent) at 5% fixed charge rates; 34.4 agorot (11.47 cents) at 7% fixed charge rate and 53.1 agorot (17.7 cents) at 10% fixed charge rate.

Negotiations are now going on with international bodies regarding the financing of the plant; no final decision has yet taken place. As this additional amount of water is not due before 1973, it is not included in our estimates. However, we shall examine later (see Chapter 4) the practicality of cultivating various crops at such water costs, which are high.

7. Water: a limiting factor?

Of all the production resources used in agriculture in Israel, water is accepted as the limiting factor in further development. But in the economic sense, this is only true if the demand from home and export markets, at prices sufficient to guarantee the minimum labour income of the farmer, is larger than the capacity of the production process. Water is the factor which will first be in short supply and prevent utilization of other resources. The predicted market capacity, as estimated in the previous chapter, was arrived at without reference to the supply of water available for agriculture. This must now be examined to see whether resources exist to enable production to expand to an output level that can be absorbed by existing and projected markets.

The estimates[20] of the National Water Authority show that the national water supply will grow from 1176 million m^3 in 1961/62 to 1931 million m^3 per year in 1970. However, 431 million m^3 will have to be returned to replenish the underground reserves which have been overutilized. The amount of water available will thus be 1500 million m^3.

The Water Authority[21] estimates that urban and industrial water needs will rise from 286 million m^3 in 1961/62 to 460 million m^3 in 1970 as shown in Table 3-6.

TABLE 3-6

Urban water requirements (according to Water Authority)

	Consumption in 1961/62 in million m^3	Consumption in 1970 in million m^3
Drinking water for towns	224	330
Industry	62	130
Total Urban Needs	286	460

The estimate is based on the assumption that industrial growth will double consumption by 1970, and that domestic needs will increase on account of the rise in population and the rising consumption per head. The estimate for domestic consumption seems to be exaggerated, particularly as efforts are now being made to reduce the average domestic consumption from 100 m^3 per head to 80 m^3 or even less. A survey shows the widely differing rates of domestic water consumption in various parts of the country. Some of the figures are given in Table 3-7.[22]

In 1961, water quotas were allocated to the local authorities with the aim of reducing water consumption to a maximum of 80 m^3 per person per year as a first stage in a more severe total reduction. The results of the year's

TABLE 3-7

Water consumption in selected towns, 1960

	m³/person/year
Ashkelon	131
Ber-sheba	70.2
Givata'im	79.5
Hadera	166
Haifa	50.2
Jerusalem	45.4
Naharia	92.6
Natania	166.7
Safed	48.8
Ramle	79.9
Tel Aviv	125.1

effort showed a fair reduction in many places. Several towns did not use more than the 80 m³, and two others, with a former rate of 166 m³ per head, reduced consumption to 130. The conclusion of the survey is that the reduction to 80 m³ per person per year can be achieved without great effort. The aim must be to reduce water consumption per head to 60 m³ per year for household, public services, business and small industry in a period of two to three years, through a strict policy of water allocation to local authorities. Such a figure means a total consumption for domestic purposes, for a population of three million, of 180 million m³ per year, a very considerable saving over present rates per head. However, in order not to anticipate results which have not yet been achieved, a norm of 90–95 m³ per person is taken to estimate the total needs of urban dwellers for water in 1972. The expected account is as follows:

	million m³
Domestic consumption	285
Industry	130
Total	415

The amount available for agriculture will therefore be 1085 million m³. Whether this is sufficient to produce the food and fibre needed for the markets of 1973 is examined in Chapter 5.

E. AGRICULTURAL TECHNOLOGY

Stages in economic progress are particularly evident in agriculture. These stages are termed subsistence-level agriculture, diversified agriculture, and specialized agriculture. These three stages can likewise be described as non-commercial, partly commercial and highly commercial. The move from an earlier to a more advanced stage is bound up with advancing national prosperity – which provides the spur to improvement – and with technology, which provides the means. Organization and management, which enable the forming and application of technological methods, are included in the term 'technology', which in the modern sense implies co-ordination of different sector occupations.

The most primitive systems of farming are based on certain techniques, but today's technology of cultivation, is connected with many techniques which depend on social organization, division of labour, flow of materials and communication of ideas. The frontier of agricultural technology is continually moving outwards with very few farms managing to keep up with it.

A farming system which is stagnant will use individual techniques, but it cannot be said to be benefitting from 'technology'. Technology is based on science and scientific methods. It implies examination and measurement of existing methods and seeks to improve them by using different and better practices drawn from the range of techniques it has at its disposal. The improvements leave the way open for others to follow. Introducing a new technique may raise production to a new stagnation level, but technology measures the improvement and uses the new level as a basis for further improvement. Thus the process is associated with factors outside the farm, with education, experiments, advice, supply of materials and organizational links which carry the improvements from their innovators to the farm. Better seeds, their selection, grading and distribution; fertilizers, their manufacture, sale and use, or any of the apparatus which the agricultural scientist may seek to apply, have indispensable connections with firms and organizations outside the farm, both before they reach it and after they are used.

1. Application of technology

There are very few tools in the whole store of modern agricultural science which the farmer can apply unaided by others outside his farm. Much of farming in the world is backward despite science being so advanced, mainly for this reason. What is lacking is not the know-how, but channels of communication between it and the backward traditional farmer. Until he reaches out from the close confines of his farm, he will not be able to take part in the advance which technology makes possible.

The relative emphasis placed on the weight given to the human resource has changed, is changing and will change even more. With modern technology, investments in people bring relatively better dividends than investments in natural resources. The modernization of cultures, the provision of education and technical knowledge becomes a predominant target for investment and is therefore acquiring increasing importance in national budgeting. Once technology enters a society, its demands for skilled and educated personnel make ignorance a limiting factor in development: the nation quickly reaches a stage at which, without mass education, development can only limp along.

There are two conditions necessary for the application of technology at the farm level: that the farmer should know and trust the processes sufficiently to use them, and that he should be connected to a supply source which enables him to obtain the materials needed for their application. These two conditions have wide implications.

Knowledge and trust are dependent on education and agricultural extension, a social framework which allows enquiry and acceptance, and instructors to whom the people will listen. It also means that new methods should be fully practical and thoroughly tried out under local conditions so that farmers will not only be able to use them but that they will work satisfactorily.

The supply of materials at the farm implies a level of institutional and commercial organization which gets the necessary materials to the right place at the right time. The farmer must also be able to buy them, so that credit and commercial facilities have to be provided. In order to return the credit, the farmer must sell produce, and therefore a suitable marketing framework has to be built and linked to centres of demand. If there should be no demand centres sufficient to absorb the products resulting from the new technology, the farmers will run into debt, the process will stagnate. Alternatively, a benign government could put raw materials at the farmers' disposal free of charge in order to raise the level of subsistence consumption, but such a huge input is an investment which does not renew itself. No country can afford to provide great sums of working capital each year which are consumed and do not return for re-circulation or investment in real capital. This procedure may be useful in certain circumstances, but only as a short-run aid to overall long-term progress.

Overall progress is so demanding in respect to skilled personnel and capital, that it is not surprising, under present conditions of immobility of labour and capital between rich and poor nations of the world, that many countries still scarcely benefit from the technological revolutions of the past decades. These, which occurred mainly in the countries of the West were largely responsible for the great increases in wealth and productive capacity which characterize them. Once the process was started, it became self-proliferating with techniques breeding techniques to form a continuous flow. Social and institutional frameworks orientated themselves to accommodate to the demands and products of the new technology.

There is ample evidence of the contribution technology has made to the advance of agriculture over the last fifty years. They fall into two categories:

A) The discovery of new materials, power sources and better breeds of crops and animals, i.e. a wider and better range of inputs for the farmer to use.

B) An understanding of the relationships of inputs to outputs, which implies a careful collection of data on the timing and integration of farm operations. The use of inputs, old and new, can then be made subject to rapidly rising productivity, with less effort resulting in more output, since it is directed in ways giving the best results. The science of crop husbandry, for example, studies and reports on the best ways of using inputs in crop growing to get the most economic results. Farm management studies the relationships within and between farm branches to arrive at their most economic and productive combination in the farm plan.

The first category listed above is largely the result of research and enterprise conducted outside the farm by national or commercial bodies. These inputs are manufactured in factories, laboratories or research stations, and must be made available to farmers through commercial or other channels. The specific contribution which some of these new or better inputs have actually given to agriculture is examined later in more detail. Their common feature is that they can be bought in a more or less standardized form and that their reliability is known and tested under particular conditions. This applies to seeds, fertilizers, tractors, insecticides and most other inputs bought off-farm.

The second category of technology is very different. It is not a set of items which can be made up and bought or sold when necessary. It consists of skill, enterprise and knowledge, applied by the farmer and his advisers to the complicated tasks of organizing and operating farms. Many farm operations can be measured, evaluated and reduced to methodical steps which can be written down and taught, a reduction which is an invaluable contribution of the scientific method to agriculture. But however precise written instructions are, and however closely they may guide one to an expected result, they cannot simply be sold and applied in the right quantities as if they were, for example, fertilizers. The way has to be well prepared in advance so that the farmer can read, understand, and follow instructions.

In some countries the farm population is becoming more responsive to the 'selling' of techniques, with many farmers highly literate. They buy books and journals on many aspects of their work, introducing improvements solely on the basis of what they read. In general they are rewarded with success. Such farmers, on their own initiative, also seek the services of extension officers, since their education and high level of business sense tells them that expert advice pays well.

On the other hand, new methods and practices are more difficult to introduce where farmers are scarcely literate. The farmers must take instruction from someone who is the bearer of the new technology. This means that the extension worker has to be first accepted in an authoritative role. Together with the teaching of new methods, he is also the bearer of new materials to the farmers. His is a key role both in countries developing their agriculture from subsistence, and in those working at a high level of output.

Research stations all over the world are undertaking the production of new techniques – work, which apart from the generally accepted fields of variety testing, seed improvement and crop ecology, should be given much more emphasis than it receives at present. Direct links must be forged between the stations and the extension personnel so that a flow of knowledge and enquiry can take place between the two. Separately, their work is of little benefit in carrying technology to the farms. Together, each learns from the other so that applied research becomes eminently practical. The institutional framework used to achieve this goal of co-ordination and communication in Israel is discussed in Chapter 8.

2. The revolution in quality

The steady increase in agricultural output, particularly over the last 50 years, cannot be ascribed only to higher levels of inputs. This increase is best known and measured in the industrial countries, where it is associated with a rapid drop in the percentage of the labour force occupied in agriculture resulting from a rising output, not only in total, but also per unit of input. The efficiency of these inputs, because of new combinations, new materials and new knowledge, has reached heights never previously conceived. Far more can now be obtained with the same total effort. Schultz[23] terms this startling change a 'revolution in quality' and holds that it has great consequences for all countries, developed and underdeveloped. He gives two examples to illustrate the change. Statistics for the United States show that between 1940 and 1958 the stock of physical capital in agriculture rose only 20%, whereas farm output increased more than twice that much. At the same time, the farm labour force fell by 30%.

Brazil, Mexico and the Argentine (before Peron), for two decades and longer, have achieved rates of increase in agricultural output that have been twice the rate of increase in conventional farm inputs. In the United States, with virtually no rise in conventional farm inputs (land, buildings, machinery, fertilizer and farm labour, combined and weighted by 1947–49 prices), farm output rose 47% between 1940 and 1958. Schultz ascribes these increases to

    a) the higher quality of labour acquired through education;

    b) the application of research results;

    c) inputs which 'act as carriers of new knowledge';

    d) additional fertilizer.

He also points out that there are strong indications that *this improvement in the quality of human effort and of reproducible physical inputs, can serve as a substitute for farm land under a wide range of conditions.* Application of the materials and methods of the new technology, now shows the way to break the problem of rising population pressing harder and harder against land which is being progressively worn out. Technology introduces new inputs which draw more out of the land and require less labour to do so. By introducing new materials and techniques, the output of all factors can be raised. This technology depends on the quality of the human factor, a quality which is developed outside the boundaries of agricultural science (see Chapter 4).

3. Predicting future yields

Prediction of Israel's crop and animal yields in 1973 must take previous yield increases into account. Such increases have resulted from intensive experimentation, research and extension, from the rise in 'quality' of farmers, of their instructors and institutional organizations, from the improvement in industries and industrial products serving agriculture and from the new skills in farm management methods and techniques of production. There has been no sign that the increase has halted; the skill and management ability of farmers keeps rising from year to year so that yields per unit of input also continues to rise. Experience shows that 'the record of today is the average of tomorrow'.

There are some villages which will be faced with severe difficulties in adapting the structure of their farming systems to higher income farming. In some cases the size of the farms will be too small, causing them to be worked too intensively to maintain soil fertility and freedom from diseases and weeds. The better understanding of irrigation farming, of fodder growing and cattle feeding, of crop growth cycles and agricultural science in general, can substitute in part for land and for labour input, but these have special requirements in regard to farm design and operation.

The Joint Planning Centre and the regional planning offices have estimated yields in regions, villages, and on individual 'guided' farms, such material giving pointers for the future. Once the elements of success in the best farms have been analysed and passed on to others, they too will reach the same level of performance. Meanwhile, the best farms will 'move on', serving again as a test laboratory for the future.

Table 3-8 illustrates recent increase in yields, comparing the extent and yield of several farm branches for 1953 and for 1959. It shows that production rose both as a result of increases in the absolute size of the branches, and also as a result of technological improvements giving rise to better yields.

TABLE 3-8

Changes in the extent and volume of production of various branches of agriculture[24]

Branch	1953		1959		Increase (index: 1953 = 100)		Increase in production due to improved know-how (%)	
	Dunams or head	Output in tons	Dunams or head	Output in tons	In extent of branch	In yields	Total over period	Annual average
Tobacco	35127	1800	44817	2400	127	133	5	0.8
Dairy-cows	37240	121800	63000	250500	169	205	21	3.5
Vegetables & potatoes	222500	258400	211615	358500	95	138	45	7.5
Cotton	3000	300	62370	7300	2079	3650	75	15.0
Sugar beet	4427	8150	30807	122100	697	1498	115	19.7

The rise in yields of the different branches of agriculture during the past few years, has often surpassed predetermined estimates.

In a seven-year plan drawn up in 1953[25], for example, planning norms (i.e. estimated yields) were fixed for 1959, which were all in fact, achieved before 1959 and have today been considerably exceeded. This applies also to yields and norms published by the Joint Planning Centre in 1955.[26] Table 3-9 shows the average yields actually attained in 1960.

The table illustrates the rate of increase of yields as compared to past estimates. By 1960 most crops had attained average yields in excess of the

TABLE 3-9

Yields and estimated yields for selected branches

Branch	Average yield in 1953*	Estimated yield for 1959 in 7-year plan[25]	Yield estimated by Joint Planning Centre in 1955[26]	Yield estimated by Joint Planning Centre for 1961[27]	Average yields in 1960*
Laying-hens (eggs per hen)	170	177	150	170	170
Dairy-cows (litres of milk per cow)	3425	3750	3600	4755	4300
Groundnuts (kg per dunam)	350[a]	300[b]	300	350[c]	330
Cotton (kg per dunam)	255[d]	220	220	–	270
Sugar beet (kg per dunam)[e]	2700	4000	3500	5000	4800
Citrus fruit (kg per dunam)	3000	3500	3000	4000	3900
Vegetables (kg per dunam-average)	1600	2000	–	3000	2000

* The data for 1953 and 1960 were calculated on the basis of the area of cultivation of various crops and the production figures of agricultural branches appearing in *Statistical Abstracts of Israel* Nos. 10 and 11.

[a] In the Sharon region only; [b] average yield, including new development zones at that time, such as the Negev; [c] in the Lakhish and the Southern regions; [d] yield in 1954; [e] yield in mature plantations (excluding abandoned orchards).

forecasts both of the seven-year plan and of the Joint Agricultural Planning Centre. Inferences can be drawn from such increase with regard to the future.

The expected yields of some of the main branches of agriculture will now be examined, confining ourselves to a detailed discussion of those which can illustrate the methods adopted, in arriving at estimates.

i. *Dairy farming*

In 1953 the annual yield per dairy cow averaged 3425 litres of milk. Expert estimates made in that year, assumed that within seven years, i.e. by 1960, the average yield would amount to 3750 litres. In fact, the average general yield in 1960 was 4290 litres, while that for controlled (herds registered in the herd book) reached about 5640 litres.

Table 3-10 shows average milk yields from 1956–1960 in the Jewish farming sector generally, and in controlled herds in particular.

TABLE 3-10

Milk production and average yields in the Jewish farming sector
(1956–1960)[28]

Year	Jewish sector total				Controlled herds only		
	No. of cows	Production in millions of litres	Average yield per cow	Percentage marketed	No. of cows	Production in millions of litres	Average yield per cow
1956	38300	164	4280	75	13600	65.8	4840
1957	42000	176	4190	77	13500	65.8	4840
1958	49500	208	4200	80	15400	76.4	4960
1959	59000	250.5	4250	82	16900	90.4	5350
1960	62000	266	4290	82	16600	93.6	5640

The table shows that while average milk production per dairy cow in the Jewish farming sector as a whole, did not increase throughout the five-year period 1956–1961, the yield of the controlled herds rose constantly and at a rapid rate. The average yield for the Jewish sector as a whole remained stable, because new farmers entered into the production cycle every year. Their lack of experience caused the national average to remain the same, despite the constant rise in yields of herds maintained by experienced dairy-farmers.

Some examples of milk yields obtained by new moshav settlements as they gained more experience in dairy-farming are given below:

Moshav A[29] is a dairy-plantation type settlement in the Central Region, founded in 1952. In 1960 the moshav was selected for an economic study because it was considered to be representative of a large group of mixed-farming settlements in the Central Region, where the various branches of agriculture had not yet been fully developed. The study showed that out of 39 dairy farms, 12 had a yield ranging between 4500–4700 litres per cow, while 20 reached an average yield of over 4700 litres per cow. The total average yield in Moshav A was 4530 litres per cow.

An increasing tendency towards specialization was noted, as illustrated by the number of cows kept by the individual settlers. A survey carried out in Moshav B, representative of the Central Region[30], showed that out of 71 settlers carrying on dairy-farming by the end of 1960, 37 had as many as 6 cows or more.

Increasing experience and know-how of new farmers, as well as the greater efficiency of all dairy-farms, resulting from more specialization and the larger scale on which they are conducted, will keep leading to higher yields. It may be assumed that at the end of the period covered by this study, all dairy farms will have become firmly established, the dairy branch of agriculture reaching the maximum production level planned for.

The total number of dairy farms is likely to fall, with only those farms remaining which are particularly suited to dairying. Hence there is reasonable ground for assuming that in the course of the development period, the yields achieved in the new settlements will equal those now commonly obtained by veteran settlements, while at the same time the yields of the latter will continue to rise. In our opinion, an average yield of about 5000–5200 litres would be a reasonable estimate. To be on the conservative side, however, we have assumed an average marketable yield of 4800 litres at the end of the period under consideration.

ii. *Citrus fruit*

Citrus yields in Israel vary considerably. This is due to the manner in which the different citrus groves have been tended, or more precisely to the neglect of certain of the groves during the Second World War and the War of Liberation. Owing to the great number of abandoned groves, by 1948, Israel's citrus harvests had declined considerably. Only after the groves had been rehabilitated did yields again rise, without yet having reached the level of the late 1930's. The present average national yield is not a reliable basis for projection, particularly as the proportion of abandoned and of old citrus groves will gradually fall. We shall accordingly base our calculations on the yields normally obtained by mature groves planted in suitable soil. Data provided by the Ministry of Agriculture indicates that the average yield of such groves exceeds 4 tons per dunam. Many groves even attain yields of 6 tons per dunam.[31] The report of the Public Commission for Investigating the State of Agriculture[32] stressed the large fluctuations in citrus yields, ranging from 60 cases per dunam (2.4 tons) to over 150 cases per dunam (over 6 tons), the lower figure relating to young or improperly cared for groves or to those being rehabilitated. In estimating the yield per dunam in ten years time, some increase in the present yield of mature citrus groves can be anticipated, although future average yields cannot be expected to reach present record yields. These record yields of 6 tons or more, are only attained by part of the groves; in the opinion of citrus experts, this will also be the case in the future. We have therefore assumed an average yield of 4.5 tons per dunam for citrus groves situated in typical citrus-growing areas. In other areas, where experience has not yet been acquired, or where the soil is heavy, we have estimated a yield of only 4 tons per dunam. It should be mentioned that the Joint Planning Centre has assumed a yield of 4 tons in all areas.[33]

Sugar beet

Sugar beet yields in Israel have been increasing continuously. Within the past 7 years, the average yield has risen from 2.4 tons to 4.8 tons per dunam, while the yield of sugar has risen from 390 kg to 800 kg per dunam. This development is illustrated by Table 3-11.

TABLE 3-11

Increase in sugar beet yields in Israel[34]

Year	Area (dunams)	Yield of beet (tons per dunam)	Yield of sugar (kg per dunam)
1953	3500	2.4	390
1954	2500	2.6	460
1955	5600	3.7	640
1956	7710	3.6	630
1957	12180	4.6	730
1958	20720	4.5	770
1959	28350	4.3	720
1960	35250	4.79	800
1961	55380	4.42	716

An examination of yields for each settlement separately shows that the record yields attained were 8 tons per dunam. These yields were obtained in various parts of the country, including the Negev, the Beit Shean Valley,

TABLE 3-12

Sugar beet yields in 1961 in selected settlements[34]

Name of settlement	Region	Net yield
Regional Council	Beit Shean	7.8
Shavei Zion	Western Galilee	7.0
Ein Hacarmel	Carmel Shore	6.6
Habonim	Carmel Shore	6.3
Or Haner	Shaar Hanegev (Northern Negev)	6.8
Nahal Oz	Shaar Hanegev (Northern Negev)	5.9
Reim	Maon	5.9
Eshbol	Nir-Shelahim	5.8
Patish	Ber-sheba	6.7
Shokeda	Ber-sheba	7.6

the Western Galilee and the Carmel coastal belt. Average yields of 5.5 tons per dunam were obtained by many farms.

A survey of sugar beet cultivation in the settlements of the Negev Region made by the Extension Authority has shown that out of 49 *settlements*, 25 attained an average yield of 5 tons or more, while 14 of these achieved a yield of over 5.6 tons. This indicates that record yields of over 6 tons were obtained even in new settlements.

In view of the large number of settlements which had sugar beet yields of over 5.5 tons per dunam, and even of more than 6 tons per dunam, it may be assumed that within 10 years the average yield will be 5.5 tons. It should be pointed out that the Joint Agricultural Planning Centre assumed an average yield of 5.5 tons per dunam for the Negev and of 5 tons per dunam for the rest of the country. In our opinion the high yields obtained in most parts of the country show that yields of 5.5 tons may be expected for all settlements growing sugar beet in suitable areas.

iii. *Cotton*

The commercial cultivation of cotton in Israel started in 1954. By 1955, about 100 kg of fibre per dunam were already being obtained (a gross yield of 290 kg per dunam). Yields were temporarily reduced, owing to an infestation of the boll weevil, but rose constantly once the pest was exterminated. In 1960 the national average yield was 106.5 kg fibre per dunam [35], but yields 110–120 kg fibre per dunam (or 300–320 kg raw cotton) were commonly obtained in various parts of the country, including new settlements in the Lakhish Region. Such yields indicated that with the right conditions and correct methods of cultivation, average yields could be considerably increased. Similar yields have been obtained in various parts of the world, where cotton is grown on irrigated land under conditions similar to Israel's. The average yield in California in 1959 was 119 kg of fibre (318 kg raw cotton) per dunam. [36] Cotton experts in Israel assume that a yield of 350 kg raw cotton per dunam may be expected in efficient farms under suitable conditions. [37]

It seems to us that this figure may be taken as the average yield at the end of ten years. However, to be on the safe side, we have assumed a yield of 330 kg per dunam (130 kg fibre).

The short experience acquired in Israel in recent years on the cultivation of cotton with partial irrigation only, indicates that yields of 160–200 kg raw cotton per dunam are obtainable. [38] The Joint Planning Centre estimated yields of 180 kg (65 kg of fibre per dunam and 115 kg of seeds) [39] for cotton cultivated under these conditions. For lack of additional data this estimate should be adopted, and the future alone will show whether these figures correspond to reality or are too low.

iv. *Groundnuts*

Considerable fluctuations, both upwards and downwards have occurred during the past few years in Israeli groundnut yields. The main reason has been the introduction of this crop to unsuitable areas. Research carried out at the National and University Institute of Agriculture, and the experience of agricultural instructors, have shown that in order to obtain yields which will ensure profitability, groundnut growing must be restricted exclusively to suitable areas, namely to light, sandy soils as in the Coastal Plain, in the Western Negev and the like. Trials carried out in experimental fields in the Western Negev have proved that yields of 400 kg per dunam or even more, can be achieved without great difficulty. This is borne out by a survey carried out by the Authority in the kibbutzim of the Negev, where average groundnut yields exceeded 400 kg per dunam. The yields of these kibbutzim for 1961 are summed up in Table 3-13.[40]

TABLE 3-13
Groundnut yields of kibbutzim in the Negev for 1961

Name of settlement	Area in dunams	Total yield in tons	Marketed yield in kg
Gevuloth	330	122	384
Magen	110	28	254
Nir Yitzhak	400	195	487
Nir Oz	150	54	360
Ze'elim	300	126	420
Mashabei Sadeh	100	41	410
Revivim	150	67	404

In the moshav sector yields were lower, owing to lack of adequate know-how and sometimes to planting in unsuitable soils. The yields of a considerable number of moshav settlements, however, came close to 400 kg. Groundnut yields for 1961 in a number of moshavim which achieved particularly good crops are given in Table 3-14.[41] In medium and heavy soils, the yields were lower (in the vicinity of 350 kg per dunam) as was the proportion fit for export.

We assume that in the future groundnut cultivation will be restricted to light soils only. The expected yield may thus be set at 400 kg per dunam.

TABLE 3-14

Groundnut yields in selected moshav settlements in 1961

Moshav	Yield per Dunam
Paamei Tashaz	372
Meliloth	406
Maslul	347
Peduyim	317
Taashur	358
Mivtahim	374

Another factor liable to increase yields is the introduction of new strains ensuring both high quality and yields. In experiments recently conducted by the National and University Institute of Agriculture, promising new strains ensuring the profitability and quality of the Israeli groundnut crop, were successfully acclimatized to local conditions.

v. *Summary of yield estimates*

The yields of all other branches of agriculture have been analysed similarly to those detailed above. Average yields were compared with those of particularly successful farms. Various studies were assessed, prospects of higher yields in the future considered, etc. Table 3-15 shows the estimated mean yields in the main branches of agriculture.

TABLE 3-15

Branch	Yield
Milk (marketed quantity only)	4800 litres per cow
Beef (from dairy cattle)	190 kg per head
Beef (from beef cattle)	390 kg per head
Citrus fruit (in typical areas in the South):	4500 kg per dunam
(in the Negev and in heavy soils):	4000 kg per dunam
Vegetables	3000 kg per dunam
Potatoes	2500 kg per dunam
Melons, watermelons and pumpkins	1800 kg per dunam
Groundnuts (only in suitable areas)	400 kg per dunam
Sugar beets	5500 kg per dunam
Cotton (fully irrigated)	330 kg per dunam
Cotton (auxiliary irrigation)	210 kg per dunam
Pulses (for grains)	120 kg per dunam
Hens	170 eggs per hen
Poultry (from lay-hens)	2.1 kg per hen
Poultry (for fattening)	1.7 kg per hen

It must again be stressed that even now it appears probable that yields of certain products will exceed the forecasts. All estimates, of course, depend on the selection of suitable areas for cultivation, on the progress of agricultural research and on greater specialization. With regard to the last-mentioned factor, it has been assumed that individual farms will tend to concentrate on certain farming activities which will accordingly be conducted on a larger scale.

4. Factors influencing yields

i. *Seeds*

The seed, by means of which plant characteristics are inherited, is the link between one plant generation and the next. Some plant characteristics are wholly desirable and important, others are negative, militating against the economic use of the plant for agriculture. The science of genetics shows that by breeding programmes, positive characteristics can be selected and concentrated, negative ones eliminated. Botanical and genetical studies also show the great variation that can occur in plants of a particular species, while crop experiments give estimates of these variations in respect to agricultural performance.

In a research programme the better plants and species are first selected, the rest rejected. Among those showing promise, breeding is carried out which concentrates the best characteristics into one variety which will be standard and reliable from one generation to the next. The work is continuous; a new variety which improves on its predecessors is examined and subjected to further selection aimed at even greater improvement and better varieties. As an example, new possibilities have been opened up with the discovery and successful commercial hybridization of maize and sorghums, perhaps one of the greatest contributions of genetical science to agriculture. Since work with hybrids is still in progress, it may produce other startling results.

The characteristics which geneticists seek to concentrate in the ultimate 'perfect seed' are varied. They include first of all reliability of performance with the seed faithfully perpetuating its original variety. The plant it produces should be high yielding, disease resistant and suited to the climatic and environmental conditions with which it has to contend. In a region tending to frost, the plant should be frost resistant. If rains come early, the plant should ripen earlier, with ripening even throughout the crop. Sometimes the farming system demands particular characteristics of the seeds and the crops produced – a high germination percentage in the seed, short stems or long stems, profuse or thin foliage, or a long or short growing season.

Markets too have special demands which may be seasonal or qualitative such as round, regular fruit of particular colour, tenderness in vegetables or thin skins in oranges. There are many other qualities for which the consumer is willing to pay, the farmer attempting to satisfy, and the geneticist in turn searching for new varieties to satisfy the farmer. The geneticist's quest is for seed which, on planting, will give high, economic returns to the farmer. Poor seed, badly treated and of unreliable performance is extremely wasteful. It leads to bare patches in the field, possibly weed problems from weed seeds being included in the seed sample, uncertainty in yield, in time of harvest and in quality of the product. The farmer also has to buy or save much more seed for sowing than is necessary when the seed is of known variety, healthy, properly cleaned and has a high germination rate. Seed improvement, both for new varieties and for better seed handling and distribution, is of vital importance. A good variety rather than a poor one may make a tremendous difference in total yield with very little extra effort on the part of the farmer. Provision of good seed is one of the most valuable tools in the hands of an authority driving towards agricultural progress.

The following example illustrates some of the effects and results obtained by Israel plant breeding stations in the production of commercial, high-yielding sorghum varieties.

A local strain of sorghum cultivated in Israel over many years of non-irrigated soil used to yield 60–75 kg per dunam. In 1950, investigations were launched on new imported strains to ascertain the best suited to local conditions and most promising of highest yields. Finally the Sooner Yellow and the Martin strains were successfully introduced, doubling and even trebling the previous yields, i.e. 205 kg per dunam in experimental plots. As a result in 1956 the national average yield of sorghum had gone up to 150 kg per dunam.

Attention was again given to sorghum in 1957, with a view to finding strains with still higher yields. It was found that two American strains, Texas 610 and Frontier 400, could be successfully acclimatized, giving record yields of 350 kg per dunam on non-irrigated land. Since these strains are a 'trade secret' of the distributors in the U.S.A., research was instituted to establish a local cross-bred strain with similar yields. Experiments proved successful with Hazera 610, equal to the best American strains. Its yields at the experimental farm reached 352 kg per dunam on non-irrigated land, in areas with 350 mm rainfall. For other regions, with more plentiful rainfall (about 500 mm per year), 500–550 kg per dunam were obtained at the experimental farm.

In 1960 the crossbred Hazera 610 strain was used in most areas where sorghum was cultivated, and the average national yield was 185 kg per dunam. A year later, the average yield was over 240 kg per dunam.

Thus the average national yield of sorghum cultivated on non-irrigated soil was multiplied several-fold in the course of ten years. On irrigated soil, Hazera 610 strain yielded 883 kg per dunam, proving that its high potential yield has not been fully realised only because of limited precipitation.

TABLE 3-16

Average sorghum yields (on non-irrigated land)[42]

Year	Average yield in kg	Common strains
1950	75	Most areas – ordinary local strains
1952	60	Most areas – ordinary local strains
1954	90	Most areas – Martin and Sooner strains
1956	150	All areas – Martin and Sooner strains
1958	190	Most areas – cross-bred strain Hazera 610
1961	240	Practically all areas – cross-bred strain Hazera 610
1963	250	Practically all areas – cross-bred strain Hazera 610

ii. *Fertilizers*

Experience has shown that although certain soil qualities must be accepted as they are found, the fertility of the soil depends very much on the way it is handled by man. Agricultural practice can ruin a soil in one generation; it can also change it unrecognizably for the good. Plants require of the soil a physical medium in which to take root and grow; they also obtain from it many of the nutrients they need to develop, flower and bear fruit. The soil may renew some nutrients of its own accord so that in some soils the original mineral stock may last for many years. Usually, however, the stock of nutrients is depleted by continuous cropping. Although careful agricultural practice can maintain a minimum level of soil fertility, without import of plant nutrients from outside the farm, almost everywhere, and on most types of soil, the addition of certain nutrients results in a response by the plants which gives higher yields and better crops. If only one nutrient is in short supply, the plant is limited in the use it can make of others, and its response to cultivation is held back by this shortage.

The main nutrients which need supplementing in the soil are nitrogen, phosphate and potash. The fertilizer industry is continually expanding research to supply agriculture with these materials in forms convenient to carry and spread, with as high a concentration of active ingredients as possible, and with fertilizers made up in particular mixes and formulae according to specific conditions of crop, soil, or climate.

Methods of fertilizer application vary from hand spreading to large and small machines behind horses or tractors, to aeroplane attachments or application through irrigation water. Agricultural engineers design and re-design fertilizer spreaders to produce a machine which spreads evenly without becoming blocked up. Fertilizer is saved by machines which place it along with the seed at sowing time, so that it is immediately available to the sprout. The aim is to ensure that the plant has the right nutrients in the right place as it needs them. Too much fertilizer scorches; too little holds back plant growth.

Research is continuing to find new and better ways by which man can supplement the store of the earth for plant use. The results achieved are striking, and have been proved all over the world, although in many countries fertilizer application is still far below the economic level.

iii. *Manures and composts*

Most traditional, intensive systems of agriculture have relied on organic manure. A balance between arable husbandry and livestock which returns manure to the land has been the hallmark of these systems, keeping soil fertility high over long periods of time. However, in modern, specialized farming, regional and climatic advantages for arable farming may be so great that livestock husbandry has no economic place. Fodder growing will give way entirely to intensive cash cropping, with artificial fertilizers adequately and efficiently used. Crops which build up soil fertility are included in the rotation. Green manuring also becomes part of the practice, the artificial fertilizers providing the nutrients, the other crops returning organic matter to the soil.

Such modern methods of crop husbandry have not been entirely accepted. Critics claim that soil structure will rapidly deteriorate without organic manure, and that the system will last only as long as original reserves of organic matter last in the soil. They maintain that animal residues are necessary to keep humus content sufficiently high.

On the other hand, the proponents of purely arable cropping systems point to the relation of animal manure produced and the area and soil depth which it is intended to fertilize. They hold that even heavy applications of organic manure can have little effect on soil structure simply because the amount of soil is so much greater and dilutes the manure application beyond effective action. Even if heavy application of manure does influence soil fertility in ways which artificial fertilizers cannot, they ask where the amounts necessary to cover the agricultural area of the world are to come from. Manure bought for one farm is a loss to the other which sells it. Even far greater numbers of farm animals than are kept at present cannot supply a fraction of the amount of organic manure needed to sustain the soil structure and fertility of most agricultural land.

Both sides offer evidence to support their claim. The one points to the high yields and improvement in soil fertility obtained on farms in many areas by intensive organic farming, while the other rightly maintains that such areas usually produce rich manures by importing feedstuffs from arable areas and asks from where these latter are to obtain their organic manure. The author toured areas in California where the soil and climatic conditions are very similar to those of Israel. In the last fifteen years, the crop yields have been continuously rising, reaching amazing heights over thousands of acres. The ceiling of the rise is not yet in sight, and each year new records are established. All this is achieved without organic manure, for the area does not sell livestock. Soil fertility is guarded and improved from year to year because of the attention paid to suitable cultivation and up-to-date fertilization by liquid application. This method, by which the plant is provided with nutrients in an immediately available form as and when it needs them, gives very good results.

In the Soviet Union too, large areas of cotton and other crops have been grown for years without manure; yields are high due to the use of fertilizers.

Such areas demonstrate the artificial fertilizer solution to shortage of animal manures, disproving the claims of those who say that without organic manure there can be no agriculture. If, indeed, systems of farming which maintain soil fertility without organic manures are proved successful, they should be wholly accepted, investigated and applied to the many areas where, for economic and other reasons, the stock-carrying capacity is insufficient to provide enough organic manure. On the other hand, where organic manure is available, it should be carefully conserved and used, together with the best cultivation and artificial fertilization processes. It is clear, however, that most areas of the world will have to be farmed without application of organic manure sufficient to affect soil structure.

In certain farming areas, compost may be a useful source of organic matter for return to the soil. The cost of making it by reducing bulky materials to something which has real value to soil fertility is in many cases prohibitive. It is the cost factor which will determine the usefulness of composting schemes.

iv. *Farming systems*

Agricultural technology depends not only on the quality of inputs, but also on their organization within a farm plan and the integration of that plan with supply and marketing facilities. Farming can be looked upon as a business whose aim is to profit from the investment of materials, labour and managerial ability, in a framework producing and selling food and fibre. To succeed, the business must preserve its capital and be run on lines ensuring long-term survival.

Farm management is now a much studied art which, in its application, leads to great benefits; it is essentially the art of reducing farm operations to a steady, flexible system geared to profit and improvement of the farm business. It cannot be haphazard. The relations between input and output in farming, the flow of materials within farms and between farms, factory and market are such that continuity within a flexible framework is essential. Each part of the system has to function smoothly in conjunction with the others so that effort is utilized productively. This is as applicable to low level, non-commercial farming as it is to high-powered farm-factories. The most primitive farmers work in a system, perhaps based only on tradition, but nevertheless a system which continues to support its operator from year to year. Harvest follows sowing, and fallow follows cropping, with appropriate reserves kept for the next year's operation. Today primitive systems can be vastly improved or replaced by others more productive. The interaction of farm branches and the need to maintain soil fertility, mean that each and every improvement must be introduced within the framework of an improved farm system. Otherwise improvement as a consistent feature cannot be maintained.

There are many other factors which influence the productivity of farming: stock breeding and feeding, machinery manufacture and use, farm building design, organization of supply, credit and marketing are among them. The methods are known; they have to be applied before they bear fruit. As mentioned earlier, the problem of how to apply them faces planners, administrators, research and extension workers all over the world, in the most underdeveloped and in the most developed countries alike.

F. AGRICULTURAL LABOUR PRODUCTIVITY

The impact of technology on agriculture results in increases in the output of labour. Technology provides materials and methods which raise farm efficiency, releasing labour for other tasks. If the farmer plants a corn variety which yields twice as much as a previous one, then he plants a smaller area of corn for his needs, growing something else with the work days on the land saved. Alternatively, the double yield with the same land and labour inputs frees him from the necessity of searching elsewhere for extra income. The result of such an improvement is that living standards rise, either through higher income or more leisure. The most modern techniques can only be applied to a farm if management constantly concerns itself with raising labour productivity. Techniques and management should be seen as the blades of a pair of scissors which work together to cut waste: separately, each of the blades is of little use.

High labour productivity is achieved by careful planning which locates the operating foci of a farm and its branches, in convenient relationship to cut unnecessary walking. It supplies tools of the right type for each job; mechanizes where machines are more economic than manual labour; avoids delays through advance ordering of materials and the planning of transport arrangements; careful management and planning also observes good maintenance of buildings and machinery; maintains orderliness in storage sheds; pays the strictest attention to operational details on which a man spends his days, and tries constantly to eliminate unnecessary movement.

The principles of the science of time-and-motion study are well-known. The results achieved in industry and farming are sometimes almost unbelievable when the cost and profit accounts are analysed before and after its application. Unfortunately, the science is not applied with sufficient frequency. Its main basis is common sense coupled with the most pedantic measurement and timing of operations as they are carried out. One of the works on this subject [43], which draws freely on actual case studies, describes in simple language the methods by which the productivity of farm labour can be raised many-fold with infinitesimal investment compared to the labour and effort saved. The book, if used by the British farmers for whom it was written, will improve the efficiency of their farms unrecognizably. Its application is no less relevant to the farms of Israel and, indeed, of any country.

In one of the case-studies quoted, a farmer walked a distance of 192 feet whenever he had to water his horses which were housed in four adjacent boxes. By installing piped water in each box – an expense entailing only the extension of the existing pipe network for a short distance – this wasteful labour of walking was completely eliminated, the farmer freed for productive work.

The same crop rotation system, as another example, with only the order of fields changed, so that similar crops are adjacent, can lead to great time-economy.

There are countless similar examples of the saving in back-breaking and inefficient work through thought and organization. One of the new settlers of Israel's Lakhish region, farming independently for only five years, described how he plans cucumber production. Each operation is worked out in detail before any work begins in the field. The row lengths and pattern are arranged while bearing in mind the placement of pickers and their containers at harvest, the access of carts to pick up the boxes and all the other steps from ploughing to the final picking. Such examples which ensure timeliness of operation with an eye to high yield, high quality and no waste of labour or materials, prove that an hour at a table with pencil and paper can save days of frustrating effort in the fields. Organization, once introduced, is always serviceable; savings on an operation are not gained on a single occasion but are obtained each time the operation is performed.

Labour productivity is highly dependent on the question of the physical arrangement of buildings and fields, the distances travelled between jobs, and on sound investment in the right tools and equipment for the work on hand. Inattention to spare-part storage and machinery maintenance can alone be responsible for days of wasted labour. Loading and unloading sacks from barns and lorries account for weeks of back-breaking work when an electric motor could do the job at a fraction of the manpower and the cost.

Attention to the farm plan is supremely important in saving labour and expense. The examples quoted above should be sufficient to make any farmer go into long sessions of deep, honest thought about the way he runs his farm. In fact, very few farmers do actually make this minimal effort, paying a high price for mental laziness by permanent physical fatigue, high costs and low profits. Many say there is no time for 'office work'. Obviously this is true. They are busy doing all the jobs which a day's office work would save them!

Farming is traditionally dependent on manual work. Where capital is available, this dependence can be broken with the range of machinery and techniques today available to the farmer since there are very few operations which cannot be economically mechanized. A small farm, however, cannot afford to buy the whole range of machinery which would be economic for it to use. The alternative to buying is contracting from either private or co-operative machinery services.

Hand labour will continue to remain essential for some operations such as picking fruit and vegetables or sugar-beet thinning, where it is usually superior to machine work. In sugar beet growing, for example, the quality of the singling work by hand has a direct influence on yield; on the other hand, 'gapping' by machines can immensely lighten the labour where attention to cultivation and good seed have given a thick enough stand to enable the machine to work well. In fruit and vegetable picking, where quality of the produce is directly related to price, hand labour – the only way of dealing with most picking – is economic. However, the efficiency of the whole operation can be raised by attention to suitably sized containers, their distribution and collection among the pickers, and adequate transport facilities for fruit and workers. Weeding in row crops accounts for a huge number of man hours. The solution is simple: by paying attention to rotational and cultivation principles, and by cutting down growth around fields, fewer weeds will sprout.

The way to the raising of labour productivity often lies outside the farm itself. In a modern farm business, the farmer is very dependent on institutions, firms, banks and government. If these do not operate efficiently, they may swallow up many precious hours of the farmer's time by late or inconvenient times of delivery, slipshod attention to orders, insufficient staff and top-heavy bureaucracy. The farmer is not obliged to accept such situations. By changing his custom to a better firm, by organizing complaints among his neighbours and by constant pressure on local councils and institutions, much can be done to gain better services. The solution may be found in co-operation to substitute for inefficient firms, or requests to private people to open new branches or start new firms.

Israel is experiencing steadily rising labour productivity in her agriculture. Average work norms for the crops and animal products are gradually being introduced, as skill improves among the new farmers, and as the quality and flow of materials and machines to the farm increase. The extension service is also making a worthwhile contribution to increasing labour efficiency, while the drive for higher living standards makes farmers more cost conscious. With the advancing specialization of our agriculture and of the individual farms, further progress is to be expected.

Table 3-17 compares the labour requirements fixed by the Joint Planning Centre in 1955[8] with those for 1960[9] and with the norms proposed in the

TABLE 3-17

Work-days according to various plans

Branch	Work-days according to 1955 plan[8]		Work-days according to 1960 plan norms[9]		Work-days according to master plan of 1961[10]	
	per dunam	per ton	per dunam	per ton	per dunam	per ton
Groundnuts	10.0	33.3	4.6[a]	13.1	4.6	11.5
Sugar beet	10.0	2.8	5.5	1.1	8.0[b]	1.6
Cotton (on irrigated land)	12.0	54.5	8.5	28.3	–	
Spring potatoes	13.0	6.5	9.5	4.2	no particulars given	
Autumn tomatoes	37.0	12.3	22.6	6.5	no particulars given	
Cucumbers	23.5	23.4	17.4	8.7	no particulars given	
Citrus	14.5	4.8	13.4	3.8	12.5	3.1

[a] The great decrease in labour requirements is due to the mechanization of threshing operations.

[b] The increased labour requirement is due to the expected rise in yields.

master plan for various types of farms in 1961.[10] These data relate to moshav-type settlements.

The figures given in the above table indicate a distinct and constant decline in labour requirements in all branches, both per unit of area and per unit of produce.

Savings which surpass the estimates and norms laid down by the Joint Planning Centre in a large number of well-run farms are likewise noticeable in the field of labour productivity. The following example will illustrate the feasibility of achieving a saving in labour[44]:

Surveys carried out in the Negev in 1962 have shown that the production costs of a ton of sugar beet vary between IL. 45 and IL. 70, the divergence depending mainly on the cost of labour. While an efficient collective farm invests a total of 6 hours per dunam in sowing, thinning and weeding of sugar beet, other collective farms require 12–16 hours per dunam, while moshav-type farms require at least 2 work days, that is 16 hours, per dunam. The relevant survey of the Extension Authority concludes that sufficient technical means are available, suited both to collective and to moshav-type farms, which would cut labour requirements to 4 hours per dunam. It cannot be doubted that labour efficiency will be enhanced by the tendency to increased specialization and by the transition to a regional structure providing adequate facilities for mechanization and improved sorting and marketing.

However, Israel still lags behind the advanced countries in labour efficiency. Although average yields are generally high – sometimes higher than in many advanced countries – the raising of labour output to the highest levels attained has yet to be accomplished.

The late S. Baumgart of the Joint Planning Centre made a comparative study of labour productivity on a trip to the U.S.A.[45] and reported: "The total labour requirements for a certain crop in Israel are higher than in the United States because farms there are much more mechanized than in Israel. However, I am referring here not to the total labour requirement for a certain crop, but rather to the labour requirements for a specific operation, such as ploughing by tractor to a certain depth, cotton picking, thinning, etc. In the output of these specific operations, Israel lags behind the United States.

In the mechanized operations the differences are less noticeable and have some justification. The fields in Israel are much smaller than those in the United States and of course the efficiency of machines is lower.

However, in manual labour, the differences in output are quite large as indicated by Table 3-18. (This table compares the norms of Israel with those of California, whose climate and crops resemble those of Israel.)"

TABLE 3-18[45]

Comparison of labour norms

	California	Israel (in kibbutzim)
Thinning of sugar beet	20 hr per acre	48 hr per acre
Hoeing and weeding sugar beets	14 hr per acre	30 hr per acre
Hoeing and weeding cotton (2 ×)	24 hr per acre	52 hr per acre
Hoeing and weeding tomatoes	20 hr per acre	36 hr per acre
Picking cotton by hand	100 kg per day (8 hr)	60 kg per day (8 hr)
Picking tomatoes for canning	1.2 tons per day (8 hr)	0.6 tons per day (8 hr)
Harvesting cabbage	1.8 tons per day (8 hr)	1.1 tons per day (8 hr)

It can be seen from the above figures that Israel generally requires twice as great a labour input as that of the U.S.A.

According to Baumgart "The difference in labour output in thinning and weeding of sugar beet can be partially explained because these operations are performed in Israel during the rainy season. Sugar beet in California is planted in the early spring, whereas in Israel it is planted in the fall. . . . However, this hampering factor does not apply to the thinning and weeding of cotton and to the weeding of carrots and tomatoes, nor does it apply to the picking of cotton and harvesting of vegetables.

These differences can be traced mainly to the fact that settlers and agricultural labourers in Israel are mostly newcomers with urban skills, who have no experience in agricultural work.

A second cause for the discrepancies in output is that in California the worker is paid for piecework (namely, according to the acre, pound or container), a method which gives the worker incentive and increases his output. In Israel's agriculture this method is uncommon. When the above-mentioned operations are carried out by hired workers in Israel, their pay is computed on a daily basis and is affected only by their goodwill and skill on the one hand, and the organizing ability of their employers on the other hand – two not always reliable factors. The largest part of the work in Israel is done by the settler himself, who is also averse to the pressure of the piecework method. In addition he has generally not yet acquired adequate working and organizing ability. Another reason for the low labour output is in some cases the use of inadequate tools and containers."

In order to calculate the labour force which will be required for agriculture in 1972, an estimate of work norms is necessary. Table 3-19 presents the figures.

TABLE 3-19

Estimated work norms for 1972/3
(in principal branches)

Branch	Yield per dunam	Work hours per dunam	Work days per dunam (for 8-hr days)
Winter tomatoes	5 tons	380	47.5
Carrots	3 tons	150	18.8
Onions	3.5 tons	102	12.7
Potatoes	1.8 tons	66	8.2
Vegetables (average)		140	17.5
Groundnuts	400 kg	38	4.7
Cotton	320 kg	60	7.5
Sugar beet	5.5 tons	43	5.4
Melons	1.8 tons	52	6.5
Maize for fodder	800 feed units	30	3.7
Lucerne	800 feed units	52	6.6
Clover	750 feed units	45	5.6
Fodder beet	2000 feed units	66	8.2
Hay	150 feed units	4	0.5
Winter crops			
Citrus	4.5 tons	100	12.5
Milk	4800 litres per cow	180	22.5

NOTES

1. GIL, N. and ROSENSAFT, S., 'Soils of Israel and their Land Use Capabilities, Part I – Summary of Soil Survey', State of Israel, Ministry of Agriculture. Publication No. 54 (Tel-Aviv 1955).
2. FAO, Yearbook of Food and Agricultural Statistics, Production, (1956), Tables 1, 3 and 4a.
3. SCHULTZ, T. W., *The Economic Organization of Agriculture* (McGraw Hill Book Co., New York 1953), Chapter VIII.
4. JOHNSON and BARLOWE, *Land Problems and Policies* (McGraw Hill Book Co., New York 1954), p. 112.
5. EVEN-ARI, M., 'Ancient Desert Agriculture in the Negev Highlands, International Farmers Convention in Israel 1959 (The Government Press, Jerusalem 1960).
6. 'Irrigation Development in Israel 1949–1959', Report submitted to the ICID Congress in Madrid, TAHAL, Water Planning for Israel (1960), (mimographed).
7. 'Report on Sprinkler Irrigation in Israel', 2nd edition (January 1960) of a report submitted in 1957 to the International Centre of Sprinkler Irrigation, Verona, Italy (mimograph).
8. BAUMGART, S. (Ed.), 'Agricultural Planning Norms', Joint Agricultural Planning Centre, (Tel-Aviv 1955), (mimograph in Hebrew).
9. BAUMGART, S. (Ed.), 'Agricultural Planning Norms', Joint Agricultural Planning Centre (Tel-Aviv 1960) (mimograph in Hebrew).
10. Joint Agricultural Planning Centre, 'Master Plan for Farm Types in Moshav Settlements' (Tel-Aviv 1961), (mimograph).
11. Extension Authority, Extension Bureau for the Negev, 'Summary of Crops for 1960/61: Sugar beets, Cotton, Groundnuts', pp. 2–3 (mimograph in Hebrew).
12. PAT, J., 'Irrigation of Citrus Trees in Israel', *Hassadeh* 39, No. 6, (1959), p. 658 (in Hebrew).
13. YARON, D., BIELORAI, H., WACHS, U. and PUTTER, J., 'Economic Analysis of Input-Output Relations in Irrigation, Trans. Fifth International Congress on Irrigation and Drainage, New Delhi, 1962.
14. MATZ, R., 'Research and Developments in Electrodialysis at the Negev Institute during 1965', The National Council for Research and Development, The Negev Institute for Arid Zone Research (Ber-sheba, March 1966), (mimograph).
15. SNYDER, A. E., 'Desalting water by freezing', *Scientific American* 207, No. 6 (1962), pp. 41–47.
16. SNYDER, A. E., op. cit., p. 47.
17. Report of the Commission for investigating water needs of Eilat 1966–1971, (March 1966), (mimograph in Hebrew).
18. BRYAN, C. J., *Fresh Water from the Sea*, reprint No. 27 (reprinted by the Courtesy of the Secretary, British Water-works Ass.), (Gard J. Weir Ltd., Glasgow 1954), p. 39.
19. HAMMOND, Philip R., 'Large Reactors may Distill Sea Water Economically, *Nucleonics* 20(21) (1962), pp. 45–49.
20. TAHAL (Water Planning for Israel), 'Basic Principles for a Development Plan for the Water Economy of Israel 1962–1970; (Tel-Aviv 1961), (mimograph).
21. Op. cit., p. 6.
22. 'Report on Agriculture Presented to the Knesset by the Minister of Agriculture', Ministry of Agriculture (Feb. 1962), (mimograph in Hebrew).
23. SCHULTZ, T. W., 'A New Era for Agriculture in Economic Growth', International Farmers' Convention in Israel (The Government Press, Jerusalem 1959), pp. 198–207.
24. WEITZ, R., 'The Role of Science in Agricultural Progress', *Economic Quarterly*, No. 33–34 (Tel-Aviv, March 1962), (in Hebrew).

25. 'Proposed Plan for the Development of Agriculture and Settlement 1953/4–1959/60'. Joint Agricultural Planning Centre (Tel-Aviv, October 1953), (mimograph in Hebrew),
26. BAUMGART, S. (Ed.), 'Agricultural Planning Norms', Joint Agricultural Planning Centre (Tel-Aviv 1955), (mimograph in Hebrew).
27. 'Master Plan for Farm-types in Moshav Settlements', Joint Agricultural Planning Centre, (Tel-Aviv, September 1961), (mimograph in Hebrew).
28. International publication of the Joint Agricultural Planning Centre: 'Basic Data for a 5 Year Plan', 7.12.60, unpublished.
29. Settlement Department of the Jewish Agency, Planning Section for the Central Region: 'Economic Analysis of a Dairy Plantation Type Moshav Settlement for 1959', p. 10.
30. Settlement Department of the Jewish Agency, Planning Section for the Central Region: 'Examination of Dairy Farming in Moshav Settlements' (1961), internal publication, not printed.
31. MENDEL, Dr. K.: Farmers' Convention in Israel 1959 (The Government Press, Jerusalem), pp. 178–182.
32. Report of the Commission for Investigating the State of Agriculture in Israel, Bank of Israel (Jerusalem 1960), pp. 85–86.
33. 'Master Plan for Farm-types in Moshav Settlements', Joint Agricultural Planning Centre (Tel-Aviv 1961), (mimograph in Hebrew).
34. PERLMAN, A.: 'Data on sugar beet areas and yields', The coordinating Committee for Harvesting and Marketing of Sugar Beet (Tel-Aviv 1960), (mimograph in Hebrew).
35. 'The Growth of Cotton Cultivation', Agriculture in Israel (Hebrew monthly) No. 6 (1961), p. 9.
36. KATZ, A.: 'National and World Production of Cotton in 1959', Hassadeh (Hebrew monthly) 40, No. 9, p. 1012.
37. See: 'Cost Calculations for Irrigated Crops', Guide for 1961/2, part III, Joint Agricultural Planning Centre (Tel-Aviv 1960), p. 3 (mimograph in Hebrew).
38. Hassadeh (Hebrew monthly) 41, No. 6 (1961), p. 660.
39. 'Master Plan for Farm-types in Moshav Settlements', Joint Agricultural Planning Centre (1961), (mimograph in Hebrew).
40. Extension Authority: 'Summary of Crops for 1960/61: sugar beet, groundnuts, cotton', published by Extension Bureau for the Negev (1960), pp. 2–3 (mimograph).
41. Extension Authority: Op. cit.
42. Figures from: Statistical Abstract of Israel No. 11, p. 155 and Statistical Abstract of Israel No. 15, p. 323.
43. PATTERSON, D. W.: Work simplification in Agriculture (Land Books, London 1960).
44. 'Extension Programme for 1962', Extension Authority Negev Bureau; (1962), p. 5 (mimograph in Hebrew).
45. BAUMGART, S.: Report presented to the International Co-operative Administration, Nov. 24th 1958 (not printed).

CHAPTER 4

THE HUMAN RESOURCE AS A FACTOR OF SUPPLY

The ability of the farmer and his willingness to use the available means to farm, assume increasing importance in an agriculture of high technological and organizational levels. Schultz,[1] in a recent study states, "the differences in the capabilities of farm people (in different countries) are most important in explaining the differences in the amount and rate of increase of agricultural production". He shows that although the quality and amount of land per capita are lower in Europe than in India, the volume of production in Europe is considerably higher. He states that a major factor responsible for this phenomenon is the quality of the farmers in terms of skill and training. Schultz concludes:[2] "The man who is bound by traditional agriculture cannot produce much food no matter how rich the land. Thrift and work are not enough to overcome the niggardliness of this type of agriculture. To produce an abundance of farm products requires that the farmer has access to and has the skill and knowledge to use what science knows about soils, plants, animals and machines. The knowledge that makes the transformation possible is a form of capital, which entails investment – investment not only in material inputs in which part of this knowledge is embedded, but importantly also investment in farm people".

Willingness to farm and ability are not simple matters, but are tied in with many factors, including the way the farmer is organized to farm, the community and cultural context in which he lives, and his level of education and know-how. These three cited factors are closely intertwined, and are manifested in the manifold problems in either 'updating' or modernizing traditional farm communities, or of founding new settlements.

The introduction of modern agricultural practices into traditional communities, or the settlement of individuals or ethnic groups, are complicated and delicate tasks, requiring understanding, knowledge and adaptation. The most careful planning for the utilization of natural resources may not bring the desired results if no account is taken of the farmer within his social context.

Israel's recent experience in handling the 'human resource' has been primarily in new settlement of ethnic groups and individuals, most of which had not previously been farmers. In the very first years of large-scale immigration after the founding of the state in 1948, sociological planning was more or less non-existent, and many mistakes were made which were costly in terms of turnover of personnel in the villages and waste of manpower, to say nothing of the wider problem of turning immigrants into productive citizens. The Settlement Department, realizing that professional help was necessary, began to use sociological concepts in its planning, and at the same time teamed up with the Department of Sociology of the Hebrew University for research, consultation and use of staff. Later, sociologists became part of the regular staff of the settlement department, while sociological instruction was introduced into courses for agricultural extension workers. When the Lakhish Region was planned, the importance of sociological factors was recognized. The final fruit of Israel planning – the composite rural structure – accepts sociological and community planning as a matter of course.

A. THE SOCIOLOGIST AND SETTLEMENT PROBLEMS

When it was first proposed that sociologists be utilized in solving crises of immigration and settlement problems, there were doubts both on the part of those directing settlement policy and on the part of professional sociologists themselves. Many of the settlement staff could not envision the utility of sociological insights on day-to-day farm problems, while some of the sociologists themselves were unclear whether they could provide the answers to questions on a practical level.[3]

There were those who maintained that the instruments of sociology or social-anthropology were as yet insufficiently developed to allow practical conclusions, which could be responsibly applied. On the other hand, there were those who held that there were, at the disposal of the sociologist, methods adequately tested to enable studies which could have a bearing on solving immediate crises and on general settlement policy. These latter felt that by use in a practical way, the methods of sociology could be perfected and widened, as is the case with research methods in other sciences.

Once co-operation began between the professional sociologists and the Settlement Department of the Jewish Agency it became clear that the use of sociology was not only useful, but also necessary, while for the field of sociology in general, the problems of settlement offered data of much variety. Studies were made on a pure research level by the research wing of the joint Hebrew University–Settlement Department, Council for Sociological Affairs, while regional sociologists became part of the settlement staff, likewise carrying out research, but advising on immediate difficulties and giving lectures to field extension workers.

Perhaps the basic problem to be solved was how to enable groups of settlers from various ethnic or cultural backgrounds to adapt to an agricultural life of modern technology and organization on the base of a more or less egalitarian, democratic and co-operative framework. Although the people who arrived in waves of mass migration after the founding of the State in 1948 were Jews, they had lived for many centuries in separate groups on separate continents and had evolved differences in custom and outlook. Many of such groups were more or less 'folk-like', people, still integrated and cohesive as well as untouched by either modern organization or technology. Other immigrants, on the other hand, arrived in Israel as individuals from broken and disintegrated modern communities.

In the first years of immigration before sociological insights were made use of, and before Israel itself had the means to avoid the situation, immigrants were placed in temporary camps or 'maabaroth' until brought to the site of permanent settlement. These temporary camps proved near disastrous, causing hardship and low morale which were added to the ordinary difficulties of adjustment once the immigrants were permanently settled. A policy was therefore instituted of 'from ship to village', avoiding the transient period.

Another mistake which became righted through experience, was the placing of individual candidates for settlement in a village without regard to background, ethnic origins and solidarity with kinsmen. The orientation of policy in the beginning of mass immigration was that of the 'melting pot' approach, or that individuals would adjust to becoming Israeli's if they were intentionally 'mixed'. The villages settled by such a system did not thrive, so that by the time Lakhish was planned, it was realized that the more homogenous the background of the settlers in one village, the better. The attempt was made to settle groups of extended families or 'Hamulas' in the same village, and in some cases, whole villages were transferred from abroad to Israel. By this time it was understood that individuals could not be separated from their cultural backgrounds, and that the general concepts of cultural change for any group applied in Israel. The villages which were settled by such insights proved to be more stable and to develop more rapidly; they are today part of Israel's modern agricultural system.

One of the studies which threw light on the importance of the sociological context on willingness to farm and on village success in farming was by A. Weingrod.[4] He compared two moshavim – one which had taken root with its settlers becoming successful farmers, and another whose settlers eked out a difficult livelihood with the aid of welfare grants and relief works, even though both moshavim had been given more or less equal means of production. The study indicated that most families in the successful village were linked by kinship, while more than half the families in the backward village were isolated from their kind and people of similar background. In the former, the family ties aroused a sense of companionship and a desire for mutual aid and co-operation. The stable traditional leadership co-operated with the settlement staff and aided the acceptance of successful extension programmes for the adoption of new cultivation methods. In the backward village, however, the settlers opposed central authority, while the leaders of the small groups were at loggerheads with one another and no one group could gain the support of a decisive number of settlers. The village institutions were ineffective and the lack of organization gave the settlers a sense of bitterness and unwillingness to work their farms. This study therefore became one of the factors influencing the change from the 'melting pot' approach to that of settling people in more or less homogeneous groups.

Once policy had become that of settling people in groups, it was then realized that sociological insights concerning cultural change could be applied, and that some of the research should be concerned with it. An example was a study by O. Shapiro[5] on changes in leadership patterns in a moshav whose settlers had come from a village in the Yemen where they had been weavers and small traders. The moshav, 'Gadish', was considered successful by the settlement team. The study indicated that one of the reasons for success was that the community had been cohesive in the country of origin. However, in the adaption to Israel, there grew an internal secular leadership, not of the elders who abroad had held primarily religious or ceremonial leadership, but of younger adults who regarded themselves as the bearers of the modernization process. This particular solution of the leadership problem guaranteed the smooth development of 'Gadish', with old patterns adapted rather than drastically changed.

An example of a sociological study throwing light on long-term political integration of immigrants, rather than on any immediate problem, was that by Dorothy Willner, one of the first social-anthropologists in the Settlement Department.[6] Her paper 'Politics and Change in Israel: The Case of Land Settlement', was concerned with the adaption of the various immigrant ethnic groups to the local Israeli political scene. Her conclusions were: "It is likely, however, that the politicization of the immigrants has been serving the welfare of the nation, however much it complicated the rational implementation of land settlement or any other programme.... Practical politics is a means through which their interests as they see them can gain representation on the policy-making levels of the country.... An ideologically-motivated partisan elite built Israel and effected its creation as a state. The decline in ideology and intense partisan commitment and their replacement by practical politics may now help to integrate the nation."

If sociology has proven useful for long-term insight, on a practical level it has led to decisions solving many crises. For example, in a village which was populated by settlers from a Middle Eastern country, an active group began to run village affairs and held a large part of the lands. The remaining settlers did not farm their plots but earned their living from outside employment. Their farms were leased by the first group so that their owners had no benefit from them. The complete control exercised by these 'managers', both over the means of production in the village and over its institutions, led to the formation of two inimical factions and the disintegration of the village community. Two proposals were made at a discussion of the problem in the regional directorate. The regional sociologist proposed that an instructor of proved stature and effectiveness be brought to the village in order to reactivate the settlers engaged in outside employment and encourage them to return to farming their plots and fight for equitable distribution of the means of production. The regional director, on the other hand, backed by the settlers' movement in question, proposed that an active group of new candidates be brought to the village. The introduction of a consolidated group of different origin would in his view, create a new class in the village, undermine the power of the 'managers', and sweep the settlers employed outside the village back into farming. After full consideration of both proposals, the Director-General decided to adopt the approach suggested by the sociologist, partly because of economic considerations, but also because it was in keeping with general policy and experience.

Another practical study led to the realization that farm branches had to be fitted to the human material, or at least their rational development would meet with resistance if it impinged on a group's traditional practices. Farmers from Kurdistan of a certain moshav refused to thin out fruit or inoculate chickens despite the oft-repeated entreaties of the settlement instructors. The study showed that thinning of the fruit and inoculation were contrary to religious tradition which the settlers were not prepared to sacrifice for the sake of economic considerations.

Altogether, Israel has gained varied experience in the last fifteen or so years since the first concerted use of sociology, in handling the human resource in agriculture. Although the experience has been primarily in problems of new settlement, many generalizations can be extended to the updating of traditional farm communities. However, the first rule is always flexibility fitted to specific situations, rather than wholesale adoption of any method solving 'human problems', even if successful in Israel or elsewhere.

B. SOCIAL CONTEXT OF THE FARMER

We stated that willingness and ability to farm were closely tied to three factors – the way of organizing the farmer, his community and cultural context, and his know-how level. Israel's experience throws some light on each one:

1. Organization of the farmer

In Chapter 7 it will be pointed out that the family farm has not been replaced by the large-scale managed farm as had been predicted, but that the great leap in agricultural production in almost all countries in the last twenty years has been within the structure of the family farm. A highly efficient agriculture can be maintained as long as either the individual farm units join forces for large operations, as in co-operative villages in Israel, or possess sufficient capital and mechanization to operate large-scale farms on their own as in the U.S.A.

One reason for the efficiency of the family farm is that the family unit is flexible in manpower. The farm as a unit of production involves many differing, relatively small tasks which have to be carried out periodically without neglect. It requires constant watch for unexpected events, such as climatic variations, and quick adjustment to crises.

The flexibility of the farm-family in terms of labour supply becomes apparent in view of the unequal distribution of labour requirements over the year. Within certain limits the family can adjust itself so that it supplies extra work-days at certain peak seasons with the help of wife and children, while in off seasons the farmer can utilize his time for maintenance and repair work.

In the planning of individual farm-units, the size, structure and capabilities of the operating family has to be taken into account along with other factors. The basis of manpower planning for two types of farms in Israeli moshavim as shown in Table 4-1, was the nuclear family with from two to four children.

As a unit of production each farm was planned for 350–400 work days per year, or a monthly average of thirty work days. Allowing for a variation from this norm of five work-days per month, there were only three months

TABLE 4-1
Work schedules of Israeli moshavim – field-crop farm and hill farm

Farm type	Total workdays	Month											
		10	11	12	1	2	3	4	5	6	7	8	9
Field-crop	355	38.7	26.5	22.7	18.4	22.3	23.4	21.6	34.7	37.0	28.0	38.8	42.9
Hill	369	12.8	15.8	21.8	15.3	20.3	26.8	25.8	29.3	59.8	57.8	45.8	37.8

of the year which were 'normal'. The fieldcrop farm had five months in which there was variation of more than eight days per month, two reaching a maximum of twelve days, while the hill farm had nine months in which the variation exceeded eight days, including five with a fifteen day variation or more. The two peak months of the hill farm – June and July – needed more than two full time workers, while during October, November and January, the farmer could well be unemployed for half of each day or even more. Hence in either farm type the variation in monthly work load made it necessary for the farm unit to be flexible, as it could be if the unit was the family. The income level to be derived from the above model farms was also calculated on the basis of the average nuclear family.

The demographic constellation of the farm-family is a decisive factor in its ability to reach the stipulated level of income and in the suitability of this income to support the family needs. However, in certain families a lack of labour supply can cause underutilized resources; in others the income may not be sufficient to cover all needs. A balance between the size and structure of the family, the work requirements and the level of income is therefore necessary.

A demographic survey of moshavim in Israel by Weintraub and associates showed that this balance was not always achieved.[7] They classified and discussed the basic types of farm-families as follows:

1) *The fully-productive childless family:* Such a family produced more than it needed. It was typical of the first period of settlement, and was, in general, a temporary stage which became marginal in the case of veteran well-established farms.

2) *The balanced Western family:* The planning of a typical moshav farm was based on the balanced western family with from two to four children. Consumption and production requirements were to be balanced so that the farm family could obtain an adequate income without overwork. Such a family unit was typical in old-established moshavim.

3) *The balanced large traditional family:* Many of the settlers who came to Israel from traditional backgrounds were in nuclear families with many children and were part of extended families or 'hamula'. The typical moshav farm could not utilize all the available labour in such families nor could supply all the consumption requirements. When balanced, such a family had members who found outside work. A balance could occur, of course, only in regions where other employment opportunities were available.

4) *The unbalanced small Western family:* Such families, survivors of war and extermination in Europe, were characterized by late marriages and consequently often consisted of elderly parents with young children. The labour capacity of the family was smaller than required and the shortage was balanced by hired labour and more mechanization than otherwise would have been necessary. In certain cases such families became fully productive when the children grew up and took over. In other cases the settlers sought outside work, using the village as residence only.

5) *The unbalanced large traditional family:* Such a family was demographically similar to type 3 described above. The consumption requirements exceeded the capacity of the farm, however. There was no excess labour force and sometimes there was a shortage because there were not enough grown children to supplement the earnings, or help on the farm. The adults were likely to be impaired by poor health, working power, etc. Such a family situation was not necessarily static and could become 'balanced' after a few years. However, when the family pattern did not change for a long time, it was liable to become a social case and a burden on the community as a whole.

The difficulties in such variation as illustrated in the case of Israel moshav settlement are obvious, especially when it is realized that an unbalanced family may become balanced by the maturing of a child, or vice versa when a secondary earner in a large family departs or starts his own family. On the other hand, unbalanced families not only cause difficulties in their own productive activities, but harm the smooth functioning of the rural community in general. Careful planning may not be able to eliminate all the imbalance in families in a village, but it can, at least, plan around the typical demographic pattern of the group which is being settled. For instance, more mechanization or hired labour may assist the unbalanced small family over the difficult period before the children grow up; larger farms may be allotted to large balanced families, etc. All these factors and possibilities should be examined during the planning process.

Organizing the farmer so that his labour is organized most productively is not only a matter of careful demographic planning of the family unit of production, but involves his participation in units with other farmers for those large-scale activities mentioned elsewhere, from ploughing, cultivating and harvesting with machinery, to marketing, obtaining finance, etc. Such matters are an integral part of any agricultural plan, and their success has a bearing on the farmer's willingness and ability to farm. In Israel the small-holder's co-operative or moshav was an answer to such necessary organization on the village level, while the moshav itself became tied into regional organization for certain activities unsuitable at the village level (see Chapter 7, pp. 311f).

Becoming part of organizations which are not based on family ties is one of the problems of modernizing 'folk' people. Family obligations can be at cross-purposes to democratic requirements of running a co-operative, for example, while family authority may not be effective in coping with requirements of modern technology, so that it may be the trained youth rather than the elders who wield power. This fitting of family structures into modern structures is one of the reasons why sociological insight is necessary in agricultural planning. However, there is an upsurge among traditional villagers all over the world to become part of the technological-scientific world of today so that such problems can be worked out if treated carefully. Family organization is part of the community and cultural complex as a whole in which the farmer exists and cannot be treated entirely apart from it.

The role of the Settlement Authority in the beginning stages of settlement was all important in introducing the settler to farming, whether he was of folk-like background or western. Out of necessity, the new 'farmer' found himself involved in a bureaucratic framework which gave him instructions he often did not understand, yet which if he did not carry out could mean lack of success in farming, and of course, meagre income. Methods and attitudes of the settlement staff were therefore all important, particularly of the village instructor who was the mediator between the authority and the farmers. A. Weingrod[8], in a case study of village development, points up the all-important role of the instructor which was one not only of handing on directives for farming and mediating between the village and the authority, in matters such as marketing the settlers' produce, etc., but who was also a model for the Israeli way of life towards which the settlers began to adapt. As has been mentioned, instruction in sociological insights became indispensable to the instructor, for it allowed him more flexibility in handling the human problems with which he had to deal in those first years. In a way, he wielded immense authority over the families under his jurisdiction. Since the whole system of the Settlement Authority was to enable the settlers to become independent of 'guided farming' as quickly as possible, handling their own affairs through elected representatives, the over-importance of the instructor's authority, lasted only for a relatively brief period. Afterwards his work became that of ordinary extension. The role of the organization which handles either new settlement or the modernizing of old farming regions is not to be underestimated and is of prime importance in any development system.

2. Community and cultural context

i. *The regional base*

Just as no doubt there are deep social-psychological reasons why an individual farming on a family basis is more likely to be a successful agriculturist rather than if he farms in a labour brigade, so is a flourishing community likely to have better farmers than one which struggles far behind standards of health, education and of living in the rest of a country. As has been learned throughout the book, Israel, in settling the large wave of immigrants which arrived after 1948, attempted to organize them in an agricultural system whose means of production would give the farmer an income equal to that of the skilled city worker. Prior to the founding of the state, the collective settlement or 'kibbutz' had constituted the main type of settlement or 51% of the total, but after 1948 the preponderance of settlement was in the 'moshav' or smallholders co-operative village. Out of 450 communities established, 258 were of this type.[9]

The village or 'moshav' co-operative, was tied into the co-operative branches of the national labour organization which also supplied the village health service. Education, welfare, and other services were provided by the government, while the moshavim became rather quickly absorbed by various political parties. Thus at the very outset of settlement, the immigrant found himself part of a democratic nation with opportunity for a standard of life on terms with that of other workers. There is no doubt that such a base greatly aided the effort of the individual immigrant to learn to become a farmer. It should also be remembered that the continued refinement of the network of villages into the composite rural structure, in an attempt to solve not only problems of transportation, marketing and services, but also human problems such as excess labour from the second generation, continues to provide a sound base for healthy communities within a region.

The question then becomes – "Although in new settlement it is possible to provide such a base, how can Israel experience throw light on the modernizing of traditional farming villages long established?" The answer lies, of course, in overall planning of regions, if not to immediately close the gap between the farmer and the industrial worker, at least to place in the hands of the farmer the means to improve his production, provide jobs for excess farm labour, and extend services.

ii. *The social background*

Most Jewish agricultural settlers prior to 1948 had come from European backgrounds with relatively high levels of education and experience in organization other than that of family. They were in general organized into socialist political parties, were trained in groups of young people for settlement, and were sustained during hardships in the pioneering years by an idealism of Zionism. Such pioneering settlements were an integral part of the Yishuv (Jewish population) as a whole, helping to set its standards. By 1948, the Yishuv was in general a modern population which utilized most of the advances in contemporary technology and which was at home in an 'urban' way of life where family ties were not necessarily the controlling factor after adulthood.

As we have mentioned on p. 141, when the wave of mass immigration arrived in Israel after 1948, it was at first expected that each settler could be thrown into a village without regard to his cultural background, and that willy nilly, he would become 'melted' into a modern Israeli, taking on all the standards and customs of the old Yishuv. When this did not happen, a realization began to grow that planning had to take into account what was known of group integration and cultural change. Settlement became planned on a basis of more or less homogeneous cultural or ethnic background of all settlers in a village with crises often sorted out by sociological help. By the time of the writing of this book, agriculture in Israel of those settled after 1948 is an important part of the total economy.

In a survey of 223 moshavim conducted in 1958, it was found possible to divide the settlers into three general groups, each with peculiar characteristics – North African, Asian, and of Western extraction. The three 'ethnic' divisions were broad, yet enough specific social and cultural differences could be distinguished to warrant a comparison.[10]

First, Settlers of North African Extraction: Although at one time the Jewish communities in North Africa had manifested the characteristics of 'folk' communities, by the 1950's their isolation and cohesiveness had been broken by the extensive influence of non-Jewish or foreign contact. They lacked autonomy, with community leadership manifested almost solely in the economic sphere. They were composed of families of small craftsmen, merchants, and a small number of large-scale tradesmen, in towns and cities. Religious leadership, having lost much of its previous influence, did not constitute an important factor in the community. Insecurity in the surrounding society was the incentive for joining the immigration organized in Israel for settlement there.

In this group as a whole, there were different levels of education, technological training, familiarity with 'westernization' or 'modernization'. In general, however, the level of education and of technological background was low. Adaptation to Israel was difficult not only because of the change from life in the towns or the back alleys of North African cities to rural life, but also because of the requirements of the modern, organized structure of the villages.

Second, Settlers of Asian Extraction: In contrast to the immigrants from North Africa, those from Asia came from folk communities, which had retained a certain integrity. Jewish community life in the Middle East and India had been characterized by a considerable degree of autonomy in the cultural and spiritual spheres, though less in the political and economic spheres. Social relations with the non-Jewish society were limited, confined primarily to economic, commercial and political contacts with the authorities.

Internal social differentiation manifested itself on the one hand in occupational and economic roles (higher status of certain trades, e.g. goldsmiths, landowners, merchants) and on the other hand in the religious sphere. Traditional religious leadership formed the central focus of the social structure and was the main source of social influence. Generally this leadership was intimately connected with the economic leadership of the community.

Immigration to Israel was principally conceived of as a mystical act, inspired by rumours of the establishment of the State. As a rule it took place within existing social frameworks under the auspices of traditional leadership, and constituted an example of a 'transfer of whole communities'.

With regards to technological training, the conditions prevailing in their countries of origin did not prepare the immigrants for modern farming, although the traditional structure of 'extended-families' (Hamula) was somewhat adaptable to co-operative patterns of settlement.

Third, Settlers of Western Extraction: Owing to the general trend of Jewish life in Central Europe, part of this group had retained its traditional quasi-autonomous social structure, while another part had completely lost touch with Jewish life. All the intervening stages of assimilation were represented as well. For the majority of these immigrants, the Nazi persecution and the ensuing events were the determining factor for emigration to Israel. The immigrants did not arrive in groups and often not even in families. Most who came had been made economically destitute although they were modern people, adapted to contemporary technology, organization and education.

The peculiarities of each group from the traditional or folk-like Asians, to the 'transitional' North Africans to the unintegrated Westerners, brought specific problems in the development of moshavim. The traditional immigrant "was thus foreign to rational economic motivation and behaviour, and hostile to the democratic government and bureaucratic administration characterizing the modern village. He lacked the complex skills and the scientific requirements of advanced farming, and the attitudes and frame of reference implied by the marketing, financing and development processes. Finally, his previous social environment had not prepared him for interaction transcending the narrow ascriptive criteria of kinship, and in most cases he could only live in homogeneous communities. Such 'organic' villages, 'transplanted' from their native milieu, could make for greater solidarity and ability to absorb change; however, they often also constituted pockets of ethnic isolation. Conversely, the more urban and 'modern' the background of the immigrant, the more he tended to reject the ideological orientation towards personal and social pioneering, substituting individualism for collectivism, 'Gesellschaft' for 'Gemeinschaft'. He shrank from agriculture as a vocation, from manual labour, from physical isolation and from cultural provincialism. Thus, one can say that the traditional section of the immigrant population was at odds with the modern social and economic elements of the moshav, while the 'Western' section was in conflict with the emphasis on colonization and the limitations placed on economic activity and entrepreneurship."[11]

3. Education and know-how level

We have stated that ability to farm is in part dependent on educational and know-how level of the farmer, and that continuous effort has to be invested, as a sort of capital, in raising the level. However, experience in Israel has proven that this factor is closely interwoven with the other two – the way the farmer is organized and his community context. When he is given satisfactory returns for effort by the way he is organized, and when he is supported and can act in a satisfactory group context, he will show more willingness to make the effort to improve his technical level.

On this score, a study by Weintraub and Lissak of certain farms in Israel showed that cultural background and ethnic origin were decisive factors in the degree of acquiring modern technology.[12] Farmers of 'western' extraction were quicker to adjust and take full responsibility of their farms than farmers of 'oriental' origin. Out of 24 villages who reached a status of consolidation, or a state of full production and autonomous government, 23 consisted of European farmers. The adaptation of oriental people with a traditional background required a longer time and a different approach in order to reach similar levels of production.

The study noted that although most new settlers had had no previous experience in farming, the acquisition of manual and mechanical skills did not present any particular difficulties. However, great variations in adaptation were noticeable in the theoretical and decision-making spheres and in the mastering of organizational problems, such as management, marketing, etc. An example cited was that of farmers in a certain village who did not sow sugar-beets at the right time, or used insufficient or excessive amounts of fertilizers, obtaining a low percentage of sugar content and consequently lower returns. An economic analysis made in that village showed that due to improper handling of farm and marketing operations the individual farmer lost on the average a potential income of IL. 1000. Table 4-2 illustrates the differences in income due to differences in efficiency.

TABLE 4-2
Production and marketing of onions in a sample of seven farms

Production and marketing of onions	Farm code number						
	1	2	3	4	5	6	7
Amount marketed (in tons)	1.30	2.00	1.90	0.60	3.10	14.90	1.80
Amount spoilt (in tons)	1.30	0.50	3.00	–	1.00	–	–
Home use (in tons)	0.05	0.03	–	0.02	0.03	0.06	0.05
Total production (in tons)	2.65	2.53	4.90	0.62	4.13	14.96	1.85
Area (in dunams)	1.50	2.00	2.00	0.50	3.00	4.00	4.50
Yield per dunam (in tons)	1.63	1.26	2.45	1.60	1.38	3.74	0.41
Income (in Israeli pounds)	145	172	164	51	285	1010	168
Average price obtained per ton marketed (in Israeli pounds)	111	86	86	85	92	68	93

C. SOCIAL BACKGROUND AND DEVELOPMENT RATE – THE ISRAEL CASE STUDY

The three important factors in willingness and ability to farm – the way the farmer is organized, his cultural background and his education and know-how level – are all closely intertwined as we have mentioned. Because of

Israel's special circumstances of settling large numbers of people from diverse backgrounds, it has been possible to make a study of development rate with the first factor constant, comparing the combination of the other two in the three main cultural or ethnic backgrounds of Israel's immigrants as described on p. 149, that is, Asian, North African and Western. A survey was made in 1958 and again in 1963 of 223 moshavim, comprising 14545 families or a population of 79529 persons.[13]

The study was divided into two parts. The first part examined four other factors or external parameters which might have influenced the development rate of the villages along with the ethnic factor. These were: proximity to the borders, housing conditions, allocation of natural production means (land and water quotas), and investments in the villages. The second part dealt with an evaluation of the development rate of the villages, according to the three ethnic groups.

The statistics of the first part of the study showed that proximity to borders, housing conditions and investment in the villages were more or less equal for all three groups, but that there was divergence in the allocation of means of production. Following is an examination of that factor:

Allocation of natural production means: From the economic view point, the terms 'dunam' of land or 'cubic metre' of water could not be used as a common denominator for different settlements. A meaningful denominator was derived from the system of farm-type models; dairy farm, field crop farm, citrus farm, hill farm. The planning of the farm-type models was based on a net income of IL. 5000 annually at 1965 prices. In 1958 the average income was IL. 3500. The actual land and water quotas for all the villages were determined and compared to the average planned quotas of the various farm-types. Accordingly the following grades were established:

1) Villages below the average range – poor (A).
2) Villages within the average range – average (B).
3) Villages above the average range – good (C).

TABLE 4-3

Comparative situation of land and water quotas in the three groups of villages

| | All villages | Villages according to settler's origin | | |
		North African	Asian	Western
Category A	21%	16%	27%	20%
Category B	32%	45%	33%	19%
Category C	47%	39%	40%	61%

Table 4-3 shows the comparative situation of land and water quotas in the three groups of villages under study in 1963.

The unequal distribution of the natural means of production, as indicated in the table, was the result of historical development, the period when the different regions were settled and the time of arrival of the various groups of immigrants. The table shows, for example, that 61% of the villages inhabited by settlers from the West enjoyed good – i.e. above the average – land and water quotas. This was an important factor which might have influenced the development rate of this ethnic group. The differences in land and water conditions between the different groups, therefore, had to be taken into account, while evaluating the effect of the settlers' origin on the development rate of their villages.

In order to compare the rate of development of the communities inhabited by settlers of the three ethnic backgrounds, it was necessary to define a 'developed' village and how a 'development' measure could be determined. First, it was assumed that 'development' could be defined by specific economic, social and political criteria. A model of an 'ideal-village' was built, measuring all the villages included in the study against the model. The model was constructed of ten indicators:

1. Percentage of outside income of the settler.
2. Acquisition of agro-technical know-how.
3. Adaptation to farm work.
4. Size of investments by the farmers in the farm.
5. Indebtedness of the village organizations.
6. Efficiency of village marketing institutions and their acceptance by the farmers.
7. General stability.
8. Stability of the village elected authorities.
9. State of village public institutions (co-operative store, consumers' and producers' co-operatives, cultural centre, etc.).
10. Co-operation of the village and its elected institutions with the national authorities.

Following are the results of the evaluation of the ten model indicators:

1. Percentage of outside income of the settler

Income from sources outside the farm in a given year was calculated as a percentage of the net income of the farm during that year.

Since the development of the agricultural means of production was implemented according to the equal income principle, the percentage of income derived by outside work indicated the degree of the settler's adaptation to the desired economic set-up of his farm.

The survey showed that outside income ranged mostly between 0% and 50%. In constructing Table 4-4 a village whose members' average outside income amounted to 50% or more of their total income, was given the minimum rating of zero points. Villages whose members had almost no outside income were given the maximum mark of ten points, as was the model village. For every 5% of outside income, one point was deducted from the maximum number of 10 points. After awarding corresponding marks to all the villages under study, they were classified into three categories and the median averages for the years 1958 and 1963 were calculated. Villages within the range of 1.5 points of the average mark were classified as category B or average, those above this range as category C or good, and those below, as category A or poor. In 1958 the average marks were 6.3, in 1963 4.7. The results of the calculations are presented in Table 4-4.

TABLE 4-4

Percentage of outside income of the settler

Villages	Average points		Category A (poor)		Category B (average)		Category C (good)	
	1958	1963	1958	1963	1958	1963	1958	1963
North African	5.4	4.1	31%	46%	62%	24%	7%	30%
Asian	6.4	5.1	14%	24%	76%	37%	10%	38%
Western	7.2	5.2	11%	27%	61%	30%	28%	43%
Total	6.3	4.7	18%	32%	66%	30%	15%	37%

Table 4-4 indicated that whereas in 1958 66% of all the villages belonged to category B, or were average in obtaining their income from sources other than the farm, in 1963 only 30% were average, with the other 70% more or less divided between the extremes of category C and A. A similar trend in the decline in the average toward the extremes was observed for many of the other ten indicators and could be explained by the process of polarization as a result of growth. The decrease in the average marks indicated that during the five-year period, the amount of outside earning in the total income increased to a certain degree, or deviated further from the model.

Altogether, the examination of outside income revealed that although the ethnic factor played a role in the first survey of 1958, with the Western villages leading the 'good' category, by 1963 the role had lessened.

2. Acquisition of agricultural know-how

The level of know-how was determined by indirectly calculating results which could be measured quantitatively, such as levels of crop yields. For example: The average vegetable yields per dunam for 1962 were calculated as a percentage of the district norm for that same year. In a similar way the percentage of deviation from the norm for the district was computed with regards to the main agricultural branches. Each of the main branches was given specific weight in accordance with the farm-type model and the composite average was calculated accordingly. The survey showed that most villages ranged between 50% to 100%. Hence, a village whose average yields amounted to 50% of the district norm received the minimum marking of zero points, and a village with an average of 100% was assigned the maximum of ten points.

TABLE 4-5

Acquisition of agricultural know-how

Villages	Average points		Category A (poor)		Category B (average)		Category C (good)	
	1958	1963	1958	1963	1958	1963	1958	1963
North African	5.9	7.1	20%	24%	73%	56%	7%	20%
Asian	6.7	7.6	13%	23%	79%	50%	8%	27%
Western	7.3	8.4	5%	11%	73%	46%	22%	43%
Total	6.7	7.7	13%	19%	75%	51%	12%	30%

The figures in Table 4-5 indicated that, as expected from the settlers' education and background, the Western villages exhibited the highest percentage of technical know-how. The other two groups, however, showed important gains in the five year period, with that of the Asians being the most notable of the three groups.

3. Adaptation to farm work

Most of the immigrants in rural settlement in Israel came from a non-agricultural background. As we have seen, the majority had been city-dwellers, engaged in commerce, trade and various other urban occupations. Even the minority hailing from rural areas was unfamiliar with the branches of agriculture practised in Israel, with the methods and techniques applied, and with the organizational structure of Israeli rural settlement. Hence, adaptation to farm work meant a decisive change in the settlers' lives.

Adaptation to farming was measured in this study by the degree of willingness the settlers showed to cultivate their fields. This was evaluated by the subdistrict officers under the supervision and control of the district directors. A scale from zero to ten was established, based on the proportion of plots cultivated individually by the settlers themselves or by neighbours, the village community or by outside bodies. The neatness of the fields was also borne in mind, the compliance with agrotechnical instructions, attitude to field work and the realization of its importance for the future of the family and community.

TABLE 4-6

Adaptation to farm work

Villages	Average points		Category A (poor)		Category B (average)		Category C (good)	
	1958	1963	1958	1963	1958	1963	1958	1963
North African	6.1	7.2	40%	21%	43%	56%	17%	23%
Asian	7.8	7.6	14%	15%	49%	54%	37%	31%
Western	7.6	7.3	18%	24%	59%	54%	23%	22%
Total	7.2	7.4	23%	20%	51%	55%	26%	25%

It is intriguing that the North African group, which in the beginning had seemed to show the greatest resistance to the social and organizational patterns in Israel, was the one to make an advance in adaptation to farm work. As to the other two groups, the figures seem to warrant a conclusion that there were individuals less inclined to regard field work as a worthy sphere of endeavour, although the communities as a whole showed willingness to adapt.

4. Size of investment by the farmers in the farms

Investment of the settlers in their own farms was computed as a percentage of the total net annual income. The results showed that most cases ranged from 0% to 20%. A village which did not invest at all was given the lowest mark of zero; one with an average investment of 20% of the total annual net income, was accorded the mark of 10. For every 2% of the net annual income invested, one point was added to the minimum mark.

Readiness to invest depends on a general awareness of the importance of investment. Unlike the Western immigrants, settleis from North Africa and Asia were used to saving for a rainy day, but not to investing their savings. This practice stemmed from the nature of the occupations in which they had

TABLE 4-7

Size of investment in the farms

Villages	Average points		Category A (poor)		Category B (average)		Category C (good)	
	1958	1963	1958	1963	1958	1963	1958	1963
North African	3.7	4.3	46%	51%	46%	20%	7%	30%
Asian	5.3	5.5	23%	38%	53%	26%	24%	36%
Western	6.7	6.4	9%	23%	49%	31%	42%	46%
Total	5.3	5.4	26%	37%	49%	26%	25%	37%

been engaged and the insecurity of their original environment. With change of economic activity under the influence of conditions in Israel, the North African and Asian settlers, according to the figures, learned some investment habits.

The recession, which took place in the rate of investment of the Western villages was most probably – this is only a conjecture – due to the fact that their effective means of production had about reached the projected scope.

5. Indebtedness of the village organization

The liabilities of the villages to third parties, over and above the amount required for the normal operation of farming activities, were computed as a percentage of the total output. The survey showed that the percentage ranged between 0% and 50%. The minimum mark of zero was thus accorded to villages with an indebtedness of 50%, a maximum mark of ten for settlements with no indebtedness. For every 5% indebtedness above zero, one point was deducted from the maximum mark.

TABLE 4-8

Indebtedness of the village organization

Villages	Average points		Category A (poor)		Category B (average)		Category C (good)	
	1958	1963	1958	1963	1958	1963	1958	1963
North African	6.2	5.1	14%	30%	73%	42%	13%	28%
Asian	6.9	6.1	9%	23%	78%	31%	13%	46%
Western	6.5	6.0	16%	27%	73%	23%	11%	50%
Total	6.6	5.8	13%	26%	75%	32%	12%	42%

The questionnaires included details which allowed us to conclude that the general improvement in becoming clear of indebtedness was due to a number of reasons, such as the increased efficiency of farm operations and the expansion of credit facilities by the settlement authority for current operations and for consolidation of outside debts. Possibly the reduced indebtedness to outsiders could also be ascribed, even if only to a modest extent, to the greater investment-mindedness noted from Table 4-8. The North African villages had not yet reached the stage of reduced indebtedness of the others, but the three groups seemed to be moving towards a common level.

6. Efficiency of village marketing institutions and their acceptance by the farmers

The Settlement Authority policy was that of the settlers marketing their produce on a co-operative basis through village co-operatives. Such marketing had a number of advantages, such as lower transport costs, centralized sorting and packaging, protection against dumping prices and the saving of time and labour since the individual did not have to look for customers. Despite these advantages, many settlers took their produce directly to the market or made arrangements with dealers on the spot. The main reason for the unorganized marketing was the wish of the settler to free himself from obligation and debt to the village co-operative. There may have been a certain temporary advantage for the settler to such unorganized marketing, but in the long run it undermined the functioning of the co-operative and the cohesiveness of the village.

In order to determine the efficiency of the marketing organizations and the degree of co-operation they received from the settlers, for each farm type one characteristic product was chosen. The products chosen carried no government subsidy since subsidies were payable only through the authorized marketing organizations. On the other hand, care was taken to choose representative items which made up a considerable proportion of the settlers' income. The quantity of the specific produce marketed through the organized channels during a given year was computed relatively to the total production of the village. This ratio was found to vary from 0% to 100%. Hence villages which marketed none of their produce through regular channels were accorded the minimum mark of zero, and villages who marketed their entire produce collectively, the maximum of ten points.

From Table 4-9 we learn that the number of 'good' villages of all three groups increased, with the Western villages having the highest percentage of good villages in 1963. The North African group was not far behind; it is the

TABLE 4-9

Efficiency of village marketing institutions

Villages	Average points		Category A (poor)		Category B (average)		Category C (good)	
	1958	1963	1958	1963	1958	1963	1958	1963
North African	6.7	6.8	28%	18%	50%	38%	23%	44%
Asian	6.4	5.4	33%	36%	51%	37%	15%	27%
Western	7.5	6.4	16%	27%	49%	24%	35%	49%
Total	6.9	6.3	26%	27%	50%	33%	24%	40%

only group whose poor category decreased, and was therefore the one to show the most improvement in acceptance of co-operative marketing. Adaptation to co-operative marketing showed less improvement in the Asian group.

7. General stability of the village

In order to estimate the general stability, the average number of families which had settled and then had left was computed for 1958, 1960, 1961 and 1962. In order to eliminate the effect of the size of the village and numerical volume of the migration, we compared the figures against the total number of families in the settlements at the end of September of each of the years. The 1963 figures are the averages for the three years 1960, 1961 and 1962.

Most villages ranged between 0% and 15%. Hence a village which had an average of zero percentage received a maximum number of 10 points, while a village with an average of 15% was accorded a minimum mark of zero. For every 1.5% above the minimum one point was deducted from the maximum mark.

TABLE 4-10

General stability of the village

Villages	Average points		Category A (poor)		Category B (average)		Category C (good)	
	1958	1963	1958	1963	1958	1963	1958	1963
North African	6.5	7.6	32%	14%	58%	73%	10%	13%
Asian	7.2	7.9	14%	17%	74%	62%	12%	22%
Western	7.5	7.1	12%	22%	81%	70%	7%	8%
Total	7.1	7.5	19%	17%	71%	68%	9%	14%

The figures show that there was an improvement in the general average, with the North African villages and the Asian showing upward trends in stability, but the Western a downward trend.

8. Stability of village councils

The number of times the elected village council was changed in a given year was ascertained, and in the majority of cases, one to three changes were recorded. A village whose council was not changed even once was accorded the top mark of 10, a village whose council was changed once received 5 points, twice – 4 points, three times – 3 points.

TABLE 4-11

Stability of the village councils

Villages	Average points		Category A (poor)		Category B (average)		Category C (good)	
	1958	1963	1958	1963	1958	1963	1958	1963
North African	6.0	9.4	28%	13%	54%	87%	18%	–
Asian	6.8	9.2	12%	14%	72%	86%	17%	–
Western	7.3	9.6	9%	7%	53%	93%	38%	–
Total	6.7	9.4	16%	11%	60%	89%	24%	–

The trend shown by the figures in Table 4-14 is that turnover of the village councils twice a year was the general rate, rather than three times or one time, as had been the rule in 1958. This was a trend, in general, towards the consolidation of the villages and an amelioration in the relations between the villagers and their elected councils.

9. State of village public institutions

The state of the various institutions – the co-operative store, the sorting and packing centre, and so forth – was evaluated by the subdistrict officers using a descriptive scale. Marks were awarded on a scale of ten to zero points, the criteria being the quality of the village services, the cleanliness of the premises, the efficiency of the management and the book-keeping.

The table shows an improvement in the state of public institutions in the North African and Asian villages with the good category increasing slightly, and the poor decreasing, the average constituting the majority. Although the Western villages had the largest category of 'good', they did not show the same improvement as the other two.

TABLE 4-12
State of village public institutions

Villages	Average points		Category A (poor)		Category B (average)		Category C (good)	
	1958	1963	1958	1963	1958	1963	1958	1963
North African	5.9	7.1	31%	28%	55%	55%	14%	17%
Asian	6.3	7.1	29%	23%	60%	64%	10%	13%
Western	7.5	7.7	7%	24%	61%	38%	32%	38%
Total	6.5	7.3	22%	25%	59%	52%	19%	22%

10. Co-operation between the village and the local government authorities

The degree of co-operation with government authorities was assessed by the subdistrict officers under the control of the district directors according to a scale ranging from zero for the lowest degree of co-operation to ten for the highest. These estimates are presumably the most subjective among the non-quantitative evaluations contained in this study. Nevertheless, the evaluations were made by persons with long-standing experience in rural settlement and development problems who belonged more or less to the same school of thought.

TABLE 4-13
Co-operation with local authorities

Villages	Average points		Category A (poor)		Category B (average)		Category C (good)	
	1958	1963	1958	1963	1958	1963	1958	1963
North African	6.4	7.5	31%	14%	55%	69%	14%	17%
Asian	7.2	7.3	15%	23%	65%	62%	19%	15%
Western	7.7	8.3	7%	7%	62%	61%	31%	32%
Total	7.1	7.7	17%	15%	61%	64%	22%	22%

The North African villages again showed the greatest improvement; the Western villages remained about the same in time periods, but constituted the highest percentage of the good villages of all three groups. The Asian villages showed a decline in willingness to co-operate with the authorities.

11. General development rate of the villages

a) The general development rate of all the villages together, and of the three groups separately was evaluated by giving an equal weight to each of the indicators examined in the previous tables and calculating their average. They were then compared against the model of ten points.

TABLE 4-14
General development rate of villages under study

Villages	Average points		Category A (poor)		Category B (average)		Category C (good)	
	1958	1963	1958	1963	1958	1963	1958	1963
North African	5.9	6.6	48%	25%	38%	63%	14%	11%
Asian	6.7	6.8	26%	19%	56%	59%	18%	22%
Western	7.3	7.3	14%	9%	45%	59%	42%	31%
Total	6.6	6.9	29%	18%	47%	61%	25%	22%

In Table 4-15 the figures for the level of development of each ethnic group in the two survey years were worked out on the basis of Table 4-14, awarding one point for each percentage of poor villages, two points for each average, and three for each good. The results were as shown in Table 4-15.

TABLE 4-15
Indices of development

Ethnic group	1958	1963
North African	166	184
Asian	192	203
Western	230	220
Total	198	206

The two tables show a trend in the North African villages and in the Asian of closing the gap between their development level and that of the Western villages. Although the settlers of Western origin had started on the highest level and remained on the highest, their development rate nonetheless declined. The progress of the Asian group reached near to the average for all groups, while, although the North African group remained on the lowest level, its progress was the most marked. It can be concluded, therefore, that the development rate was greater the lower the starting point, and that the rate tended to decline once a certain standard had been reached.

b) In Part I of the study, external parameters were examined and only one, the allocation of natural production means, or the land and water quotas, was found to be significantly different for the three groups of villages. This factor is introduced in Table 4-16.

TABLE 4-16

Rate of development under varying allocations of natural production means

Villages	Deficient conditions A		Average conditions B		Good conditions C	
	1958	1963	1958	1963	1958	1963
North African	100	183	171	184	191	189
Asian	162	179	197	200	208	222
Western	162	193	221	223	250	229
Total	146	186	189	195	223	218

From Table 4-16 the following may be noted:

1) When land and water conditions were equal, the Western villages were on the highest level of development, after them the Asian villages, and lastly the North African.

2) Under average and deficient conditions, all groups showed development in the five years between 1958 and 1963, but when the conditions were 'good', only the Asian group progressed with the Western declining noticeably.

Hence, when land and water conditions were taken into account, the same conclusions were reached as above, that once a level of development had been attained the rate of development decreased, and in the case of the Western villages, became negative. Again it is seen that the Asian group was nearing the level on which the Western villages had begun, while the North African group, although progressing, was behind. Table 4-16 agrees with the results of the other tables, that each ethnic group exhibited its specific rate of development, but at the end of the five-year period, the level of development was tending to be the same.

12. Significance of the Israel Case Study

From the foregoing study of the influence of the ethnic factor on the development rate of the 223 moshavim, several conclusions can be drawn:

1) That uprooted people with no previous experience in farming could be settled into agriculturists provided they were carefully organized and guided.

2) That the ethnic background was an influential factor in adjustment to farming and the level attained, but that given proper guidance, it was a factor

which became less important through time. The Westerners, the group with the highest level of technical and modern background, were the quicker to adjust, exhibiting the highest level of development in both survey years. At the end of the five-year period, however, the level was slightly declining, while that of the two other groups tended to catch up to it.

3) That 'traditional' or 'folk' people could be removed from their old background and settled into a village farming life requiring adaptation to modern farming techniques, co-operation, democratic way of life and other aspects of a modern state, provided they were given equal conditions and provided attention was paid to rules for settlement which will be discussed below.

In addition to the above conclusions, the difference between the development rate of the Asian group and the North African group may have pointed to the difference in success in settlement and development of a cohesive non-modern people such as the Asian, who settled in family groups with cultural traditions more or less intact, as against those more 'broken' groups or more motley such as the North African. From the results of the study, the importance of keeping ethnic groups 'whole' would seem to be borne out, but the problem deserves further research.

D. SOME RULES OF THUMB FOR CULTURAL CHANGES

Out of experience with the problems of settlement in Israel has come the following rules for handling the social context of the farmer. Though they were worked out from experience in new settlement, they can be extended to the updating of traditional farm communities:

1. Study the social background

Professional methods of either sociology or social-anthropology should be utilized.

We have already seen how important was sociological insight in Israeli settlement. Another example of a clash between the concepts of the settlement staff and that of traditional villagers, which was solved once the background was understood, was as follows[14]:

The village of 'Lebanon' was settled by Jews from the Yemen who were a folk-like group which had never farmed. The group social structure was a patrilineal kinship group or hamula, composed of fifty households. The old leadership of the elders was maintained, adapting itself to economic requirements of the moshav, becoming familiar with various marketing and accounting activities and adapting the hamula to co-operative enterprise. Two scales of social prestige developed in the moshav, the one of age, lineage and religious learning being the more important, agricultural or administrative achievement the lesser. No genuine democratic organization developed and the council was answerable to 'hamula' or extended-family leadership. The solidarity of the village was intense, with the hamula taking good care of its social cases even though the State would have undertaken their burden, settling them elsewhere.

Despite the positive aspects of such a situation, the settlement staff tried to introduce democratic ways, and wanted to have the village council staffed by younger people (as had occurred in the previous example of 'Gadish'). The village absolutely refused to co-operate. Solution to the clash was a removal of the settlement staff, and studies instituted in various villages which brought out the implications of familialism in settlement.

Weintraub, with 'Lebanon' in mind and other moshavim of traditional or folk people wrote: "Familialism as a value and as a pattern of social organization not only continued to exist after immigration to Israel, fitting 'naturally' into the small community pattern, but was even reinforced by the crisis of immigration".

Understanding the background of communities where development programmes are to take place will help those responsible to avoid unnecessary clashes with existing traditions and will moreover make it possible to often use those traditions to activate people toward the acceptance of necessary changes.

Knowledge of particular social problems which may arise in the process of development enables policy-makers and planners to evaluate prevailing social customs in terms of their possible effects on the development programme. Certain customs need not, or should not, be interfered with, either because they have no effect on the adjustment to modern agriculture, or because any interference may raise strong opposition and jeopardize the process of development as a whole. Other customs may be utilized to push new ideas. There will always be found certain social patterns which must be discarded, gradually and carefully, because they are not compatible with the aims of development. Still, it must be taken into account that sometimes the community itself will adapt innovations to its own concepts.

2. Respect the customs and social pattern

When working out a plan for either modernizing or settlement, the cultural background should be utilized as far as possible, and changes instituted only in those areas of life necessary for modern farming. A simple and obvious example is the lack of marketing activity on the Sabbath in Israel moshavim, which would have clashed with religious custom.

A more complicated example is the fitting of kinship structures into an Israel co-operative village framework, even though the result was not 'classical co-operative organization'. Actually, in one case, when the settlement staff tried to introduce democratic decision in a moshav which had been flourishing under traditional leadership, strife developed and the village disintegrated.

More can be expected in new settlement in the way of modification of the culture than in the modernization of traditional communities in the early stages, particularly if settlement is within a modern framework as has been the case in Israel. This is because the fact of migration from an old background is already a disruption which necessitates change. New settlement must have more guidance towards change, however.

In recent years, the applied social science of community development has worked out techniques which are often useful in instigating cultural change without breaking down the traditional backgrounds of groups. Use is made of group dynamics where leadership is activated and people made aware of their needs, and ways of obtaining them.

An example of community development techniques which helped motivate people who had never been farmers to take up farming, is as follows[15]:

Another moshav of Yemenite Jews in Israel, had been settled for ten years, yet produced almost nothing. In 1959 a community development worker found the people apathetic towards agriculture and the agricultural staff convinced they were lazy or their 'mentality was wrong'. Three quarters of the village was composed of one extended family or hamula which dominated the other hamula. The head of the main hamula was a belligerent and dominating man who had been elected to the village council and controlled it. Unusual for such a traditional leader was his pocketing communal money for the health service so that the village had no doctor. Yet when the villagers first learned of his corruption, they were too dominated to act against him.

The community development worker began by having talks with the women who came to the nurse. Finding out that they were worried about their children's poor diet, she suggested they grow vegetables. They agreed and the activation of the village was thus started. Slowly, through two years of effort, patience and of getting the staff of the various outside agencies to work on a team (nurse, teacher, agricultural instructor, home economics instructor), the village began to come alive.

Although the team tried to engage the co-operation of the clan leader, he refused to be involved and even attempted to break up whatever had been started. The confidence of the other members of the village council was built up by helping them learn their duties as councillors. Finally, the leader, finding his power had disappeared, left the village with his strongest followers. The other men of the village gradually took up farming so that by 1963 it was a normal moshav, existing on agriculture.

Examples of adapting agricultural extension work to a culture occurred in several villages in Israel which were divided by their kinship or hamula structure. It was learned that it was necessary to establish a demonstration plot for each kinship group separately. When this was done, positive results were obtained.

3. Introduce changes gradually – evolution rather than revolution

The introduction of simple innovations, which are easily understood, easily carried out and which bring results within a short period of time, may convince people to later accept more drastic changes which they might otherwise resist. For instance, introducing a better variety of seed has often been a first innovating step. As yields increase, the farmers become less suspicious and more open to other suggestions. Also, a suitable demonstration plot or farm may be more effective than any compulsory regulation.

In new settlement, the fact that people are uprooted from their original environment in itself represents a drastic change which makes for difficulties in adaptation not only to new methods of farming, but also to a new way of life, new social contacts and sometimes even to a new profession, as was the case in Israel. On the other hand, this same break may facilitate the adjustment to new technology because many old customs do not apply to the new place, and resistance is likely to be weaker. This is especially the case for settlers who have had no previous experience in farming. The person who learns a new trade has nothing in his past experience about which to be conservative and is therefore more willing to learn new and better methods. Weintraub revealed this in a comparative survey of a Yemenite village, where the settlers did not engage in farming prior to their arrival in Israel, as opposed to a village settled by Kurdish immigrants, stemming from an agricultural community. The agricultural branches in the latter group's new village were the same as in their old one, but the methods and technology were obviously different. The settlers often found it difficult to understand the reasons for this difference, claiming that the old practices were good enough and required less effort. The Yemenite settlers on the other hand proved to be more adaptable. Their lack of farming experience was one of the contributing factors.

In its settlement of people who had not been either modern farmers or farmers at all in their previous lives, the settlement staff had to develop careful and flexible extension techniques to guide the settlers and to hand to them advances in technology and scientific research. As mentioned on p. 140, extension workers were alerted to take into account the cultural backgrounds of the settlers by lectures and other training. A method of management was also devised which allowed a temporary system of centralized management until the settlers were able to take over their farms completely.

During this period the settlement authorities or their agents were responsible for the management of all the farms in the village as one unit, while the settlers were gradually learning the trade of farming and how to operate a modern farming enterprise both on the individual and the community level. The length of time required for the transfer of the farms from the settlement authority to the farmers proved to depend both on the capabilities of the farmers and on the suitability of the methods of transmitting know-how to the particular group under question. The cultural background or ethnic factor entered into the picture since it affected the methods which had to be adopted in each case.

As can be seen from the study previously described, these extension and management methods allowed settlers of both folk and modern backgrounds to become agriculturists, with the cultural background reflected primarily in the length of time it took to adjust.

4. Settle groups of homogeneous background, or integrated group structure

An advantage of homogeneous group settlement in each moshav has been that the group retains a structure, providing the individual farmer with a sense of security which is particularly important in the first stages of adjustment. It is true that communities of heterogeneous settlers which do not disintegrate, will in time develop a structure. However, particularly among people of non-modern background where family ties are so supportive, heterogeneous communities are likely to prove extremely costly in terms of turnover and wasted effort.

It was found successful in many cases of transfer of immigrants to Israel, to transplant whole villages, including those individuals who would not be able to support themselves, such as the aged and the disabled. In any event these latter would have constituted a drain on the State, but by not separating them from their group, their function in social cohesion was not eliminated, to say nothing of human considerations.

It was found in Israel experience that in the settlement of people from modern or western background, group integration was less important, although the rule of homogeneity could be applied on the level of national groups and educational level. Knowledge of cultural background was less important, nor did there have to be as much effort in demonstration and guidance. On the other hand, if such groups had not had previous training and were not integrated, more effort had to be expended to develop leadership.

This brings up the important question of selection of candidates for settlement. There are two approaches – one that selection should be based on the attributes and capabilities of each individual, disregarding his group context. Such an approach selects on qualifications of age, educational level and other criteria. Underlying it is the concept that selected individuals are best equipped to engage in modern farming and that successful farmers make for a successful community. People who may prove difficult to adapt, are thus eliminated in advance.

The second approach is that which was found best suited to Israel's agricultural system – the transfer of whole communities. Experience has proven that the cohesiveness of the community and the sense of belonging to the same familiar social group, make the individual feel the impact of change to a lesser degree, and more able to cope with it. The approach is that a strong community makes for successful farmers.

Concluding the chapter on the Human Resource, it may be stated again that development planners must take into account the family and cultural contexts to which the individual farmer belongs. Family context includes the demographic constellation. Whether the farmers belong to a 'western' modern society or to a traditional society, whether they are part of old-established communities or of new settlements, the background is important. Willingness to farm and the acquisition of know-how are so woven through the intricacies of the cultural context, that ignoring it can bring disappointment and failure in the realization of the farm plan.

In order to take the cultural context into consideration for the development plan, the services of sociologists or social anthropologists should be called on, while extension workers, who work with the farmers to realize the plan, must have sociological training. The techniques of applied sociology and community development should be utilized to activate the farm community towards the goals of development.

Israel's experience has shown that modern people can be handled in a slightly different way from traditional people; more effort should be spent on knowledge of the latter's culture. Both groups, however, are capable of absorbing modern concepts of farming, though each in its own way. In the selection of candidates for settlement, group integration and cultural wholeness seem to be more important than cultural background or educational level.

Finally from Israel's experience it has been learned that, given the means of production and a viable organizational frame, even people who have never been farmers or known rural life can be settled into villages as modern farmers. Ethnic differences are influential particularly in the initial stages of development and should determine the methods of guidance and training to be applied. With proper material conditions and extension methods, and given the necessary period of time, all ethnic groups are capable of attaining economic and social progress.

NOTES

1. SCHULTZ, Theodor W.: *Transforming Traditional Agriculture* (Yale University Press, New Haven and London 1964).
2. *Ibid.*, p. 205.
3. WEITZ, Raanan: 'The Role of the Sociologist in the Rural Development Process in Israel', *Transaction of the Fifth World Congress of Sociology* 1 (Sept. 1962).
4. WEINGROD, A.: 'Stability and Instability in Immigrant Moshavim'.
5. SHAPIRO, Ovadia: Internal Report of the Settlement Department, Jewish Agency.
6. WILLNER, Dorothy: 'Politics and Change in Israel: The Case of Land Settlement', *Human Organization* 24, No. 1 (Spring 1965), p. 72.
7. WEINTRAUB, D.: 'Demographic Structure and Development in Israel' Hebrew University Manuscript (Jerusalem, March 1966).
8. WEINGROD, A.: *Reluctant Pioneers* (Cornell University Press, Ithaca 1966), p. 133.
9. WEITZ, Raanan: 'The Ethnic Factor in the Development of Rural Settlement in Israel' *Sociologia Ruralis*, 2 No. 2, 1967.
10. WEITZ, *ibid.*
11. WEINTRAUB, D. and SHAPIRO, O. 'The Tale of Region Organization in Rural Development in Israel – Some Preliminary Remarks', manuscript of the Settlement Department of the Jewish Agency, (Jerusalem 1966), p. 7.
12. WEINTRAUB, D. and LISSAK, M.: 'Physical and material conditions in new moshav' in UNESCO, Arid Zone Research XXIII (1964).
13. WEITZ, *ibid.*
14. WEINTRAUB, D.: 'Rural Cooperation and Social Structure', Hebrew Univ. Manuscript, (Jerusalem, March 1966) p. 45.
15. DEMBO, Miriam: 'Activating the Community', in *International Seminar on the Role of Voluntary Agencies in the Development of the Community*, State of Israel, Ministry of Foreign Affairs (Jerusalem 1963), pp. 105 ff.

CHAPTER 5

THE PLAN

The present chapter sums up the planning data for Israel set out in Chapters 2 and 3, and combines them to form a plan of operation for the years until 1972/3. This combination of resources to meet targets is the second stage of planning – preparing the blueprint. The subsequent stages – implementation and follow-up – will be dealt with in later chapters.

A. RESOURCE COMBINATION: ART AND SCIENCE

Production targets, for various agricultural commodities should be established in the light of domestic nutritional considerations and export possibilities. These targets should be attained within the required period by the agricultural population, using the resources which it has at its disposal. In fixing targets, the planners must use national data on land use possibilities for the different categories of land – fertile or marginal – with or without particular climatic and regional advantages. Detailed planning has to translate and apply the targets to land, water and labour factors, and combine these factors so that their effort coincides with the targets at that value of output which is in the best interests of long-term agricultural and national considerations. The planner does not strive simply to attain the maximum value of agricultural output (since absolute and alternative costs are of great importance), but rather strives for that output resulting from the interaction of scarce factors – transferable to alternative pursuits – which endows them with a marginal output value not less than they would earn in other uses.

Agriculture within the national economy should operate at that output which, at the margin, has a value neither below nor above that of the marginal output of other sectors of the economy. In other words, agricultural output should be so adjusted that the economy would lose on total output if resources were to be transferred into or out of agriculture, from or to other industries.

Similarly, within agriculture, once its relative contribution to the total economy has been determined, the combination of resources which gives an equal marginal product value to each farm branch, must be sought. It is impossible to achieve this combination exactly, because of continually changing costs and market prices, and the operation of factors both internal and external to agriculture, which closely affect its effort. The task of the planner is to approximate the goal as nearly as possible, by organizing and directing production so that there is no waste in any particular branch either from over- or undersupply. When production in one branch increases over that market demand at which it is profitable relative to other branches, policy must aim at transferring resources from that branch to others, in which high profits are being made because of underproduction. When the factors are balanced, one branch is as profitable as any other; a transfer of resources from one to the other would lower the output value of agriculture as a whole.

It is this balance which the planner seeks. He is limited in his search by his inability to forecast market price conditions caused by factors operating in the general economy, in which movements of the general price level markedly and rapidly affect the demand for farm products at any particular price. The agricultural price level, which establishes the basic framework within which the planner works, fluctuates according to the movements in the general price level. Instability in the latter severely affects the reliability of demand predictions made by the planner. As economists gain a clearer understanding of the workings of the general economy and its relations with those of international trade – and trade cycles are accordingly more closely controlled and dampened – planning can be more certain in its estimates. Predictions become more reliable and the execution of programmes can be more confidently undertaken. The difficulties of predicting the future and the immobility of fixed resources make planning an attempt to arrive at the best compromise between ideal and reality.

A plan for agriculture must necessarily be related to a specific period, and certain of the basic factors for the period in question must be closely and accurately estimated. The basic production resources – labour force, land, water and capital availability – must be combined both at the beginning and during the period so that they as nearly as possible meet the market conditions predicted for the conclusion of the period. Investments in agriculture take time to mature; each product has its own requirements, and direction of resources must take this into account in the timing of investment.

The science of linear programming, in which the several production and marketing factors relating to each main branch of agriculture are appropriately weighted and mathematically calculated to produce an integrated programme, needs for its operation data of a nature and accuracy which, in many economies, are not at the present stage available. This is the situation in most of the developing countries which lack basic data on the means of production, on the various branches, on various import items, etc., a lack which does not allow the planner to make use of the linear planning method.

In developed countries, much research is now in progress on the application of linear programming to agricultural planning. It is likely that it will be some time before this method supercedes others, although it can aid greatly in improving their accuracy.

1. Budgetary planning

The planning method proposed here, and used in the design of the macroplan presented for Israel over a ten-year period, is a method of budgeting by calculated trial and error, to reach the balance previously described. It begins with an examination of demand by investigating a proposed food basket for the end of the period, in accordance with predicted incomes for that date, and export markets for Israel's agricultural products. The food basket per capita when multiplied by the predicted population gives the total food needs; these, when added to the export market requirements, give the commodity targets.

The production resources, which have been kept in mind in their overall dimensions during the estimation of the targets, are next examined in more detail to establish their actual production potential. The stage of resource combination for producing the targets has then been reached. The first two stages which were covered in earlier chapters, are less difficult to explain than the third. Planning is an art as well as a science; the planner is faced by an infinite number of possible resource combinations. While he is guided by certain scientific rules, his art lies in anticipating the situation in which the rules will find their most satisfactory application. Though the only method is one of trial and error, the field can be narrowed by preliminary work so that the trials are not made at random.

Factors of production are scarce and vary in type and quality. The value of any one factor depends on the demand for it and the income which its use can generate in combination with other factors. The value of land differs according to its quality and location and the possibility of using other resources in combination with it. Land of high quality too far from any market is of lower value than poor agricultural land near a large city. Water channelled into a pipe line for irrigation has a value and is a resource for production; water running to the sea is a free good. In planning, decisions have to be made on the uses to which land and water are to be put and the degree of combination between them and their combination with labour. In Israel, for example, labour and water are scarce – scarcer than land which, without irrigation, is of low value. In the south of the country, where rainfall is slight, land alone is of very little value, while labour is also not readily available. The decisions to be made, then, will be concerned with the most advantageous combinations of labour and water with land.

The value of labour's marginal output cannot fall below a minimum, as labour will immediately be attracted to town industry: agriculture must provide the farmer and his workers with a living as good as that offered by employment in town with, probably, prospects of a relatively better standard of living to compensate for the risk of farming and the harder work. Capital in agriculture, apart from first long-term cheap establishment loans, must earn for its owner a rate of interest equal to that which it could earn in other industries; as with labour, the value of marginal output for capital cannot fall below the interest rate prevailing in the economy.

For the planner in Israel, as elsewhere, the return to labour and capital in agriculture cannot be less than that in industry. Investment in agriculture must give a return in the long run equal to that which would accrue from investment in industry; even if, for national settlement purposes, cheap long-term credit is granted to agriculture, the factors used in current production must yield a competitive return.

For water, agriculture is the main user. The cities do exert a considerable demand, but is absolute in its quantity and rates of increase. Agriculture cannot compete with cities for scarce water supply; urban consumption is an automatic deduction from a nation's total water resources, with the rest left for agriculture and industry.

2. Land-water combinations

There are degrees of land and water combination – from no irrigation to auxiliary irrigation of part of the land to full irrigation. Fully irrigated land is more costly, but its production possibilities for high value crops are greater. The criteria for determining the extent of irrigation works, and the intensity of water use per land unit, are considered below.

The first possibility – no irrigation at all – is not feasible. It only allows the cultivation of extensive grain crops on large areas of land at a high drought risk, leaving other foods, which need irrigation for their production in short supply and therefore at a very high price. This high price encourages the extension of irrigation for the production of those foods until the marginal product value of an additional area of land irrigated is below that which the investment and labour could bring elsewhere, or until all of the annual water supply is utilized. In the latter case, the intensivity of use of the existing irrigated lands will increase, their value rising in response to the rising price of foods. The demand for foods in such a case will become beyond the capacity of home agriculture so that imports of agricultural products in competition with those of the irrigated land will set a ceiling on the price rise.

It is, therefore, the market demand and value of the various agricultural products which determine the extent and intensity of irrigation within total water availability. Land will not be used for low value products when, with irrigation, it can produce high value products for which a market exists (by 'high value' we mean crops which will assure the farmer a high return for his labour and capital investment).

Fully irrigated land can grow all crops which the climate will permit, and with a high degree of certainty. The number of crops which will justify the cost of the irrigation is, however, limited. At the same time, most of such crops can be grown only on irrigated land, and their value is such that, within market limits, they must claim preference over other crops in their use of this land. It is part of planning, through a price structure as discussed later, to aim at an output of the higher value crops which ensures their continued production at a regular price, balanced by long-term supply de-mand considerations. Because of the greater skill and larger input of labour and capital necessary for their production, these will tend to remain high value crops. They will therefore have the first claim on irrigated land. Because of rotational practice, a larger area of irrigated land will have to be allocated to these crops than that which they actually use each year. At the same time a variety of crops has to be grown in order to conserve soil and limit disease and weeds, not all of which would otherwise justify their place on high cost land.

There are crops which can be grown successfully both on irrigated or on partly irrigated land, the irrigated giving a higher yield at higher cost than the partly irrigated. These, particularly cotton, and to a lesser degree sugar beet, can be transferred from irrigated land to land with auxiliary irrigation when and if a rising market demand for 'high value' crops increases the pressure on, and value of, fully irrigated land and scarce water resources. Auxiliary irrigation, by using land extensively, effects a saving in water.

The market for the products of Israel's agriculture is limited in size and in the range of goods it can absorb by the population size of the home market and its income, and by the costs of transportation and keen competition from other countries of the export market.

Because Israel is restricted by the amount of water at her disposal, in expanding the irrigation network over all land, high value crops such as vegetables and fruit will, to the limit of the demand for them, lay first claim. Lower value crops, such as cotton and sugar beet, will use the remainder. Any water still available when the demand for these has been met, will be used for auxiliary irrigation in supplementing more extensive farming of certain crops. Unirrigated crops, mainly grain, will be grown where rainfall and soil permit. When and if the demand for high value crops increases, the irrigation network will be extended, and existing full irrigation used more for these crops. Other crops, such as cotton will be grown less and less on fully irrigated land and more under auxiliary irrigation. Should circumstances arise in which home consumption and export requirements for high value crops can only be met if all the water is used for producing them on land under full irrigation, there will be no water available for lower value crops, which will have to be imported.

The situation in Israel today is that water supply is becoming critical, and is a first consideration in long-term planning of new agricultural settlement and in the expansion of existing settlements. For the present, even with the prevailing demand for 'higher value' fruits and vegetables, there is enough irrigated land to provide the country's needs in such crops as cotton. As the home demand generated by population increase will rise, and export markets for existing and new products will expand, the low relative value of cotton will likely force it off the irrigated land onto areas irrigated by an auxiliary pipe network. The water used per dunam will be much less than under full irrigation while the yields per dunam will not be on the same scale. However, since land can be used as a substitute for water in cotton growing, the same quantity of cotton can be grown by extensive use of land with a minimum of water.

In planning by the budgeting method, it is therefore the market requirement which determines the use to be made of the limiting resource factor, which in Israel's case is water. The development plan for Israel's agriculture thus begins with calculation of the market potentialities, then of the resources available, and finally of the allocation of these resources to meet the market demand. An estimate of the amount of imported products needed to supplement the possible home production is also made.

In other countries the conditions are in all probability different to those prevailing in Israel, but similar reasoning in planning for agriculture on a national scale is valid.

On the basis of national data, predictions are made for the end of the planning period (best fixed at between five and ten years) for population size and composition, work force, resource potential and rise in income per capita. The role of agriculture in the economy and its future size in terms of output and factor use, must be determined in conjunction with target-setting on the basis of home and export market considerations. This is followed by detailed investigation of the country's natural resources which are to serve agriculture; in overall terms, these resources have to be combined in a plan to produce that variety of goods fixed by the targets. It is obvious that in order to set the targets, there must be considerable knowledge of the country's agricultural resources and their possibilities: the national planners who determine the role of agriculture in the general economy have need only of round figures; the men who translate the targets into a plan of action through resource combination at the regional and farm level need the minute survey detail.

The planning of agriculture is an art; the application of economic princi-ples must always be weighed and timed according to other factors of varying importance, and never are two sets of conditions alike. In this book attempts are made to lay down only the general, logical lines of thought to be followed in budgetary planning. In all economies there are one or more factors limiting economic progress; labour, land, water or capital are limited in their po-tential by the factor in shortest supply, in terms of which the others have to be combined to the best advantage. The first stage is thus to determine the limiting factor, and, according to market demand, to utilize it first of all in satisfying the demand for higher value products. The remainder of the limiting factor is then used with the other resources for producing lower value products until its full use has been established, according to the relative profitability of the products. The aim is maximum profitability over the whole national farm. If land shortage is the limiting factor and labour plentiful, then labour is used extravagantly to get the most out of the land; if labour is scarce relative to land, then extensive farming is undertaken to economize on labour. This, however, is obvious; deeper analysis is necessary in order that the planner estimate the degree of intensity and of activity that his planning makes necessary.

3. Developing subsistence economies

The planning aims described above (i.e. the direction of scarce production factors to the most advantageous economic channels) are valid for Israel and for other societies with established markets and with some degree of sector-specialization. In many developing nations, where subsistence-farming prevails, the situation is very different: the major part of the population works the land, the minor part in the towns constitutes the effective market. In such circumstances the farmer will produce mainly for subsistence, for the market will be small, probably distant, and oversupplied. The farmer's income for exchange will be negligible, so that he will not be able to employ the specialized services of others or buy their goods. Manufacture and services in such a society will be at a minimum; there will be little division of labour and people will be very poor.

The task of agricultural planning under such conditions is to initiate economic activity – to build markets, set exchange and division of labour in motion, increase the production of primary products to provide a saleable surplus, and release workers for other employment.

In brief, although macro or national agricultural planning lacks scope unless a market exists, it can play an important part in developing markets. Surveys of agricultural resources will show their productive potential in terms of saleable commodities; agricultural planning together with agrarian reform will point the way for use of these resources to realize their potential. Market surveys will indicate the possible outlets for commodities and direct the setting up of a distribution network to collect the products and send them to markets as the plan comes into execution. Such comprehensive action can be undertaken in settlement schemes on new land or in the reorganization of an existing agriculture to produce for a market.

There are, however, two things which planning cannot achieve: it cannot, through specialization, produce agricultural goods for a market if the farmers have also to grow their own food; their food must be provided by other farms. Secondly, distributing, processing and marketing functions cannot be implemented unless sufficient people are released from producing their own basic necessities so that they can concentrate their work in specialized activity. If specialization is undertaken in one area of a country, it demands specialization in others and labour which has to be fed. A market therefore becomes created not only for food and fibre products, but also for labour. This means that subsistence farming becomes reorganized, giving its surplus labour to the new enterprises and using the remaining agricultural population efficiently enough to provide surplus food. New settlements will sell their products, with the proceeds buying their food and other goods; the established farmers will sell their surplus, exerting a demand for the services of town industry. This leads to the building of towns within regional plans, until slowly the snowball of economic progress will begin to roll. Farming regions will develop gradually according to natural advantage, trading between them and the towns increasing.

The process cannot be completed in a year but is long and continuous. However, planning in these days of advanced technical, economic and social knowledge can lead to a many-fold increase in the rate of development. An economy, once it begins its upward trend, must advance on all fronts: in education, in social organization, in health, in the application of techniques to agriculture and industry – so that the general movement is not blocked by inadequacy in one field of operation. All these factors are important in planning.

B. PAST AGRICULTURAL PLANNING IN ISRAEL

Until the creation of the State of Israel little was done in the way of long-range national planning. The late Professor I. Volcani, founder of the Jewish Agency's (later the Government) Agricultural Research Station, made a forecast of the structure of the individual farm unit and a plan for the general development of Jewish settlement.[1] He regarded the Palestine of that time as an independent economic unit – a daring point of view, since the Jewish authorities were not then responsible for the country's economy, and agricultural planning was limited to forecasting the characteristics of an individual farm unit: i.e. farm type, range of farm branches, the water schedule, crop rotation, details of fodder feeding, etc. The Jewish agricultural authorities limited their activities to professional problems, such as improving stock, the acclimatization of varieties, adjustment of machinery to specific soil features and farm management. A national agricultural plan was non-existent. In 1946, shortly before the partition of Palestine, two separate plans were submitted to an Anglo-American commission, one by Mr. L. Samuel and the other by Mr. L. Ga'aton. They dealt with general economic development of the country, but both plans were incomplete owing to lack of necessary data.

The need for comprehensive agricultural planning became obvious with the establishment of the State, when the government was faced with the task of agricultural development in the context of general development. In 1949 the Economic Research Department of the Prime Minister's Office prepared a general development plan for Israel which included plans for agricultural development. This plan, published in 1950, estimated that within the four years, from 1949/50 until 1952/53, the population would grow by 200000 per year to reach 1.8 million in 1953; the percentage of farmers (among total employed workers) would increase from 13.1% in 1949 to 21.6% in 1953; agricultural exports would increase rapidly, and by 1953 would cover 47% of imports, compared to 15% in 1949.

In the same year, the Joint Planning Committee for Agriculture and Settlement of the Ministry of Agriculture and the Settlement Department of the Jewish Agency prepared a detailed programme, entitled Agricultural Settlement Plan.[2]

This plan forecast the settlement of 66000 new farm units in the period 1949–1953, including the 16000 which had been already set up since the establishment of the State. The farm population would be 26% of the total population, which, it was estimated, would reach two million by the end of 1955. The investments necessary for implementing the programme were estimated at $55 million, of which $26 million would be in foreign currency.

For the final stage of development, home agriculture was to supply most of the country's food, except wheat and meat. The food supply was determined on the basis of the farm types then accepted, especially the 'mixed farm' with its largely dairy basis. The richest type of food basket was thus proposed, plentiful in animal protein. It was estimated that per capita consumption of milk, for example, would be 300 kg, which is twice today's consumption. For meat, 24 kg per person was estimated, which is higher than the present quantity consumed. No attempt whatsoever was made to assess foreign currency expenditure, an omission which had important consequences in the execution of the plan. The irrigated area required for the programme was 330 thousand hectares, with 1560 million cubic metres of water.

In the same year (1950)[3], the writer also prepared a development programme which corresponded with the Government-Jewish Agency plan in many respects, but differed markedly on certain points. The creation of 57000 farm units was proposed for the four-year period 1950–1954. The main agricultural branches were also of livestock, but every farmer was to have 2–4 milk cows and 100–200 laying hens. The proposed food basket was changed accordingly; the per capita consumption of milk was estimated at 262 litres annually and meat consumption at an average 14 kg per person.

The writer's proposed programme forecast that on its completion, the country's agriculture would supply 75–100% of food needs (with the exception of grains) for a population of 2 million. It attempted to evaluate the balance of payments accordingly, and estimated that the value of agricultural exports would be $3 million, and the value of agricultural imports

$3.4 million. This was the first time that such an estimate had been included in an agricultural development programme. Generally speaking, however, neither development programme paid enough attention to the basic governing facts which must be taken into account when preparing such a plan; nor was sufficient consideration given to the costs involved in the plan's execution, or whether the State could affort to maintain so high a level of nutrition. The cost of executing the plan was estimated at $40 million, with an additional $9 million for developing further regional engineering work, and installations to supply 600 million cubic metres of water annually.

Both of the plans included not only the planning of types of settlement and farm branches, but also farming methods. They encouraged the use of work animals on farms in order to reduce mechanization to those tasks which animals could not perform economically. Both plans had basically the same aims and advocated the same farming methods. The gap between plans and execution widened, however, when the predicted immigration failed to materialize: in 1952, the rural population constituted only 14% of the total population. Similarly, the natural increase of the rural population fell short of the prediction.

Other plans were prepared, but these as well were guided by basic policies and motives which did not take long-term considerations sufficiently into account. The three main considerations were:

a) For security reasons, the more settlements established, the more secure would be the country.

b) The shortage of food and the extremely speculative prices, created a seemingly endless boom in agriculture which dulled the planner's consciousness of the need to calculate a ceiling on agricultural production.

c) Most did not account for the impact they would make on the foreign currency needed for current consumption and investment. No calculation was made to assess the investment costs for pumping, tractors, etc.

d) Price structures consequent on increased production and its effect on farm incomes was not fully investigated.

Slowly an understanding of these factors began to take hold. As a result, two schools of thought emerged in the Planning Centre: one stressed the continuation of orthodox policy; the other demanded a change in agricultural policy to avoid overproduction for the local market and to shift the production means of the existing agricultural population to new goals – i.e. to industrial and export crops such as citrus, fresh fruit and new crops (groundnuts, sugar-beet, cotton, etc.). The clash between the two schools was seen in the alternative plans that emerged. As a result, efforts were made to rethink policy and to achieve balanced planning, taking into account all the factors influencing the course of agricultural development.

In 1952, the Joint Planning Centre was constituted as a permanent institution. Its function was national and regional planning, including individual farm planning for both veteran and new settlements. In 1953 it prepared a seven-year plan for the years 1953–1960. The plan attempted a close evaluation of actual conditions, and, assuming a population of 2 million by 1959/60, its aims were:

1) A cultivated area of 3.6–3.7 million dunams and 1300 million cubic metres of water for irrigation;

2) A dairy herd increase from 39000 to 72000 milkers, including the import of 10000 milk cows. This would give 3–4 milk cows per farm unit;

3) The addition of 41300 farm units (from 42300 farm units in existence in 1953 to 83600 farm units);

4) 3.6 million laying hens;

5) The production of industrial crops.

Publication of this seven-year plan gave rise to heated discussion in agricultural circles; severe criticism was made of the proposed food basket, the role of livestock and the suggested farm types. The writer opposed the seven-year plan and published his comments, which stressed the importance of changing current thought and methods of evaluation.[4] He especially questioned the role of livestock in the agricultural economy, maintaining that Israel's primary aim must be *economic* independence. In his view, the mixed farm impeded progress towards this goal and if new farms were developed as in the past, the goal would not be achieved. He suggested a new farm type, called the 'field crop farm', based on industrial crops. Another local economist stressed the need for basic re-appraisal of the country's agricultural development. He also suggested building intensively cultivated industrial crop farms, and a change in emphasis from production of animal protein to vegetable protein. At a later stage Dr. A.G. Black, an adviser sent to Israel by FAO, supported these suggestions and proposed a number of other changes in the family farm unit (discussed below).

The Joint Planning Centre accordingly revised the development plan, and the plan eventually received official approval.[5] Its aims were:

1) A dairy herd of only 62000 head, reached by natural increase;

2) 2.25 million laying hens;

3) Less animal protein from milk and eggs, but more from sheep and fish;

4) Maximum production from the livestock branches;

5) Efficient use of water resources and an increase in irrigated land;

6) A supply of oils, fibres and sugar sufficient for local needs, together with a drastic reduction in imported foodstuffs.

This third alternative showed substantial improvement on previous plans, but nevertheless lacked balance in several respects. Its main importance was in the intention to build new farm units dependent on field and industrial crops. In its execution, however, more than the planned emphasis was placed on livestock and poultry, leading to further imbalance in these branches.

The natural sequel to the seven-year plan was the four-year plan prepared by the Joint Planning Centre in 1955 [6] which is in principle no different from the seven-year plan. The programme, however, included details of execution according to the rate of development and the regional water schemes. Table 5-1 shows a comparison of the seven-year and the four-year plans for 1959/60.

The writer published a second plan in 1958 [7] in which he tried to direct national policy to the goal of closing the foreign payments gap through holding down the standard of living. He suggested a food basket with little animal protein, since, he maintained, the country could not afford the rich food standard it was demanding. However, the norms he predicted were low in relation to the rapid increase which the new settlements, only just on their feet, actually achieved in the succeeding years. It was then too early to predict the full impact of the economic development which has since taken place on such a wide scale in recent years, and that was only getting under way in 1958. This plan was therefore soon out of date.

The plan presented below represents an attempt to learn from previous planning in order to make more accurate predictions for the future.

C. CHOOSING THE BEST ALTERNATIVE

The first decision of macro or national planning is to determine the place to be occupied by the future development of agriculture in the overall development of the country. This decision refers to the allocation of capital and sometimes also to manpower. In Chapter 1 the general factors which affect this decision were mentioned; a detailed technical discussion is beyond the scope of this book, though the question is referred to below when an attempt is made to evaluate the capital/output ratio in agriculture.

Once the role of agriculture in the general development plan is determined, the planner must examine the various alternatives of resource combination, compare them, and suggest the combination best suited to the aims of agricultural development in the country. In this respect, it must be emphasized that the choice between various alternatives is not entirely economic and

TABLE 5-1

Development forecast for 1959/60 according to the seven-year plan of 1953 and the four-year plan of 1955

Branch	Forecasts for 1959/60	
	according to the 7-year plan[5]	according to the 4-year plan[6]
Total water supply for agriculture	1 300 million m³	1 040 million m³
Total irrigable area	1 854 000 dunams	1 770 000 dunams
Unirrigated area	1 796 000 dunams	1 695 000 dunams
Irrigated orchards	353 000 dunams	419 000 dunams
Irrigated field crops	1 462 000 dunams	1 313 000 dunams
Fish ponds	39 000 dunams	34 000 dunams
Livestock:		
Milk cows in the Jewish sector	62 000 head	56 000 head
Milk cows in the Arab sector	15 000 head	20 000 head
Beef cattle	25 000 head	17 000 head
Laying hens	2 250 000 head	3 800 000 head
Sheep and goats (Jewish sector)	100 000 head	147 000 head
Food basket:		
Wheat and flours	150 kg	185 kg
Potatoes	60 kg	39 kg
Legumes	10 kg	5.2 kg
Sugar	24 kg	22 kg
Oils	14 kg	18 kg
Fruit (including melons and watermelons)	108 kg	103 kg
Vegetables	120 kg	116 kg
Wheat	9 kg	11 kg
Eggs	200 eggs	259 eggs
Fish	15 kg	15 kg
Milk	143 litres	145 litres
Value of agricultural exports	$ 50.7 million	$ 58.7 million
Value of food imports	$ 16.8 million	$ 39.7 million

sometimes the decisive factors are non-economic; social, political and security factors play their part in the choice. Their specific role in such crucial decision varies from country to country and from period to period, according to a vast complex of factors which cannot be analysed or even clarified.

The trial-and-error method used in determining the best resource combination from the economic point of view, means that the planner must construct many alternatives by altering one factor at a time. The final conclusions of each alternative must be calculated if it is to be of any significance in this method of evaluation. This implies a search for alternative modes of combining and estimating the extent of the different crops, taking into account climatic and soil conditions and the need, as far as field crops are concerned, of ensuring normal crop rotation. This search for the most favourable size and configuration of the agricultural branches within the national economy thus makes it necessary to test a large number of possibilities in order to attain a combination which on the one hand gives the highest and most desirable returns, both to the state and to the farmer, and on the other hand, meets agrotechnical requirements insofar as it is based on the availability of suitable soil and on climatic conditions. Each macro plan resulting from each of the alternatives is considered by the planner, taking into consideration the factors mentioned above. This process though both laborious and prolonged has not yet been replaced by any scientific system. Undoubtedly, linear programming will enormously facilitate the task of the agricultural planner, but will not be able to supplant the method of trial and error entirely, because of the difficulty in measuring non-economic factors. These are particularly significant in the non-western countries and in those with a heterogeneous population so that the possibility of introducing streamlined mathematical systems to national planning is slight. There are, however, some basic rules to guide the planner in his choice of alternatives. These fall into three categories. Firstly, the evaluation of the relative profitability of the different branches with reference to units of the basic resources; secondly, the system of supply based on marginal profitability; and thirdly, the presuppositions deriving from general policy.

D. DESCRIPTION OF THE PLAN

1. Estimation of the relative and marginal values

An attempt was made to establish whether the present kind of profitability among the main agricultural branches in Israel was the same for both the national economy and for the individual farmer. The aim was to test whether the interest of the one accorded with the interest of the other, and to rank the branches according to the most profitable use of water. The calculations were put in tables presented below. They were made on a number of alternatives showing the changing profitability of the various branches as efficiency changes, and on varying basic assumptions. For citrus, figures are given for three possibilities: the first is based on the assumption that the crop will be 4000 kg per dunam, the present average; the price given is the one prevailing in recent years. The second set of figures assumes the present yield (4 tons per dunam), but with lower prices (20%). The third possibility is a 20% drop in price, as forecast by economists, with yields increasing to 4500 tons per dunam – also predicted by experts.

The calculations for sugar-beet were based on two possibilities: the present yield (5000 kg per dunam), with 8 work-days per dunam, as found in many *moshav* farms; the second assumption is that yield will be 5500 kg per dunam, while work-days needed will drop to 5.5 days a year per dunam – an achievement which, in the view of experts is attainable during the period considered by this projection.

For export tomatoes, there were also two assumptions: a yield of only 4000 kg per dunam with 28 work-days per dunam p.a.; or, on the basis of recent experiments, a yield 50% higher of 6000 kg per dunam – which has been shown to be perfectly possible: in this case, however, the number of work-days necessary would rise to 50 per dunam, because of the high yield and the special treatment it requires.

For groundnuts, calculations were made on two possibilities, reflecting expert opinion on yield and water requirements.

Tables 5-2 and 5-3 show that under certain conditions and with proper organization of the various branches, it is possible to combine profitability to the farmer with that of the national economy at the same time, taking into account the requirements of the internal market and export prospects. At the head of the profitability scale are citrus fruit and export tomatoes, the income per work-day being the highest for citrus fruit and per dunam for export tomatoes. The income per work-day (at 1964 prices) is IL. 35.4 for citrus fruit, while according to the projected price for 1972/73 it will be IL. 27–31. The citrus industry is also close to the head of the list for the amount of added dollar value per m^3 of water and per dunam of land.

On the same high if not higher level of profitability are tomatoes for export. From the national standpoint (using added dollar value per water unit as the criterion), they rank at the top of the list while for the farmer they are also highly profitable. Accepting the view of the Negev Regional Office that 60% of the harvest will be of exportable quality, the profitability of tomatoes is actually greater than that shown on the table. All such calculations apply exclusively to export tomatoes – grown in particularly suitable areas, as for example the Jordan Valley or the Western Negev, which alone offer climatic conditions allowing the export fruit to ripen at the right season and to have the desired quality.

Cow milk ranks higher than citrus in the profitability scale per dunam due to the prevailing policy of government support and subsidies. The stability and future development of dairy farming depends on the adoption of the

TABLE 5-2

Income of selected branches of agriculture per dunam per work-day and per 1000 m³ water [8]

Branch	kg per dunam	Work-days per dunam	m³ water per dunam	Income in IL. at 1964 prices after deduction of direct expenses		
				per dunam	per work-day	per 1000 m³ water
Citrus fruit – present price	4000	12	700	425	35.4	607
Citrus fruit – estimated price [a]	4000	12	700	329	27.4	470
Citrus fruit – estimated price	4500	12	700	380	31.6	547
Cow milk per head	4700	34 [d]	1420	940	27.6	
Sugar beet	5000	8	500	179	22.3	358
Sugar beet	5500	5.5	400	190	34.5	475
Export tomatoes [b]	4000	28	500	720	25.7	1440
Export tomatoes [c]	6000	50	800	1200	24.0	1330
Groundnuts	400	4.5	550	147	32.6	267
Groundnuts (Western Negev)	450	4.5	680	170	27.5	250
Cotton (Acala)	330	9	500	165	18.3	330
Cotton (Acala)	330	3.5 [e]	500	165	42.8	366
Spring potatoes	2800	8	400	205	25.6	512
Miscellaneous vegetables	2000	17	500	225	13.2	500

[a] The estimated price of citrus fruit is 80% of 1960/61 prices, i.e. $105 per ton as against $130 per ton in 1960/61 and $120 in 1964.
[b] It is assumed that the proportion exportable will be 25%, i.e. 1 out of 4 tons per dunam, the average price per ton to the farmer being IL. 220.
[c] This calculation is based on an estimate of the Negev Regional Office of the Settlement Department, which sets the proportion exportable at 60% while assuming a larger water and work factor.
[d] Work-days per head includes work in the growing of fodder for the cows.
[e] Mechanical picking.

price policy outlined elsewhere, while its expansion is restricted by the limited capacity of the internal market – i.e. by production quotas. There is no doubt but that with practical new alternatives to dairy farming, its importance within the general agricultural structure will decline, although the rate of such decline cannot be foreseen. Our projections for dairy farming have therefore been based more on anticipated policy than on purely economic considerations.

Further down the profitability scale are the industrial crops – cotton and groundnuts, with projected yields which are sometimes higher than those obtained at present. Income per unit of water is a little higher for cotton

TABLE 5-3

The amount of dollars saved or earned per m³ of water, per dunam of land and per day of work are presented in the following table: Added (or saved) dollars per m³ of water, per dunam of land and per work-day[9]

Branch	kg per dunam	Work-days per dunam	m³ water per dunam	Added or Saved Dollars		
				per 1000 m³ water	per dunam	per work day
Export tomatoes[a]	4000	28	500	870	145	16.1
Export tomatoes[b]	6000	50	800	790	522	13.0
Citrus fruit – present prices	4000	12	700	700	490	40.8
Citrus fruit – expected prices	4000	12	650	677	440	36.6
Spring potatoes	2800	8	400	237	95	11.9
Groundnuts	400	4.5	550	147	81	18
Cotton	330	9	500	136	67.5	7.5
Sugar beet	5500	5.5	400	160	64.3	11.7
Sugar beet	5000	5.5	500	111	55.5	10.0

[a] An exportable proportion of 25% is assumed, i.e. 1 out of 4 tons; the f.o.b. price per ton is 200 dollars.
[b] According to the estimate of the Negev Regional Office (which sets the exportable proportion at 60% while assuming a larger water and work factor).

than for groundnuts, though the added dollar value per dunam is slightly higher for the latter. It must be stressed, however, that light soil suitable for groundnut cultivation, where yields of 400–415 kg are obtainable, are of limited extent, and that future expansion of citrus groves will further curtail them.

For the sake of comparison, we have included the income derived from other non-export commodities, such as milk, vegetables, etc. The farmer's income per work-day and per dunam for vegetables and potatoes is also found to be adequate.

The table also shows that the prices of the various commodities determining the level of the farmer's income, can well be made to conform to national requirements, so that the needs of both the farmer and the country can be satisfied. The national agricultural production system and the types of farm here proposed, were drawn up with a view to achieving this satisfactory balance, which is essential to the success of the programme. In other words, the plan is designed to achieve an adequate level of income for the farmer, while at the same time making a substantial contribution towards the consolidation of the national economy and closing the gap in the balance of payments. The figures also indicate alternative economically profitable uses of water. These are reflected in the expansion forecasts of the different branches, and hence in the agricultural development of various parts of the country.

Another calculation to be carried out before determining the crop structure is the computation of the lowest possible export dollar price of various products which would cover their production costs. This was done on the basis of future market prospects and an analysis of the water factor, the main limiting factor in Israel agriculture. In Chapter 3 it was learned that desalination of sea water has come within the range of feasibility, so that the limitation on water is its cost rather than its physical absence. Calculations were made therefore on the future cost of water obtained partially by desalination. Although the rise in cost would likely not take place before the end of the period of the present plan, the rise had to be taken into account because the development of export production had to be already begun in the present period. The prices obtained in the calculations can serve as a valuable guide to the planning of the various export branches.

Nine export crops were examined: winter potatoes, autumn carrots, early-spring-season onions, cotton, groundnuts, peppers, winter tomatoes, gladioli and citrus fruit. The calculation was made both at 1964/65 and at 1974/75 prices and conditions for a projected price range from 10 Agora per m³ of water to a maximum of 40 Agora per m³ of desalinated water. The value of unexportable produce sold in Israel was deducted from the total of each crop, and the resulting difference was divided by the dollar returns for the exported produce. The result obtained thus represented the dollar value – in Israeli pounds – required to cover the production costs. The calculation was drawn up separately for the required dollar value at the farm gate and at the foreign export market.

It was seen that at 1964 conditions the dollar rates (IL. per dollar) required to cover cultivation costs, i.e. at the farm gate, were as follows:

Crop	Cost of m³ water	
	10 Agorot	40 Agorot
Gladioli	IL. 2.06	IL. 2.30
Tomatoes	IL. 1.89	IL. 2.66
Peppers	IL. 2.99	IL. 3.96
Carrots	IL. 2.41	IL. 4.38
Citrus fruit	IL. 2.69	IL. 4.75
Groundnuts	IL. 2.59	IL. 6.23

From the table it is seen that there are certain crops whose entire cost, including marketing costs and freight, is not overly affected by the higher water rates. The dollar rates required to cover the entire costs, including delivery at the foreign port of destination were found to be as follows:

Crop	Cost of m^3 water	
	10 Agorot	40 Agorot
Gladioli	IL. 2.50	IL. 2.63
Tomatoes	IL. 2.52	IL. 2.86
Peppers	IL. 2.98	IL. 3.31
Citrus fruit	IL. 2.84	IL. 3.84
Carrots	IL. 2.89	IL. 3.37
Groundnuts	IL. 2.73	IL. 5.17

It may thus be stated that taking the entire production process into account, all the crops examined, except groundnuts, remain profitable even at the highest projected water cost, provided the agrotechnical conditions and the organizational structure remain the same as in 1964/65.

The calculation of the projected export dollar rate for 1974 was based on a number of assumptions regarding the future course of development, viz.:

a) A certain decrease in the labour input of the various crops.

b) A 50% increase in daily wages.

c) An average 10–20% reduction in water consumption.

d) A 20–25% increase in yields.

e) An increase in the share of the exportable output.

f) A certain rise in input prices, such as fertilizers, insecticides, etc.

Hence, following the projected rationalization of farm production, the results of the computation will be different in 1974 than in 1964. The general trend is towards a reduction in the relative share of water in the aggregate production costs, so that water will become a factor of secondary importance in the development of export crops.

In Table 5-4 it is seen that the dollar rates required to cover the cultivation costs and the c.i.f. expenditure (marketing costs, freight, insurance, etc.) will be lower in 1975 than under existing conditions.

The figures show that under the projected technological conditions, all the crops examined are likely to be profitable even considering solely the cultivation costs and leaving aside the subsequent marketing stage. Groundnuts will fetch the highest added dollar price of IL. 3.78, even at a water rate of

TABLE 5-4

Dollar rates for production costs

	Dollar rate (in IL.) required to cover cultivation costs		Dollar rate (in IL.) required to cover total production costs (c.i.f.)	
	Water at 10 Ag.	Water at 40 Ag.	Water at 10 Ag.	Water at 40 Ag.
Gladioli	1.6	1.75	2.08	2.17
Tomatoes	1.73	2.02	2.35	2.49
Peppers	2.59	3.04	2.70	2.87
Citrus fruit	2.18	3.54	2.57	3.29
Carrots	2.21	3.43	2.66	3.00
Groundnuts	1.86	3.78	2.21	3.55

40 Agorot per m^3, considering cultivation costs only, and slightly less – IL. 3.55 – taking into account all the production costs.

With respect to the crops examined, the price of water does not seriously affect and certainly does not limit future export prospects, even at a cost of 40 Agorot per m^3. Altogether export farming is expected to constitute an ever-growing proportion of Israel's agriculture, which is, of course, considered in setting up alternative production possibilities in the agricultural plan.

Many products which could well be produced in Israel are partly or wholly imported from abroad, since local production is not always economically worthwhile. A good example is furnished by local vegetable-oil production. During the first years of the State, agricultural planners thought that the entire oil requirements of the population could be economically produced locally; experience has shown this assumption to have been erroneous, since oil made from locally grown groundnuts cannot compete with imported oil. Local groundnuts, therefore, served mainly as an export crop, with only unexportable surpluses directed to oil production.

The production of sugar provides another example. Sugar manufactured from Israeli beet is at present considerably more expensive than imported sugar. The marginal profitability of sugar production should therefore be weighed against other alternatives. As indicated above, the cultivation of sugar beet should be restricted exclusively to those areas providing high yields and in which the costs of processing can be reduced: these conditions will determine the extent of sugar beet cultivation in the future.

The marginal profitability of the various branches should therefore be taken into consideration when selecting the best of several production alternatives.

2. Supply combinations

Investigation of the marginal value of agricultural products is therefore one of the principal factors determining the structure of the farm economy's branches and the various supply alternatives; in accordance with its results, we can draw up the various supply combinations – i.e. the types of product and the quantity of each type which farmers can profitably produce for marketing. We have already noted that if the profitability of various products, such as food oil and sugar is marginal, it is worthwhile neither to farmers nor to the country to produce them since they can be more profitably imported. This is true in the case of a free market and free import: the position may change if the government adopts a policy which favours local products – even if they are more expensive – either by administrative controls which make import impossible, or by a farm-support policy by means of high subsidies. A typical example is provided by milk production: the subsidy on milk and the limitation of dairy imports have made the branch extremely profitable for farmers. Continuation of this policy is likely to lead to a constantly increasing milk supply because of the high returns that dairy production offers farmers. Abolition or curtailment of the subsidy, or competitive import, would radically change the position, in view of the much lower costs of dairy production in many European countries (Holland and Denmark, for example): local producers could not compete with the imported product.

In this case, the policy of the Israel Government has been determined by social rather than economic factors. Many farmers in Israel depend on this branch, and in the absence of suitable water and soil conditions, they lack production alternatives. Free milk import or cutting of the subsidy would be a severe blow to the thousands of families dependent on dairy farming: in order to avoid the resultant stresses, Government policy in this field remains unchanged.

In working out the possible alternatives, we have relied on three guiding principles which have determined the framework of discussion and limited its scope:

a) The size of the livestock branch generally and of dairy farming in particular is determined exclusively by the capacity of the internal market (apart from a quantity of eggs for export, which is insufficient to affect any of the following remarks). As long as the price policy discussed previously remains in force, the size of these branches (as set out in Table 5-5) will remain unchanged. The quantity and composition of animal feeds are determined by two main factors: the know-how available regarding the best diet for the animals in question, and the optimum combination of the foodstuffs available to provide the cheapest food. These considerations have guided us in the compilation of the animal feed tables. The area required for fodder, straw and other forage crops has been determined according to these figures, in conjunction with the expected crop rotation of the different types of farms.

b) The scale of vegetable production is determined by the level of local and of foreign demand. We have already seen that to the extent that the obstacles now hampering the export of Israel vegetables can be overcome, vegetables can be one of our most profitable crops and their cultivation will expand to the limits determined by the availability of suitable areas: the limitations are therefore mainly of an agrotechnical and organizational nature. We have assumed export quotas which at the moment appear to be the most reasonable estimates, but are nonetheless to a certain extent arbitrary. It is however clear that if advances in know-how and in organization are made at a more rapid pace than envisaged at present, vegetables will achieve absolute priority over all other irrigated field crops – as shown by the two preceding tables. The extent of vegetable cultivation for the internal market, on the other hand, is determined and limited by the monthly supply quotas, according to which our figures have been calculated.

c) The extent of the citrus industry has been determined mainly by an analysis of the future market capacity, bearing in mind its high ranking in the priority scale.

It follows that particular difficulties are encountered by the planner in assessing the possible alternatives to irrigated crops and irrigated industrial crops. The reasoning which has decided us in favour of the structure set out in the following sections has already been explained. Nevertheless, we must stress that further systematic research is required to achieve adequate projections of the national agricultural structure – a field in which the linear programming method might become applicable. For the present, we shall use the estimates available, notwithstanding their defects and their generality.

3. Projections of the various branches

In working out alternatives for the most profitable future size of the various agricultural branches, we have borne in mind the following considerations:

TABLE 5-5

Agricultural production, area and quantities at end of projection period

Crop	Area (dunams)	Production[a] (tons)
Vegetables and potatoes	215000	620000
Cotton (irrigated) - fibre	200000	26000
Cotton (auxiliary irrigation)	120000	8300
Cotton (unirrigated)	60000	3500
Sugar beet	90000	495000
Groundnuts	90000	36000
Winter grain (irrigated)	150000	48000
Winter grain (unirrigated)	750000	187000
Summer grain (irrigated)	40000	28000
Summer grain (unirrigated)	100000	30000
Pulses and oil seeds (irrigated)	25000	3500
Pulses and oil seeds (unirrigated)	70000	6500
Fodder crops, perennial	75000	75[b]
Fodder crops, annual (irrigated)	200000	160[b]
Fodder crops, annual (unirrigated; silage)	100000	12[b]
Hay (irrigated)	180000	29[b]
Hay (unirrigated)	350000	46[b]
Melons and watermelons (irrigated)	25000	50000
Melons and watermelons (unirrigated)	30000	25000
Tobacco (unirrigated)	60000	3000
Miscellaneous (unirrigated)	100000	–
Citrus	500000	2100000
Subtropical fruit, miscellaneous	47000	47000
Deciduous fruit	85000	155000
Olives (irrigated)	10000	8000
Olives (unirrigated)	125000	25000
Table grapes (irrigated)	50000	60000
Table grapes (unirrigated)	35000	18000
Wine grapes (irrigated)	20000	20000
Wine grapes (unirrigated)	36000	21000
Bananas	30000	60000
Carobs (unirrigated)	40000	20000
Green manure (irrigated)	50000	–
Green manure (unirrigated)	115000	–
Total cultivated area, irrigated and unirrigated	4203000	
Total cultivated area, irrigated	2202000	
Total cultivated area, unirrigated	2001000	

[a] At full maturity of the plantations, expected about 1980. In 1973 production will be less.
[b] Millions of feed units.

TABLE 5-6

Livestock and its production

Type of animal	Number	Milk (million litres)	Meat (tons liveweight)	Eggs (millions)
Milk cows	83 000	398	15 900	
Calves for fattening	50 000	–	19 500	
Cows for meat	40 000	–	14 000	
Local Arab cows for meat	25 000	–	2 500	
Sheep	150 000	28	3 700	
Pedigree goats	30 000	22	400	
Arab flocks (sheep and goats)	200 000	17	3 000	
Other animals for meat			6 000	
Total		465	65 000	
Laying hens	8 300 000		17 000	1 420
Poultry for fattening	31 000 000		53 000	
Total poultry			70 000	1 420
Grand Total		465	135 000	1 420

the demand for the various commodities on the internal market in relation to the price system; the average disposable income; the food basket and predicted changes in its composition; the projected external market; the various alternative products of the agricultural sector; and the relative profitability of the various branches of agriculture as regards labour input, the balance of payments (which will to a considerable extent also affect government policy and the degree of its intervention in price regulation), and the restrictive factor, water. The alternatives worked out and selected as the best available are presented in Tables 5-5 and 5-6.

The areas for the various kinds of fodder (green fodder, hay, sown pasture, etc.) were determined according to the extent of the livestock branches proposed in the preceding table and the accepted nutrition norms as determined by animal nutrition experts. The areas allotted to seasonal and perennial green fodder, hay and silage can provide the full requirements of livestock for these types of feed. Production of grains and protein concentrates (generally obtained as by-products, such as oilseed cakes) account for only part of the livestock requirements; the remainder will be covered by imports.

Tables 5-7, 5-8 and 5-9 detail the livestock feed balance: the first shows consumption of the various types of feed; the second feed production, whether primarily feeds or by-products of other agricultural branches; while the third summarizes the production and consumption balance of animal feeds and the import requirements.

TABLE 5-7

Feed requirements of livestock at end of projection period

Type of animal	Number	Feed requirements per head (feed units)						Total feed requirements (thousands of feed units)					
		Protein concentrates	Carbohydrate concentrates	Fresh feed	Natural pasture and stubble	Hay and straw	Total feed units per head	Protein concentrates	Carbohydrate concentrates	Fresh feed	Natural pasture and stubble	Hay and straw	Total feed units (1000s)
Milk cows	83000	1400	920	2470	–	410	5200	116000	76000	205000	–	34000	431000
Calves	50000	800	600	80	–	170	1650	40000	30000	4000	–	8500	82500
Cows for meat	40000	300	200	100	1800	500	2900	12000	8000	4000	72000	20000	116000
Sheep	150000	50	50	40	360	45	545	7500	7500	6000	54000	6750	81750
Goats	30000	130	100	400	–	100	730	3900	3000	12000	–	3000	21900
Arab sheep and goats	200000	50	50	40	320	40	500	10000	10000	8000	64000	8000	100000
Arab cows	25000	600	400	800	800	500	3100	15000	10000	20000	20000	12500	77500
Laying hens	8300000	13	34	–	–	–	47	108000	282000	–	–	–	390000
Poultry for meat	31000000	2	2.7	–	–	–	4.7	62000	83700	–	–	–	145000
Work animals	50000	200	1000	200	–	600	2000	10000	50000	10000	–	30000	100000
Total								384400	560200	269000	210000	122750	1545000

TABLE 5-8

Production of animal feed at end of projection period

Crop or branch	Area (1000 dunams)	Green fodder and silage	Production in 1000 feed units			
			Hay and straw	Pasture and stubble	Carbo-hydrates	Pro-teins
Seasonal fodder (irrigated)	200	160000				
Seasonal fodder (unirrigated)	100	12000				
Perennial fodder (irrigated)	75	75000				
Hay (irrigated)	180		29000			
Hay (unirrigated)	350		46000			
Groundnut stubble	90		7500			
Straw from summer and winter grains	1040		33000			
Straw from pulses (irrigated and unirrigated)	155		6500			
Winter grains – barley					93000	
Summer grains – sorghum and maize	140				58000	
Stubble from grains	1000			30000		
Carobs	20				20000	
Cotton oil-seed cakes						31000
Other oil-seed cakes						7000
Cakes from imported soybeans						81000
Sugar beet – leaves and head [a]	90	10800				
Sugar beet – pulp [b]					25000	
Citrus peel		15000				
Chaff from local and imported wheat						50000
Natural pasture	2500			180000		
Carobs from forests and pasture [c]					25000	
Dried potato leavings and other peelings					29000	
Total		272800	122000	210000	250000	169000

[a] 120 feed units per dunam.
[b] 16.6 kg wet pulp = 1 feed unit; there will be a total of 381000 tons wet pulp.
[c] Not included in cultivated area.

TABLE 5-9

Animal feed balance sheet (in thousands of feed units)

	Protein concen-trates	Carbo-hydrates concen-trates	Fresh feed	Hay and straw	Natural pasture and stubble
Total requirement	384400	560200	269000	122750	210000
Total production	169000	250000	272800	122000	210000
Deficit to be covered by imports	215000	310000	–	–	–

4. Evaluation

i. *The balance of trade*

In an analysis of the possibilities for agricultural export in Chapter 2, citrus and tomatoes were used as examples. Citrus is already Israel's principal export, while export of tomatoes is in its infancy; however, examination of the economic and agrotechnical aspects indicates a good possibility for a large increase in tomato export. We have made similar examinations for all other products, establishing a scale of export. Determining the extent of each agricultural branch in the national agricultural economy and the proportion of each destined for export, makes it possible to establish the agricultural trade balance.

Although agricultural exports can earn the country large sums of foreign currency (in our plan $156 million), all agricultural production – whether for export or local consumption – involves a fixed annual foreign currency outlay. Local production of agricultural production materials is indeed increasing steadily: chemical fertilizers, for example, once imported, are now locally produced and even exported in quantity; this is also true of insecticides. However, production of these items is also tied to certain foreign currency outlays. The trade balance for agriculture is therefore determined both by the export projections and by the development of the service branches – fertilizers, insecticides, packaging materials, implements, etc. The trade balance for agriculture is therefore determined on the basis of existing forecasts in this field.

This balance is detailed in the following tables: in Table 5-10, agricultural export and its value; and Table 5-11, the foreign currency outlays for agricultural production materials.

From the tables it is seen that the agricultural trade balance is positive, exports not only covering the foreign currency outlay needed for agriculture, but leaving a surplus of 54 million dollars, such a sum is a significant and valuable contribution to the economy of the country as a whole.

ii. *Production value and its meaning*

We shall now examine to what extent the value of agricultural production, expressed in yields and harvests, will be increased by expansion of the cultivated areas under irrigation and by improved agricultural know-how.

According to data supplied by the Central Bureau of Statistics, the value of the total agricultural output in 1963/4 was about IL. 1248 million at current prices.[10] We envisaged an output value of about IL. 1830 million at the same

TABLE 5-10

Value of agricultural export at end of projection period

Product	Quantity	Expected price in 1972/3 ($)	Total ($)
Tomatoes	15000	200	3000000
Eggplants and peppers	3000	280	740000
Potatoes	15000	85	1275000
Carrots	5000	70	350000
Onions and garlic	15000	65	975000
Processed vegetables	37000	50	1850000
Total vegetables	90000		8190000
Durum wheat	15000	125	1875000
Groundnuts	25000	270	6750000
Bananas	12000	120	1440000
Table grapes	7000	180	1260000
Melons and watermelons	17000	100	1700000
Olive oil	2500	500	1250000
Wine	4000	550	2200000
Eggs (1000s)	200000	2.7 cent	5400000
Cotton	19000	710	13500000
Tobacco	1000	1200	1200000
Various fruits	5500	250	1375000
Flowers, strawberries, bulbs, fish miscellaneous			3800000
Total, excluding citrus			48140000
Citrus (fresh)	1000000	100	100000000
Citrus (processed)	150000	40	6000000
Total agricultural export			155940000

prices (see Table 5-12) for 1972/73. This means that the extension of agriculture and the development of know-how will raise the value of agricultural production by IL. 582 million over the 1963 level, an increase of some 48%. If we add to this a further IL. 150 million of produce from citrus areas planted during the development period but due to bear fruit only afterwards, we arrive at a total production value of IL. 1980 million, i.e. 60% more than the value of 1963 production.

The share of agriculture in the national product and the national income at the end of the development period, in relation to present levels, must now be considered.

In 1963/4 the gross value of agricultural production at current prices was IL. 1248 million[10] without taking into account fixed assets (the value of perennial plantations and unslaughtered livestock, soil conservation, irrigation networks, etc.). The national income derived from agriculture (the net agricultural product) in that year amounted to IL. 652.2 million[11], i.e. the nation-

TABLE 5-11

Estimated foreign currency balance sheet at end of projection period

	In millions of dollars
Irrigation and irrigation equipment	
1100 million m³ at 1.7 cents	18.7
Work of tractors and implements	
2 million dunams unirrigated land at 1.5 work hours	
$= 3$ million work hours	
2.2 million dunams irrigated land at 4 work hours	
$= 8.8$ million work hours	
$\overline{11.8}$	
Less work hours of 50000 work animals at 100 work days $= 100$	
tractor work hours 5.0 million work hours	
Total tractor work hours $\overline{6.8}$	
6.8 million tractor work hours at 75 cents	5.1
6.8 million work hours of tractor-drawn implements at 30 cents	2.0
Pesticides, sprays and fertilizers	
All these are produced locally, their foreign currency component	
is estimated at	3.0
Packing materials	
25 million cases for citrus at $1	25.0
Packing material for other export products	5.0
Miscellaneous for animals (excluding feed)	
For cattle, sheep and goats: buildings and equipment, spare parts	
and depreciation	1.5
For poultry; depreciation of equipment and buildings	1.5
Seeds and miscellaneous	2.0
Fishing materials	
5000 tons of fish in lakes and ponds at $50	0.2
16000 tons of fish in coastal and deep water at $100	1.6
Total expenditure on production materials	$\overline{65.6}$
Annual expenditure on fodder for animals	
Carbohydrates – 310000 tons grain at $61	19.0
Proteins – 215000 tons oil cakes, etc. at $80	17.0
Total expenditure on feed for animals	$\overline{36.0}$
Total exports and substitutes for imports	
(excluding feed)	
Value of agricultural export (f.o.b.)	156.0
Less imports of agricultural production materials 65.6	
Less imports of feed 36.0	
Total imports of feed and production materials	$\overline{101.6}$
Foreign currency balance: Surplus of agricultural exports over expenditure in foreign currency	54.5

TABLE 5-12

Agricultural production and value at end of projection period (1963/4 prices)

Branch	1964		End of projection period	
	Quantity- tons	Value IL. 1000	Tons	IL. 1000
Wheat	126500	33532	125000	33125
Barley	116600	22504	110000	21240
Summer grains	86800	17800	58000	11600
Pulses	7000	5639	10000	8050
Green fodder hay and sown pasture	–	51845		65000
Total grains and fodder		131320		139015
Groundnuts	9500	7125	36000	27000
Cotton-fibre	15650	36103	38000	87628
Cotton-seed	24500	4410	65000	11700
Tobacco	980	2889	3000	8800
Sugar beet	256000	16209	495000	31185
Other industrial crops		5733		13000
Total industrial crops		72469		179313
Melons, watermelons and miscellaneous	–	24320		45000
Vegetables (incl. export vegetables)	317300	91568	475000	158000
Potatoes	106800	23596	145000	31900
Total vegetables and potatoes		139484		234900
Citrus	838900	182061	1394000	319000
Table grapes	49300	17736	67000	24100
Wine grapes	34000	12056	40000	14160
Olives	21000	13464	32000	20480
Bananas	49800	19581	60000	23600
Carobs	1750	231	20000	260
Deciduous fruit	105125	81154	155000	132000
Other fruit	11427	6977	40000	26000
Total fruit	1111425	333260		559600
Cow milk	322200	110571	398000	136514
Goat and sheep milk	43550	16215	67000	26800
Eggs (thousands of units)	1278500	130520	1420000	144840
Honey	1220	3354	1200	3354
Beef	35250	92463	51900	136135
Mutton and goat meat	6950	16528	7100	17000
Poultry meat	74500	126721	70000	120000
Other meat	4000	11070	6000	16580
Fish	18722	32527	25000	37500
Total animal products		539969		638723
Miscellaneous and produce consumed on farms		31535		80000
Grand total		1248037		1829951

al income derived from agriculture constituted 52% of the value of the total agricultural product.

We have assumed 15% greater efficiency of agriculture in the future through increased yields and output. As a result of the rise in yields and of reduced water requirements, the net national income from agriculture may constitute a larger share of the value of the agricultural product; according to calculations that have been made, this enhanced efficiency will increase the share of the national income derived from agriculture to 60% of the value of agricultural production.

If the gross value of the agricultural product by the end of the development period amounts to IL. 1830 million, at 1963 prices, the net product thence derived will be IL. 1098 million, while we estimated the total net domestic product envisaged for the end of the development period at IL. 12 800 million at 1963 prices (Chapter 3, Section B). Agriculture will thus constitute 8.5 of the net domestic product as against 10.3% in 1963. It must be pointed out that simultaneously with this decrease in the share of agriculture in the national income the number of persons employed in agriculture will relatively decline by 2.5%. The decline in the share of agriculture in the national income thus corresponds to the projected decline in the percentage of persons employed in agriculture.

We shall now examine the allocation of the added agricultural product to exports and to the local market.

Of the IL. 582 million total added agricultural product beyond that of 1963, produce valued at IL. 275 million (at 1963 prices) is destined for export in terms of our plan:

TABLE 5-13
Value of additional export production above that of 1963

Commodity	Quantity (tons)	Value (IL. 1000's)
Hard wheat	5 000	1 500
Potatoes and vegetables	83 000	30 000
Groundnuts	19 000	14 200
Citrus fruit	800 000	168 000
Table grapes	6 700	3 000
Deciduous fruit	3 000	3 000
Miscellaneous fruit	2 500	1 500
Melons	14 000	2 800
Cotton	17 000	39 000
Miscellaneous products (wine, tobacco, fish, olive oil, flowers, tubers, etc.)		12 000
Total value of added exports		275 000

The sum of IL. 275000 constitutes 74.2% of the total added production at the end of the development period. If it is borne in mind that the local population is expected to increase during the development period by nearly 50%, and will have to be supplied with all its requirements in food and fibre, it is obvious with the available means of production, our programme provides for maximum diversion of produce to export.

iii. *Capital-output ratio*

Given the above addition to agricultural output, we are able to calculate the profitability of investments in agriculture, i.e. the ratio between investments and the net value of the added product.

For this purpose, it is at the outset necessary to calculate the investments required for future development and the investments in young plantations planted several years ago, due to begin producing during the development period.

TABLE 5-14

Investment requirements of the development programme (at 1963 prices)

Kind of investment	Total in IL. 1000's
550000 dunams irrigated land at IL. 180 per dunam	99000
Planting and cultivation of 65000 additional dunams of citrus groves at IL. 1000 per dunam	65000
Cultivation of existing young citrus plantations	50000
Planting and cultivation of 20000 dunams of other plantations at IL. 800 per dunam	16000
15000 dairy cows at IL. 1500 per head	22500
5000 head of sheep and goats at IL. 160 per head (the rest will accrue from natural increase without investments)	800
20000 meat cattle	15000
5000 work animals at IL. 1000 per head	5000
1.0 million lay hens at IL. 4 per head	4000
Tools and machinery at the rate of IL. 35 per dunam of irrigated land	19200
Supplementary tools for existing areas	
Farm structures for 4500 new farming units at IL. 6500 per unit	29000
Working capital for 4500 new farming units at IL. 4000 per unit	18000
Supplementary farm structures for existing farms	30000
Working capital for settlers in existing settlements	10000
Miscellaneous	10000
Total direct investments	393500
Water works:	
Total share of investment in development of water resources for the added irrigated area	260000
Total investments	653500

The investment of IL. 653 million will increase the value of output to IL. 1830 million. We have calculated the output of the orchards on the assumption that they will have all reached a stage of full production, to compensate for the fact that we have included in total investments during the period in question orchards which will only begin producing after the end of the development period. The value of added output as compared with that of 1963 thus amounts to IL. 582 million.

As we have seen, the value of the net added output will constitute about 60% of the total gross product, i.e. IL. 349 million. The total investment of IL. 653 million will thus result in a net annual output of IL. 349 million, i.e. an annual return of IL. 534 for every IL. 1000 invested. It should, however, be borne in mind that part of the additional product is a result of increased yields and not of new investments; this additional product would have been achieved even without further investments, or with investments only in agricultural research and extension. The real return for investment should therefore probably be less than in the above calculation.

In a similar calculation made previously, (R. Weitz, *Agriculture and Settlement*), the author estimated a net output of IL. 450 per IL. 1000 invested. We would also point out that erroneous calculations of the investment to output ratio are often made in that non-productive consumer investments, (investments in the development of the infra-structure such as housing for settlers, public buildings, electricity supply, access roads, etc.) are sometimes included in calculations of agricultural investments. All these items, although included in the budget received by the settler as a settlement loan, cannot be considered means of production. In comparing agriculture with other branches of the economy, a common denominator must be found. Investments in other branches of the economy, such as industry, do not include these items; only the actual direct investments in the particular manufacturing plant are taken into account. Costs for the erection of workmen's housing or the building of access roads are excluded from the calculations; such investments cannot therefore be included for agriculture.

NOTES

1. VOLCANI, I.: *The Planning of Agriculture* (Rehovot 1937), in (Hebrew).
2. 'Agricultural Settlement Plan', The Joint Planning Committee for Agriculture and Settlement (August 1950), (mimograph in Hebrew with English Summary).
3. WEITZ, R.: 'Settlement of Israel', The Jewish Agency Settlement Department (July 1950), (mimograph in Hebrew).
4. WEITZ, R., 'Agricultural Development in Israel' Jewish Agency, Settlement Dept. 1953 (mimograph in Hebrew).

5. 'Proposed Plan for the Development of Agriculture and Settlement', Joint Planning Centre (October 1953), (mimograph in Hebrew).
6. 'Four Year Irrigation and Agricultural Development Program', Government of Israel (April 1956), (102 pp. mimograph).
7. WEITZ, R.: *Agriculture and Settlement* (Am-Oved Ltd. Tel Aviv 1958), pp. 109–123 (in Hebrew).
8. The IL. returns for citrus, cow milk, groundnuts, cotton, sugar beet, spring tomatoes are based on data of the Agricultural Planning Centre in 1964. Changes have been introduced in view of the yields, work days and water requirement estimated by us. The IL. returns for various vegetables are taken from 'Profitability of *Moshav* Settlements' published (mimograph in Hebrew) by the Joint Planning Centre, with certain changes in view of divergent yield estimates.
9. The added dollar estimates are based on 'Comparative Investigation of Production Costs of Agricultural Commodities', Joint Agricultural Planning Centre, August 1960 (mimograph in Hebrew).
10. 'Five Year Plan for the Development of Agriculture 1966/7–1970/71', Table A1; Agriculture and Settlement Planning Centre; (Tel Aviv, June 1965), (mimograph in Hebrew).
11. State of Israel, Ministry of Agriculture, 'Report on Agriculture submitted to the Knesset by the Minister of Agriculture', (February 1965), p. 10.

CHAPTER 6

ORGANIZATION OF THE UNIT OF PRODUCTION

The previous chapter dealt with building and implementing the planning framework on the national level, or macro-planning. This chapter is concerned with the 'cells of the organism' or with micro-planning, the production units and their organization including the reciprocity between the farmer and his farm at the village level. The intermediate level, that of the region, will be the subject in the chapter following.

A. FAMILY OR LARGE-SCALE FARMS

1. Theories of family and large-scale farming

Recently, certain eminent economists who have investigated the make-up of land settlement projects and the structure of agriculture in Europe, South America and Asia, have posed the basic question of whether agricultural resources are not better exploited through large administered farms or managed co-operatives than through family farms. An opinion often expressed is that the pattern of land settlement should closely follow the course of industrial developments in this century, especially in its scale of production and use of mass production techniques. The seemingly logical conclusion has been drawn that the family farm, both generally and for new settlement in particular, impedes the promotion of the technological revolution which modern agriculture demands. The large industrial farms of California and Arizona, and the communes of Soviet Russia have been given as examples of farming virtually directed by one or a few men employing a labour force divorced from direct interest in the enterprise.

The concept of the superiority of the industrial farm over the family farm is far from new. In the early days of mass production in industry (at the turn of the century), several American economists predicted a continuing development in agriculture on 'industrial' lines. Similar predictions motivated Russian experts to amalgamate small farms into the *kolchoz* and *sovchoz* units. The more recent Chinese experiments in the large communes and agro-cities followed the same thinking. They were based, apart from political considerations, on the assumption that large-scale farming is more efficient.

Such dooming of the family farm has not taken place unless forcefully pushed through. Although the type and scale of family farm production has altered radically, the majority of the gross agricultural product in the U.S.A., for example, is still produced by family farms. Bachman, in discussing changes of scale in commercial farming in the United States notes that there is still considerable confusion with respect to contemporary changes in size and number of farms in the United States. "Trends in acreage and output have been subject to many interpretations, partly because of the lack of data for an intelligent evaluation.... The number of 'full-time' farms has actually been decreasing much more rapidly than is apparent from over-all statistics. The modern era of mechanical power and equipment in agriculture has greatly increased the size of our business units."[1] Available evidence suggests, however, that the increases are of no greater extent than is consistent with man's ability to care for a larger business through technological developments that have broadened and deepened his command of resources.

From the standpoint of strengthening the family farm, changes of scale arise more from the adjustments necessary to incorporate modern technology, than from attempts to prevent the supposed trend toward factory farming. "Contrary to popular conceptions," continues Bachman, "operators of large-scale farms frequently have chosen to substitute capital for labour rather than to increase their size of business consistent with the increased availability of labour compared to other resources. The middle groups of commercial farms have been characterized by the most rapid growth in acreage and output. This has occurred largely by adding land and capital to a relatively fixed labour supply centred around the farm family. The smaller business units in agriculture have faced more difficult problems of adjustment in meeting the challenge inherent in modern technological developments. Their relative importance has increased, and in the current setting they delineate an important sector of inefficiency in the use of farm resources." Later he adds: "But the emphasis on the essential difference between agriculture and industry, with respect to suitability for large-scale production, has apparently been well justified. The greatly increased scale of operation on our commercial farms has not meant any significant tendency toward a general development of an industrial type of organization in agriculture. Estimated labour requirements per commercial farm for 1950 are only slightly more than 400 man-days per farm. Estimates of annual average employment of hired labour per commercial farm varied from a little over 0.5 of a worker in 1930 and 1945 to a little less than 0.6 of a worker in 1940 and 1950."

Looking at the family farm issue from another point of view, Parsons and Owen state:

"So far as we are aware there is no clear evidence that any way has yet been found to operate the general farms of America that is more efficient than the usual type of family farming. Consequently, the services of professional managers used even more intensively than in the past, may simply strengthen the competitive position of farm families as farm operators. Perhaps the issue can be put more clearly by turning the statement a little: if the survival of the family type farm is to be seriously challenged it is not likely to be on the open field of operating efficiency." [2]

In Russia and other East European countries, attempts at doing away with family farm units have not been entirely successful, although social and political as well as economic considerations had led to these attempts. On the *kolchoz*, the peasants give as much time as they can to their own small plot of land and livestock. Russian agriculture has not kept pace with industrial expansion, and the family farm structure is being carefully re-examined. The coming years are likely to witness a strong reaction in favour of the family farm in both Russia and China.

The family farm, as opposed to the large administered farm, is advocated by economists such as R. Schickele[3], Joseph Ackerman[4] and others, who have thoroughly analysed the family farm basis of U.S. agriculture. They conclude that the family farm is in fact the only satisfactory type of agrarian structure. Whether this conclusion is valid for other countries will become clearer in the course of this chapter. First, however, the views of these investigators must be examined.

Schickele[5] points out that U.S. agriculture has been based on the family farm from the days of Jefferson onwards. He suggests that farm tenure problems in the United States can best be understood in the light of two major theories: 'The family farm theory of tenure' and 'the farm business theory of tenure'. Although both are rooted in the general ideology of humanitarian-democratic values, they differ in the ranking assigned to various sets of more specific beliefs. In Schickele's formation the 'family farm theory of tenure' consists of the propositions that welfare of the nation and of the rural community is best served if:

a) Farmers own and operate their farms as independent entrepreneurs;

b) Farm units are large enough to yield farm families an acceptable standard of living;

c) Farm units are not larger than that which the farmer and his family can operate without depending upon a substantial year-round hired labour force, so as to provide a wide dispersion of land ownership among farmers;

d) Farm families are secure in the occupancy of their land...

"Family farmers are self-employed proprietors not subject to being hired and fired, and develop entrepreneurship with its attendant qualities of self-discipline, responsibility and self-assurance. They are craftsmen with a close personal relation to their work, which is variegated and requires every day a series of managerial and technical decisions, adapting the various tasks to the weather, seasons and growth processes of plants and animals, and to price and market conditions. Family farm communities are spared the tensions and class conflicts that separation of management and labour has brought to so many industrial communities. The theory in modern terms is that, for these and related reasons, family farm tenure conditions are more conducive to democratic ways of life than are tenancy and large-estate conditions."

On the farm business theory of tenure, he notes that the welfare of the nation is best served if:

a) The business of farming is conducted along the financial, organizational and managerial principles applied in any other non-farm businesses;

b) The free market forces are allowed to determine the tenure status, size of farm, and family income for each farmer, agricultural worker or employer according to his individual ability to take advantage of the market;

c) Farmers are not accorded any legislative protection and aid that are not given other producers elsewhere in the economy.

"Here, let us recognize the first essential difference between the two theories. Clearly, the farm business theory emphasizes the production efficiency aspect of economic welfare; the family farm theory stresses the distributive equity, social status, and security aspects of community welfare."[6] Dr. Schickele examines these two theories and concludes that each, within limitations set by the other, is correct. Ensuring that family farm units have sufficient resources for efficient operation meets the requirements of both land tenure theories: social and welfare criteria are satisfied within a framework guaranteeing high production efficiency. He points out that there is no conclusive evidence that management-labour separation and large-scale industrial organization of agricultural production processes yield in the aggregate a superior production performance, which could not be achieved under family farm tenure conditions, except perhaps in a few highly specialized types of farming.

Two further points should be noted: firstly, that the family farm theory does not call for the largest possible number of farmers regardless of their standard of living, but provides for efficiency criteria by proposing that family farms be large enough to yield the family an acceptable living; secondly, the farm business theory does not call exclusively for large-scale factories-in-the-field or concentration of management in professional agencies, but demands only that resources should be used as efficiently as possible. As family farmers become more efficient producers, their tenure status is compatible with the farm business tenure theory. In examining the land tenure systems in countries still under the feudal system of tenure, Dr. Schickele finds that the issues are distribution of income, social status and power, rather than maximum production.

Joseph Ackerman[7] notes the main factors which make successful operation of the family farm easier than that of the large administered farm: mechanization has been developed primarily to meet the requirements of the smaller and medium-sized farms, making their output more competitive with that of large farms; secondly, the lower level of predicted world agricultural prices has deterred investment in large-scale farm enterprises while favourable loan facilities have furthered the advance of the family farm; thirdly, wages and social security payments for the agricultural labour force are lower than for industrial workers, causing a decrease in the agricultural labour force and serious labour problems for large farms. The family farm is independent of hired labour and of rising labour costs.

Ackerman points out the importance of mutual aid and the need for establishing co-operatively linked family farm units in order to facilitate use of improved technology and to permit use of modern mechanization of smallholdings through co-operation. He warns against the danger of land fragmentation and stresses the importance of planning units of a size adequate to provide for two families, the father and the adult son, and suggests, moreover, vocational guidance programmes to train the remaining sons in other professions, to avoid the danger of subdividing farm units through inheritance.

Turning to Israel, the question of the family unit has been an important one. Dr. A. G. Black, the FAO representative to Israel in 1957, praised the technological advances made in Israel agriculture, but pointed out that such advances combined with the existing small farm structure of the moshav would only lead to difficulties.[8] These small farms were best fitted to the production of those commodities – milk, eggs and vegetables – which were already appearing in surplus, while the need was for staple industrial crops – cotton, sugar beet, groundnuts, feed and bread grains – which are all major items of world trade and most cheaply produced by large scale cultivation. The small farms, he predicted, would not be able to support their operators in keeping with Israeli standards for long.

At that time the new villages in the Lakhish Region were organized as large administered farms for the production of industrial crops without individual parcellation, as an interim measure to introduce the new settlers to agriculture. Dr. Black expressed the hope that the land would not be divided, but worked on a system of large-scale farming until the farmers themselves decided that this was the best system for their villages. In his criticism, Dr. Black was hitting at the basic weakness in the family farm structure then existing in Israel exemplified by the traditional moshav structure.

Before describing the structure used to adapt the moshav to efficient industrial crop farming, it is well to note the advantages of the large-scale administered farm, which can be stated as follows[9]:

1) Both the farm manager and the labour force can specialize in a limited number of branches on a relatively large scale of operation, thus raising productivity.

2) Large farms can choose production branches which offer true comparative locational advantages (optimum use of land, water and climatic conditions).

3) Maximum suitable mechanization for cultivation, harvesting and grading can be applied.

4) Cost can be lowered by mechanization to raise labour productivity, improve production quality, grade standards and methods.

5) Farm improvements can be rationally applied in accordance with ever increasing advances of modern technology, without hindrance from land fragmentation or surplus farm population.

6) In view of decreasing transport costs and their relatively small part in farm production costs, farm labour can be housed in neighbouring towns rather than in expensive accommodation specific to each farm where amenities will necessarily be scanty and poorly organized. The housing of farm labour in towns can, under certain circumstances, bridge the gap between town and country.

2. The family farm: a competitive unit?[10]

The traditional Israeli smallholders' village or moshav was until recently a closed economic unit, operating a self-contained serving system and can be used as an example of the disadvantages of family farms not organized for today's technological level. Each farm as a production unit had its land located in proximity to the farmstead, with the parcellation and irrigation network isolating each holding from its neighbour. Inherent in this structure of village organization, particularly in the first settlements built after establishment of the State in 1948, were the following limitations:

1) The possibilities of large-scale cultivation through partnership or co-operative effort were severely limited due to the isolation.

2) Overhead expenditure was high because of the small scale of individual enterprise, hence insufficient use of productive investment.

3) The extended farm layout and isolation of the separate units, often meant clumsy village design, long distances and no natural centre for educational, recreational and other services. This was not conducive to good social atmosphere. For the same reason, comprehensive rural extension activity and the corresponding technical change were hampered.

4) The conditions for farm specialization were adverse; labour productivity was low.

5) The land pattern and plot allocation were fixed and could only be changed in scale at loss of the original investment. This inflexibility would in the future militate against natural tendencies towards an increase in average farm size.

The result of this traditional organization was the difficulty entailed in increasing production or altering production patterns to raise the living standards of the farm population. Because of their physical layout, these villages could not adjust easily to new practices and innovations made necessary by advancing techniques and changing market demands. They remained static within their original rigid production framework, while their operators fell behind the general rising income levels in other spheres of the economy. Such farms faced the danger of becoming relatively too small both for labour and new capital input. Capital investment was a poor economic substitute for inflexibility in physical layout. This was in contrast to the flexibility of resource use in administered farms.

Before discussing the advantages of the family farm in practice, not only theory, differentiation must be made between it and the small farm. The family farm is characterized by family labour utilizing the resources at its disposal within the production unit. If these resources are insufficient to afford the family a satisfactory income, the farm is too small in scale and scope. The disadvantages of the too-small farm are not necessarily those of the family farm. Although the problems are closely interrelated, the too-small farm and the family farm have to be considered separately.

In its various forms, the family farm which has retained its position as the basis of farming for a very long time in many countries and under the most varied conditions, has attributes lacking in other farm patterns:

1) It provides its owner-operator with not only an income but also a way

of life. The continuity of rural communities based on the system of family farming has resulted in national stability of great social and economic importance.

2) The very nature of the agricultural production process with its seasonal fluctuations and dependence on climatic conditions, gives the family farmer both scope for initiative and a direct incentive to complete the job.

3) The members of the family farm are not bound by an eight hour day: the main criterion of their labour input is the timing of operations and the resulting higher returns. Certain products, suited to the labour organization of family farms, are uneconomic on a large scale and independent of the growing scarcity of hired labour which is attracted to industry because of its unstable ties to the land and the lower wages prevailing in agriculture.

4) Not all farm operations are economic on a large scale. Picking, for example, can be done more efficiently on a shorter row to prevent unnecessary haulage. On large-scale administered farms, the sum of the advantages of large-scale operations outweighs the disadvantages of the necessary small-scale operations. If the essentially large-scale operations (ploughing, sowing, cultivation, etc.) can be organized along with the family set-up, however, the family farm gains on the small-scale operations.

5) The difficulties of management and control limit the economic size of many farm branches and decrease the benefits which can be obtained from concentrating upon a limited number of processes on a large scale. The family farm unit can find its own optimum size of operation, dispensing with hired labour and the problems of supervision in the field.

The most serious criticisms of the family farm have been based on its production inefficiency. It is clear that many family farms are small, outdated, undercapitalized and inefficient, but they do not have to be. Lack of adaptation to economic realities rather than the family farm idea is the cause of poverty; abolishing the family farm would not abolish the poverty of its operators. Within a suitable economic framework, the family farm structure can hold its own, as will be seen in the next section on the changes introduced in Israel settlement. In his study of production efficiency, the Danish scholar Hooker found that family farms in Denmark have a production efficiency significantly higher than the administered farms. Co-operative mechanization and marketing virtually cancelled out the advantages formerly enjoyed by the large farm.

A prominent feature of the Western World is the widening social and economic gap between the family farm and other sectors of the economy. This ever-growing gap, both social and economic, has cast doubts on the ability of family farms to compete in an affluent society. In a modern state, the family farm can only continue to exist if it provides its operator with a standard of living equal to that which his labour can earn in other employment. Today's economy can neither afford underemployment on farms nor the wasted resources and loss of agricultural production which this represents. The central question in all discussions of the subject is whether the satisfactory system of rural life provided by the family farm with all its social attributes can be consistent with flexibility and efficiency in the agricultural process, so that output and rural standards of living can rise according to the needs and norms of the general economy.

The Executive Committee of the FAO European Commission on Agriculture at its fifteenth session (March, 1960) passed a resolution calling for study of "regional models of farm organization and management systems adapted to the physical and economic conditions prevailing in the different parts of Europe". It was clearly recognized that "the connections between agriculture and forestry as well as the problem of combining agricultural and non-agricultural activities... should be considered as aspects of a regional economic policy to alleviate the problems of small farms under prevailing conditions".

The European organization had arrived therefore at the concept of a regional economic policy offering solution for converting family farms into efficient agricultural units without discarding their positive aspects. Israel had arrived at similar conclusions, proposing concrete solutions, one of which is discussed below.

3. Planning for scope

To overcome the problem of small-scale operation, its irrational investments and negative social consequences, agricultural planners in Israel have attempted to combine the production advantages of the large-scale administered farm with the socio-economic advantages of the family farm, by adopting a new approach to rural structure and organization. The characteristics of this structure are:

1) Individual plots have been parcellated within greater land blocks in order to adapt them to a larger scale of mechanization. This system preserves individual tenure rights while enabling voluntary large-scale co-operative efforts in mechanization. A special sprinkler irrigation network has been introduced, locating permanent water installations where they do not hinder the free cultivation of the plots comprising one land block. The permanent underground pipes with their connecting upright installations are located at

the ends of the extended land block. The physical conditions for co-operative cultivation are thus provided, so that ploughing, planting and spraying operations can be mechanized. Participation of the farmer in co-operative cultivation is voluntary. The growing of the crops and their irrigation remain at all times the responsibility of each farmer on his plot. Since the land is kept in unbroken blocks, the houses must be in one small perimeter allowing for compact village planning and the architectural expression of a sound social environment.

2) This form of parcellation, through which specialized equipment can be introduced, has enabled the initiation of specialized agriculture, in which a limited number of crops are chosen, suited to the climatic and soil conditions, the farm layout and the farmer's ability (the human factor is of great importance in the choice of farm types). In this way, the farmer specializes and acquires full knowledge of his branches, enabling him to keep up with new techniques.

3) Large block cultivation is the justification for centralized economic services based on a number of villages. For instance, all the tractor services, aerial spraying and despatch in industrial crops to the processing industry can be centrally organized. Heavy mechanization is costly. The land area of one village is insufficient for full utilization of specialized equipment. There is a definite cost advantage, and the possibility of far wider equipment range, if investments for several villages are made collectively. A tractor station built close to several villages provides rational services for all. Similar cost advantages are gained by a concentration of marketing, purchasing and banking services.

4) Social considerations in rural areas demand more and more attention. Centralized educational, health, recreational, cultural and shopping services must be provided on a high standard, yet at low cost. A village population alone, however, is too small to support them.* Centralized services of a number of villages can supply the needs.

5) The establishment of a service centre leads to the creation of a suitable location for the integration of industry to meet the needs of agriculture. New inter-relationships between agriculture and industry can therefore be formed. The organization of concentrated resources in rural areas can be extended to include industrial development: crops can be contracted for industry, and agriculture can be co-ordinated with urban planning to supply the particular needs of agricultural industries.

---

* An Israel farming village is based on 80–100 families.

This policy of agriculture linked to industry acts through many spheres of co-operation and social connections, building an interdependence between the two. The farmer has his immediate market close by; the industries need his production for their operation. They offer him an outlet for his investment through financial partnership, employment for those of his sons who cannot be absorbed on the farm, and thus reduce the evils of rural underemployment.

6) This structure of extended land blocks and centralized services offers the flexibility necessary for present needs and future eventualities. Changes can be introduced in resource allocation to enable the rural family to attain a higher living standard, while avoiding the drastic changes in living patterns so often caused by economic change. Industry can be located at the rural centre, providing employment for surplus farm population. This is particularly important should trends necessitate enlargement of farm size and displacement of some operators, who will thus be able to continue living in the village and working in the centre. The necessary land reparcellation is very easily achieved. The whole structure of rural life and agrarian stability is maintained. This comprehensive farm, service and industrial pattern has been called the composite rural structure, analysed in Chapter 7.

To summarize: Israel experience has shown that the following four factors are basic to the success of the family farm:

a) The family farm must be planned in such a way that it remains a unit within a co-ordinated rural structure based on principles of mutual aid and co-operation. This refers to co-operative mechanization facilities, marketing and other services.

b) The planning of the family farm must be related to actual market demand and production possibilities of the region, labour and capital.

c) The size of the farm unit, its land and its water resources, must be large enough to guarantee the settler an income no lower than the national average wage. Capital availability must be sufficient to allow for intensification of farm-unit production and shifts to higher value products so that the farmer's income can rise concurrently with the income of those who buy from him.

d) The farm layout must be planned to allow for simple and full mechanization, both on the individual farm and in co-operation with other farm units in the village. Both large- and small-scale mechanization are thus allowed for.

Today Israel accepts that the family farm meets the social, political and security needs in rural areas and that the economic advantages of the family farm will become evident once the new settler is integrated in his new environment. In spite of the great strides that have been made in the sphere of agricultural mechanization, the individual initiative and management of the farmer remains the essential factor determining his production capacity and success in farming. The four basic principles above provide the framework within which the farmer can best use his personal initiative to exploit the advantages of modern technology and co-operation, and keep up with the rising incomes of non-farm people.

The individual and co-operative farm structures in the new moshavim described above differ radically from the physical organization of the old moshavim which were so severely criticized by Dr. Black, and whose advantages were set down on p. 212. In the design of her new villages, Israel believes a positive answer to the critics of family farms has been found.

Israel settlement patterns can be viewed in an additional perspective, that of village organization, each of which contains a certain number of family units consuming the raw materials and producing the goods of the agricultural process. Within these villages, the families are organized in ways which vary according to the way in which the land is farmed. It may be divided into equal parts, one part for each family, with perhaps some pasture or grain land worked collectively. This is the co-operative smallholders' village, or the moshav.

The lands, on the other hand, may form one large unit with the families organized as a single labour force having commercial facilities and a communal way of life, except for living quarters. This is the collective village, the kibbutz.

These two forms of settlement represent the Israeli expression of small-scale (moshav) and of large-scale (kibbutz) farming – in other words, of family farming within the single labour unit of one family, and the very special case of family farming in the kibbutz with the total of family labour units organized as a single unit.

These two types of village organization have provided material for comparison of the question of large- or small-scale farming. Studies of comparative profitability per family have not shown a significant advantage of one over the other. Similar resources, whether organized in the separate production units of the moshav, or in the single production unit of the kibbutz, can return the same profit per family.

B. THE DIVERSIFIED FARMING PATTERN IN ISRAEL[11]

The historical development of agricultural organization from subsistence, through mixed farming to specialized farming has been discussed previously.

Most countries with a developed agriculture have passed through each phase. The planning problem in a country working for higher output from its farmers, is usually concerned with graduating from one stage in order to reach the next. This matter has been an important issue in Israel's agricultural planning.

1. Evolution of the diversified farming pattern in Israel

The Jewish farming pattern at the end of the nineteenth century consciously departed from the centuries-old *fellaheen* or peasant farm pattern of Palestine, which sustained most of the population through dry farming producing mostly grain, supplemented by some fruit, vegetable and pulse crops.[12] The settlements established under the aegis of Baron Edmund de Rothschild were based exclusively on fruit or vine plantations; the second wave of immigrants, at the turn of the century, turned to extensive grain farming, hoping to avoid the recourse to hired labour made necessary by the seasonal labour peaks of the plantations and the resultant blurring of the image of the self-sufficient Jewish farmer. In fact, neither the plantations nor the grain farms, which failed to provide the settlers with a minimal income, succeeded in creating an atmosphere conducive to the national cultural revival which was the mainspring of the Jewish land settlement movement.

The subsequent immigrant waves, at the beginning of this century, gave attention to the problem on both the ideological and economic levels. A. D. Gordon, whose writings embodied the concept of self-labour and the 'conquest of labour' by the Jewish settlers, wrote in 1904 that as long as "every piece of land, which came to us with so much effort, is worked by others, it is certain that we are yet far from a true national life". The intensive search for an economic solution to the problem led to the birth of a new idea – the mixed farm, or diversified farm.

The mixed farm would include the advantages of the fruit farms with those of grain farming, and include livestock to widen the base of the farm and to stabilize its income and its labour requirements over the years. The first village founded on the mixed pattern was the kibbutz Degania (1909) in the Jordan Valley. Because maximum employment had to be ensured for its members, and because it was against kibbutz ideology (following Gordon), to make use of hired labour, the mixed pattern was suitable. Fruit orchards and grain farming were planned to obtain a balanced labour schedule, while milk production provided a needed cash turnover, with the manure from the stable enabling the fertilization of lands.

The advantages of such balanced farm practice were its good income prospects and the benefits of providing home-grown produce for consumption on the farm. Dr. A. Ruppin advanced other points in favour of the mixed farm; the interlocking of the different branches could balance labour needs throughout the year, strengthen the security of the settler and above all, enable the maximum to be obtained from the land. It would thus lead to a reduction in the size of the farm which had been necessary for grain production alone. Even at that early stage before World War I, it was recognized that the land at the disposal of the settlers was limited, while the necessity for settling many people was great.

The mixed or diversified farm thus became the basis of new settlement for many years in both kibbutzim and moshavim. In one sense, Jewish settlement had, by adopting diversification, come full circle back to traditional Arab agricultural practice, though avoiding the landlord system, and producing for cash as well as subsistence.

In 1910 a research station was established at Ben Shemen, headed by Professor Volcani. Various combinations in mixed farming were tried. The characteristics of all these were family labour, a balanced labour schedule, home-grown fodder for the livestock on the farm and rigid adherence to crop rotation principles. At first the mixed farm was extensive in structure with 130–140 dunams in the final plan.[13] An example is as follows:

  85 dunams – providing fodder for 6 head of cattle and their offspring
  20 dunams – providing fodder for work animals
  10 dunams – grain production mainly for home needs
  10 dunams – fruit orchards
   5 dunams – farm house and farm yard
  ――――
130 dunams – Total farm size.

Smaller mixed farms were established in irrigation areas (Beth She'an Valley, Jordan Valley):

  20 dunams – providing fodder for cows
   5 dunams – vegetables
  ――――
25 dunams – Total farm size.

At the same time, a methodical research programme was undertaken at Ben Shemen to investigate the potentialities of the various branches which could form the basis of the mixed farm unit. The Ben Shemen farm enquired into the structure, management and principles of the new system. Friesian cattle were introduced, the theory and practice of crop and animal husbandry were worked out, and crop rotation examined; different breeds of livestock and various types of vegetables were tested. From the experience of Degania and Ben Shemen it was proposed that agriculture in Israel be based on certain main farm types – extensive mixed farms; fruit farms; and intensive irrigated farms including fruit, fodder, dairy and intensive vegetable production.

In the years of the First World War and after, a problem arose with seasonal workers in the citrus groves who had no income except in the harvest season. There was need to find ways of establishing independent smallholding farmers. A number of workers' villages was therefore founded with a plot of land attached to each house to provide the worker with a supplementary income and part of his food needs. The size of the plot differed from place to place and was usually less than ten dunams. The general approach was, however, the same – to fill out the farm worker's income and give him a stake in the soil. Many opposed this system, proposing instead that villages of independent farmers be established on the land of the Jewish National Fund, working on their own mixed farm units. As mentioned in a previous chapter, the founders of this new system, the *moshav ovdim* (*workers' moshav*), wanted the farm family to be self-sufficient in food provision.

During the twenties, the mixed farming system was consolidated, although definite farm types were not fixed. The size of farms differed as did the make-up of the farm production. Some had high incomes, while others did not have sufficient resources for minimum living needs. In order to find a common denominator in the allocation of production resources, the Zionist Movement formed a committee in 1929 to draw up 'keys' for the farm structure of new settlement in the different areas of the country. This committee set itself the task of determining the following seven points[14]:

a) A suitable living standard;

b) Secure sources of income according to the state of agriculture in the country;

c) The farm size necessary to maintain family by family labour;

d) The full complement of production resources necessary to equip the farm completely;

e) The minimal equipment necessary for partial establishment;

f) The length of the necessary first transitional stage;

g) The minimal resources for this transition period.

The committee was not free to work as it wanted. There was no area of land available large enough for extensive settlement; neither were water resources sufficiently developed for a large project. It was faced with existing villages which had been built without any definite system or principle. The immediate practical application of the committee's work was severely limited by these restrictions, but it tried nevertheless to introduce system into agricultural planning and settlement.

2. Achievements of the diversified farm

The diversified farm, during its evolution from extensive grain farming to intensive dairy, fodder and vegetable farming, fulfilled the functions it was designed for. It succeeded in forming a stable tie between the settler, his farm and his land, and built up a firm rural life securely based on the soil of the traditional homeland. It made the worker independent of hired labour and ensured him an income all the year round.[15]

In addition to the agricultural and social factors, this farm type had great political importance. Agricultural villages could be settled supporting as many farmers as possible. Such settlement re-established the claim to the soil of Israel, with footholds in far-flung regions, each one separate and often a large distance from the next. The diversity of these villages and their structure as independent economic units with regard to all necessary services enabled them to hold out during periods of rioting which became a regular feature of the 20's and 30's, even when they were cut off from the rest of the country completely. The farm not only provided the family with its immediate food needs, but in the context of the village it formed part of a self-sufficient cultural, social and community unit for all the services – educational, health and economic, including grading stations, packing houses, dairy and so on.

The pattern of separate communities which developed was tied together by a co-operative network for buying and selling extending from the northernmost to the southernmost village. Such co-operation was given expression in the forming of buying and selling institutions, '*Hamashbir*' and '*T'nuva*' respectively. Farm produce was transported by the co-operative to the marketing centres which were sometimes very distant. Similarly, raw materials and consumption goods were supplied by the consumers' co-operative to the villages irrespective of the distance involved.

On the other hand, the ties of each village with its neighbours, as far as economic and social activity were concerned, were loose. Often, distances prevented any ties or common activities at all. Even where villages were closely settled, the great efforts needed for establishment, meant that little inter-village life was developed.

During this period, the Mandatory Government imported non-perishable foods, such as sugar, oils, flour, grains and fibres. The mixed farms therefore concentrated on producing fresh produce such as vegetables, milk, eggs and fruit, which could not be imported. During the disturbances of the thirties, the mixed farms supplied the Jewish population with all of its food needs when other supplies were cut off.

The mixed farm therefore, was a product of the times, supplying the struggling town and country settlers with their food. It was well-adapted to the security position of those troublous years, and was the backbone of a new rural people. Certainly it was a great improvement on the traditional subsistence farming of the country; its establishment represented an important economic advance and a turning point in agricultural development.

With statehood, there was need for another turning point. Conditions changed at once; the economic, security and social patterns of the country became completely different and the functions of agricultural settlement and production were dictated by the new needs. In the aftermath of the war and the flood of immigration, this was not at first realized. Many new villages were built on the same pattern as the old, with the mixed farm still the basis. Doubts were however expressed by some planners about the ability of the system to meet the new economic needs.

3. The trend towards specialized farming

Some five years after the country gained its independence, planners re-evaluated existing concepts of farm types. The needs of the national economy called for a change from the traditional 'mixed farm', which could not contribute substantially to import reduction or to greater exports. In 1953, the writer, in an article entitled 'Self Examination'[16], stressed that: "mixed farming has misled the agricultural economy, and its expansion cannot be continued. The aim must be economic independence as soon as possible. If mixed farms are developed at the same pace as in recent years, a surplus supply of animal protein – eggs, milk and meat – is inevitable, while import of essential foods, such as sugar, grains and vegetable oils will increase."

The writer in proposing a food basket differing somewhat from that accepted at the time[17], included a reduction in the consumption of animal protein and an increase in vegetable protein, the curtailment of fodder production for animal feeding and the increased production of grains and industrial crops, such crops being cheaper to produce, and with lower water requirements than fodder crops. He pointed out that crops produced for direct human consumption were cheaper than those consumed as animal protein through animal feeding. He likewise urged the production of industrial crops to save or even earn foreign currency as an important step towards economic independence. Others stressed that while the volume of agricultural production must increase, the output must be qualitatively more in line with the population's nutritional requirements in the austerity period, and also be directed towards narrowing the trade gap.

The writer urged the planning of a new farm type, called the field crop farm[18], which "should become the future basis of agriculture and settlement in our country, complementing in its output that of the dairy and citrus farms".

The plan he proposed for the field crop farm was originally: 2–3 dunams for the farmyard and house;

40–42 dunams for field crop farming, mainly industrial crops under irrigation;

10–12 head of sheep and 100 laying hens.

A sound rotation of winter and summer industrial crops was planned. Where possible, 5 dunams of citrus groves were to be included. The resource allocation was to be decided on the basis of equal income opportunity, so that profitability in the light of prevailing price levels would be the same as in the dairy farm type. Prices for the main field crop farm products would be set high in the initial years, until yields and efficiency in their production increased, when they could be reduced.

Heavy farm mechanization would be necessary only for the basic agricultural operations, lighter cultivation could be done with draught animals. In contrast to the planning of mixed farms where lands were concentrated around the farm house, lands of field crop farms could be separate from the farm yard, so that optimal cultivation and irrigation methods could be rationally applied, co-operatively over the whole village where necessary, as, for instance, in the aerial spraying of cotton. The quantity of irrigation water required would be no higher than that of the smaller mixed farm due to the particular choice of field crops to be produced. The labour schedule was admittedly expected to be less balanced over all the months of the year, than in the case of the mixed farm plan. However, the principle of self-labour through the work of the farmer and his family was maintained.

The proposed introduction of the field crop farm caused a storm of opposition in some agricultural circles. Many veterans rejected the plan outright.

Schutzberg, a member of kibbutz Merhavia, and also a board member of the Agricultural Centre of the Labour Federation, expressed his opposition in the following way[19]:

"The mixed farm with its emphasis on dairy farming is the structure that has been predominant in our agriculture for over 20 years. It became the method of farming as a result of years of trial, error and experimentation, starting with a dry crop farm type consisting of grain production on the one hand, and the intensive fruit farm type on the other. Mixed farming is in our experience the best method of solving the problem of yield level and land fertility. If new farm types are established against the accepted principles of organic mixed farming, there can be little doubt that resources and man-power will be wasted without accomplishing our national aims."

Others questioned the yield figures taken for calculating the projected incomes of the field crop farm types as being too high – for instance 300 kg of groundnuts or 4 tons of sugar beet per dunam. The Agricultural Centre also expressed its support for the conventional mixed farm. In its discussions in 1953, it recommended that mixed farming with its traditional emphasis on dairying, supported by the usual auxiliary branches, should be continued with additional new crops such as irrigated grains and sugar beet.

The Joint Agricultural Planning Centre, however, continued to examine the proposed field crop farm type in an endeavour to suggest practical ways of implementing the proposals. The present Minister of Agriculture, Mr. Haim Gvati, wrote that "without neglecting the basis of mixed farming it is now necessary to institute the intermediate planning of new farm types, based on crop cultivation in place of animal husbandry."[20] After discussing the reasons for changing planning directives, he stated: "This development of our agriculture in the coming 5 years will effect some changes in the economy. Limitations must be imposed on the expansion of dairying, so that the branch will only grow by natural increase. Poultry farming cannot be expanded further and new farms can only be established on the assumption that they will not develop a dairy or poultry branch. On the other hand, special emphasis must be laid on the production of industrial crops and irrigated grains."

Eventually, the new concept was adopted by the Agricultural Planning Board, the body responsible for laying down general policy directives in settlement and agriculture. From the many discussions in various forums and in the press, the field farm eventually became a reality. A turning point had been reached in the change-over from diversified to specialized farming.

The pros and cons of the two systems are as follows:

Diversified farming

a) The labour schedule is evenly distributed throughout the year;
b) An economic sense of security is obtained by cutting risk;
c) Land fertility is maintained;

d) Since a large part of the produce is for home consumption and the goods sold are varied, there is relative independence of market fluctuations: if one product fetches a poor market price, it is likely that another will make up for it;

e) The farm does not have to invest heavily in production resources or engage in large-scale buying and selling procedures;

f) The income is balanced over the year.

Specialized farming

a) Farm branches most suited to the regional conditions can be developed and other particular advantages can be utilized for efficient agriculture. In this way, regional specialization connected with appropriate commercial frameworks can obtain the maximum from agricultural resources;

b) The farmer can concentrate his efforts on a limited number of branches, become expert in them, and raise his standards of production;

c) A large scale of output in a product enables quality standardization, cheaper packaging and transport, and more competitive prices;

d) Farms, villages and regions can be planned for efficient production, marketing and transportation of the bulk output;

e) Investments can be geared rationally to the specific branch so that specialized equipment becomes worthwhile.

An advocate of specialized farming has been the Harvard economist J. D. Black[21], who holds that farms being engaged in many branches of agriculture indicate a degree of backwardness in a country. He maintains that diversified farming where the farmer supplies his own needs, is an indication of primitive distribution channels or of fear of outside competition, whereas specialized farming is achieved through improved transport facilities, increased market opportunities and improved agrotechnical methods.

After 1948, under new conditions of Israeli statehood, it became clear that increasing emphasis had to be placed on specialization – in new planned settlements, certainly, but also in the older settlements. The market needed new products; some newly acquired areas were unsuited to livestock production; the market capacity for fresh products was being exceeded; while some products were totally unsuited to combination with livestock on the same farm. There were areas – notably the hills and the south, unsuited to the mixed farming pattern, even though it had been tried in them on the conviction that it was the only sound system.

By 1953, diversified agriculture had already undergone a reorientation as testified by the results of a survey conducted in the older moshavim by

Dr. Lowe.[22] He found that balanced mixed farming of livestock, fruit and vegetables had given way to farms with some specialization which he classified in four categories: 'mixed'; 'dairy–poultry'; 'poultry–fruit'; 'poultry–vegetables'.

1954 saw the first planning of a whole region of new settlement – the Lakhish area. It was based on industrial crops, and as we shall see in another chapter, planned for services, industry, and for the handling of diverse ethnic elements. In the early 1960's, the planning of the B'sor region further south was begun on the basis of export crops. The investment, lay-out and organization were to be for low-cost, high efficiency, high quality production suited to very specialized markets (see p. 232).

The general change in Israel's agricultural economy towards increased specialization has been speeded by several factors: the establishment of numerous new settlements which necessitated solving surpluses in some commodities; the altering of the profitability of various agricultural branches; concern over the foreign payments gap; the security situation calling for a stable, efficient economy. In addition, the short distances and the natural variations in climate and soil being great, have allowed farm types to be developed for specific districts. This has lowered costs, so that prices have become competitive on the home market and footholds have been gained on world markets. Specialization has become the key to higher income for the farmer and has opened the door to efficient agriculture.

Planning principles for specialized farming in new settlements in Israel are[23]:

a) Resource allocation must guarantee equal income opportunity. Whether farms are established in the north or the south, in the hill region or on the plains, there should be no significant difference in the planned average income.

b) Farm enterprises should be few, while permitting a labour schedule within the capacity of the farmer and his family without recourse to hired labour.

c) The absence of a main livestock branch in certain farm types necessitates care in crop rotation, improved cultivation methods, and a careful use of artificial fertilizers. The introduction of a form of livestock complementing the work schedule and able to utilize surplus, low grade and residue products, should be part of policy.

d) Since the specialized farm is more vulnerable to climatic and other natural hazards, a system of crop insurance must be introduced to protect the farmer from crop failure and at least partially guarantee his income. The accepted method is yield insurance. Recent surveys have shown that losses resulting from crop disasters are by far outweighed by yields and incomes during normal or good years, and therefore crop insurance has a good possibility of commercial success.

e) The specialized farms must be established on a regional plan. Choice of farm type must be determined by the resources and economic advantages of the region. The dairy farm type, for instance, has been established mainly in districts near towns. Central milk refrigeration plants, feed storage and milk processing plants serve the particular needs of the dairy farms and milk consumers. The citrus farm type has been established in areas where soil, climate and water availability provide excellent conditions for the trees, and, in the typical citrus belt, regional packing houses and grading plants have been built. Industrial field crop farm types have been established in the northwestern districts of the Negev, where climatic and soil conditions particularly favour the production of cotton, sugar beet and groundnuts. Cotton gins and mills, a sugar refinery, groundnut grading and potato cold storage plants have been constructed to serve the region. Such services make rational marketing and processing possible, reducing production costs.

C. FARM TYPES

Agricultural micro-planning in Israel is based on a number of theoretical models of different farm types. Rural development and settlement has so far been based on three main types, while a fourth is now being introduced: they are discussed in sections 1–4 below. Obviously, the more refined adaptation of these basic types to the different regions necessitates the use of a much wider range of models, and the Joint Planning Centre has recently been using some twenty. For the purposes of our present work we shall, however, deal only with the basic types.

1. The dairy farm

The farms specializing in dairy production are located mainly on the fertile plains close to the large population centres of the country. A unit comprises 28–30 dunams of arable land, of which 26–28 are irrigated, 5 dairy cows with their young and 5 dunams of fruit trees. The crop rotation includes cash crops, the residues of which are suitable for cattle feeding. If, by private investment and natural herd increase, the farmer increases the size of his herd, more of the land is devoted to fodder.

Each village has its own milk collection centre where the milk is collected in churns or by tanker and taken to the towns for pasteurization and distribution. Regional organization of milk collection, artificial insemination, veterinary inspection, etc. are undertaken when the proximity of several villages of the same farm type make it possible.

The evolution of the dairy farm is of interest in view of the preceding consideration of the transition to specialized farming. Its beginnings, as we have seen, lay in the 'diversified organic farm', the basis for farming for several decades in Israel, and given scientific expression by Prof. I. E. Volcani in 1935. The development out of the mixed farm to the specialized dairy farm where the farmer produces almost nothing for his own needs, is illustrated by Table 6–1.

TABLE 6-1

The development of the dairy farm

Structure and composition of the farm	Organic mixed farm proposed in 1935 by Prof. Volcani[24]	Mixed dairy farm as established between 1949–1955[25]	Dairy farm as proposed by the author in 1953 and officially approved in 1959[26]	Specialized dairy farm as proposed for the development period
Total land holding	23 dunams	28 dunams	28–30 dunams	28–30 dunams
Total water quota, m³	No data available	15–17	16–17	17
Distribution of crops (cultivated area)				
Citrus groves	3 dunams	4 dunams	5–6 dunams	5–6 dunams
Vegetables (market-garden)	4 dunams	4 dunams	0–3 dunams	–
Hay and green fodder	11 dunams	14 dunams	17 dunams	28 dunams
Cereals (grain)	20 dunams	7.2 dunams	2 dunams	–
Industrial crops	–	3.6 dunams	4–8 dunams	–
Livestock				
Dairy cows	3	3	5	8
Lay-hens	60	100	200	–
Gross production (in IL. 1964 prices	7000	9700	15000	25000
Net income from the farm at constant prices of 1964	3200	3600–3900	5000	6650
Net income as % of gross production	45.7	39	30	26.5
Percentage of production for self-consumption	15–20	10–12	3–5	1–2

As can be scanned from the table, at the end of the 1972/73 projection period, the dairy farm will be based on one main branch (milk production and one auxiliary branch (citrus groves). The income will be derived mainly from the larger number of dairy cows and from the increased efficiency of all production processes, from field work down to the despatch of produce. Although no increase in the allocation of land and water during the development period will be required, the income of this farm type will reach the level planned for all farm types in 1973.

An economic analysis of established family farms in Israel, carried out by the Falk Project under the supervision of Dr. Y. Mundlak[27], contains the conclusion that the value of land as a factor of production in the farms surveyed is extremely low. The report of this investigation concludes "A striking result is the zero elasticity of production with respect to land;... irrigated land accounts for only 3.2% of total output. The meaning of this finding is that marginal increases in the land holdings of the farms would not have resulted in any significant change in the overall value of their outputs.

These findings, arrived at by economic analysis, are explained by the fact that the livestock branches were relatively more profitable than other alternative branches, and that the proportion of the main coarse feeds in the total feed consumption declined while feed concentrates and purchased feeds were relatively cheap. The zero elasticity of production of the land may further be explained by the constant rise in recent years in the yields of fodder crops. This investigation assumed therefore that the low marginal productivity of land was a reflection of the fact that a shift of resources from livestock and poultry branches into the field-crop branches would not have resulted in any substantial gain in the value of production."

The dairy farm structure we have proposed is therefore consistant with Mr. Mundlak's economic analysis.

2. The citrus farm type

Citrus farming in Israel was initiated by private farmers under the patronage of Baron de Rothschild in the last decades of the 19th century. In 1929, the Settlement Department of the Jewish Agency offered loans to farmers planting citrus groves on mixed farm units. Agricultural day-labourers were also given smallholdings with citrus for supplementary income. In the early thirties, 'middle-class' settlements were planned, based on 5–9 dunams of citrus groves and poultry farming. During World War II citrus plantings ceased. After the War of Independence, citrus farms were planned as an alternative to the dairy and field crop farm types.

The citrus farm type consists of 28 dunams, of which 10 are citrus groves and 16 field crops. About 1500 units of this type have been established, mainly in the Central and the Southern Coastal Plain. The investment and water requirements of this farm type are similar to those of the dairy farm type in the original plan, but once again intensification and larger citrus plantings must be planned and carried out to enable these farms to keep up with the national income average for 1972/3.

In land holdings and water norms, this type of farm does not differ essentially from the type of citrus farm evolved in the past. The difference lies in the internal distribution between the different branches. While the citrus farms established a number of years ago had 8–10 dunams of citrus groves and about 16 dunams of field crops, the citrus farm proposed for the future will have 20 dunams under citrus trees. Most of income will therefore be derived from citrus, at the same time achieving more efficient utilization of equipment, enhanced productivity of labour and greater specialization of the settlers. Eight dunams will be set aside for field crops, mainly groundnuts and vegetables. The proposed livestock will consist of 200 hens. This change in the configuration of the different branches, and the greatly enhanced efficiency obtainable as a result of the high degree of specialization, will enable this type of farm to achieve the required income within the period of development, i.e. IL. 6500 p.a. net.

3. The field crop farm type

Most of the *moshavim* established since 1953, especially in the south of the country, are of the field crop farm type, including most of the settlements in the Lakhish Region. Yield levels of industrial crops and grain have achieved or surpassed the original forecasts. The irrigated fields have been planned so that large block-cultivation is possible. As a result of developing the field farm, the national production of industrial crops and grains has greatly increased; the area under sugar beet has undergone considerable expansion, reducing imports of sugar, while local cotton production is supplying most of the country's needs, some being exported.

Field crop farms have lately undergone certain changes, although less far-reaching than those in dairy farms, because at the time of their establishment, the shift from the mixed farm to the specialized has already been in progress. In the first proposal put forward by the author in 1953[28], and worked out in detail by the planning authorities, the unit of land per farm was fixed at 38 dunams and the quantity of water at 19000 m³ in the Negev, and at 15000 m³ in the Lakhish Region. The livestock included 100 chickens,

7 head of sheep and 2 goats. The irrigated area was intended for vegetable and industrial crops, and in addition, the farm included 3–4 dunams of fruit (wine grapes in the Negev, and citrus fruit in Lakhish). The income was to be derived partly from livestock and partly from field crops.

The field crop farm in our present proposal is more highly specialized, the livestock branch comprising only 300 chickens, or in grazing areas, sheep for meat and wool. On the other hand, larger areas are to be devoted to field crops (industrial crops and vegetables). During the development period this farm should reach a net cultivated area of 55 dunams with a water requirement of 19300 m³ in order to reach the desired income of IL. 6000 net p.a.

4. The export farm type

There are foreign – especially European – markets for a range of Israeli produce, mainly fresh vegetables, potatoes, melons and perhaps also for table grapes. The period of high prices for each product is short, with quality demand high. Producing for such a market is a specialist operation, demanding skill, regional organization and swift transport arrangements. The exploitation of this potentially valuable market depends on the planning and building of farm units geared to its needs, organized on a regional basis to facilitate rapid grading, packing and transport. The requisite climatic conditions are found in certain parts of Israel, such as in the Jordan and Beth She'an Valleys, and in parts of the Negev. This farm type has been tentatively planned on the pattern of the field crop farm. About half the land will be under vegetable cultivation for those seasons when export is profitable, the labour schedule will be partially balanced by a smaller area in citrus groves for winter work.

A recently opened area suitable for export crops is the B'sor Region of the Western Negev[29], comprising 733000 dunams, of which 683000 are arable. Some 240000 dunams have been allocated to existing settlements, while 443000 dunams remain to be distributed among new settlements.

The soil of the B'sor Region may be divided into two main types:

1) Various kinds of *loess*, consisting of medium to light soil, with only a small part medium to heavy.

2) Sandy soils and sand above *loess*, particularly suitable for crops such as citrus fruit, groundnuts, etc.

Topographical conditions in the B'sor Region are also particularly favourable for the cultivation of irrigated field crops. There are no sudden sharp transitions from one area to another. The sandy southwestern part of the

B'sor Region is level and there are few hills. To the west, towards Ber-sheba, hills become more numerous.

In addition to the favourable topographic and soil conditions of the B'sor Region, its climate also has special qualities. It may be considered as a continuation of the climate prevailing in the southern Coastal Region (Ashkelon Region) except that the amount of precipitation gradually decreases: rainfall varies between 360 mm in the northern part and 160 mm in the southern part. In the Western and Central B'sor Region, the climate is temperate without any extreme fluctuations in temperature between day and night. The maximum mean temperature in summer is 32° C and the minimum temperature 20° C. In winter the mean day temperature is 19° C falling to 7° C at night. Frost during winter is extremely rare. The entire region is distinguished by heavy dew during the summer season, which is of great importance in reducing the water requirements of the crops cultivated there, and by a high rate of sunshine important for winter crops.

Since in the eastern part of the B'sor Region (known as the Ber-sheba Hills) the climate is more continental, with differences in temperature between day and night greater and with frost of frequent occurrence in winter, main attention will be devoted to development of the western and central area. Considerable experience has been gained through research carried out by the National and University Institute of Agriculture, as well as on irrigated plots of settlements such as Mivtahim, Nir Yitzhak and others.

The prospects of the various crops may be summarized as follows:

Citrus: The quality of lemons in the B'sor Region is high, exceeding the national average; grapefruit is of good quality, equal to that in the rest of the country. Shamuti orange trees, planted recently, are developing well and show promise of high quality fruit. The northwestern part of the Besor Region has the typical light sandy soil suitable for citrus groves.

Groundnuts: The groundnut yields exceed the national average. In observation and experimental plots, yields exceeding 400 kg per dunam were obtained without difficulty. The quality of the groundnuts is excellent, exceeding that obtained in other parts of the country. Experiments at Mivtahim carried out in 1960 have shown the water requirement of groundnuts to be 680 m³ per dunam.[30]

The yields at Nir Yitzhak were 487 kg per dunam, at Zeelim 420 kg per dunam.[31]

Winter potatoes: This crop has succeeded beyond all expectations. The average yield of potatoes is two tons per dunam of a quality suitable for export. Actually, the main share of potatoes exported by Israel comes from other Negev settlements where the climate is similar to that of the B'sor Region. In the winter of 1961, Negev settlements supplied about 2000 tons of potatoes for the export market.

Winter tomatoes: Experiments in growing export tomatoes in the Mivtahim area carried out by the National and University Institute of Agriculture have shown that the yields of the Moneymaker strain in the experimental plots were within the range of 6.3–6.4 tons per dunam.[32] Experiments conducted in 1961 on the extermination of tomato early blight, a major cause of low yields and defective fruit, were, to quote the investigators' report, "on the whole successful since they have proved that it is possible to grow tomatoes and reach yields of up to 8 tons per dunam".

The main problem encountered by growers of export tomatoes has been the low percentage of tomatoes exportable (6–7%), due to irregularity in shape and hollowness of the fruit. The research of the National and University Institute of Agriculture on this problem has, as mentioned above, reached an advanced stage. A strain has been developed which assures exportable fruit of only slight hollowness, which gives high yields, exceeding those of the Moneymaker strain.[33]

The average yield of this strain – Potentate – in the experimental plots at Mivtahim was 9.3 tons per dunam.

Early (spring) onions: The experience of the last two years has been promising: quality and yields have been high (about 4 tons per dunam in the observation plots).

Other crops, such as beans, carrots, marrow and sweet maize were also investigated and showed good yields and high quality. In the light of the above, we propose that special farms be established, based on export crops, namely citrus fruit and field crops.

The proposed holding per farm in the B'sor Region is 52 dunams, of which 10 dunams will be devoted to citrus groves and 40 dunams to field crops (such as groundnuts, export tomatoes, export onions, melons, peas for canning and carrots). The water requirement per farming unit is about 26000 m³ per year and its expected net annual income is about IL. 6000.

5. Equal income principle

In the past, because of the prevailing ideology, mixed farms were set up in parts of the country totally unsuited to livestock farming. Their economic position was weak because of high costs, low yields and inadequate services. Equal physical size of the farm unit was considered the key to equality between farms in different areas.

Land and water are not in themselves indicators of equality; there is no importance to be attached to the term 'dunam of land' or 'cubic metre of water' unless the overall conditions in which the land or water are used are precisely defined. A dunam of land in the hills is not equal to a dunam in the rich coastal plain, while the proximity of a town gives a completely new importance to land. A cubic metre of water in the north of Israel is more productive than one in the south, where evaporation is higher. Physical units alone cannot serve as a basis of comparison between farms in various regions unless other conditions are similar – which they rarely are.

Adoption of specialized farming policy has led to the establishment of farms which exhibit great variation in size and in the composition of their branches. The key to the allocation of resources in all these farms has been the social ideal of equal income opportunity: in other words, whether a farmer is settled in the north, in the hills, or in the south, he must be assured an income not less than that he would have earned had he settled in a town. With additional investment, harder work or high ability, he is able to earn more, but the system sets a minimum living standard in rural areas. However, the resources allocated to villages in difficult or dangerous border areas may be above this minimum, for the settler in them not only deserves a higher income than a townsman, but better resources are perhaps necessary in order to induce the best farming elements to settle there.

Table 6-2 shows the key to resource allocation proposed in 1959.[34] Our projections for 1972/73 differ from these in important respects, as will be discussed.

If the orchard (citrus) farm type in the coastal area is compared to the field crop farm in Ta'anach, it is seen that the first has 28 dunams and 15000 m^3 of water, while the second has almost twice the land and 12000 m^3. The income for both at full development is the same, about IL. 4200 at 1959 prices, IL. 5000 at 1963 prices, which is less than in our plan.

In order to enable the farms laid out according to the settlement key to keep up with rising incomes, some changes must take place. For the field farm, the future land allocation will have to be raised, and for the dairy farm the number of cows and the average yield per cow must rise. More citrus for the citrus farm type will be needed if it is to keep up with the national average income. Table 6-3 sums up the proposed model farms based on the common denominator – equal income opportunities.

These proposed farm types are designed as sound production units, able to carry out the tasks which a modern state demands of its agriculture, in keeping with the social principles of a people which recognizes the right of rural

TABLE 6-2

Resource allocation proposed in 1959

Region	Farm Type	Land allocation (Full irrigation)				Total irrig. fruit	Auxil. irrig.	m³ water /year	Live-stock
		Total land	Field crops	Citrus fruit	Other fruit				
Negev	Field crops – orchard (vine)	40	35	–	5	40	–	19 000	beef 2–3
	Field crops – orchard (citrus)	40	32	5	–	37	–	19 200	beef 2–3
	Milk – orchard (vine)	40	35	–	5	40	–	19 300	5 milk cows
Lakhish[1]	Field crops – orchard (citrus)	45	35	4	–	39	–	15 800	beef 2–3
South	Milk – orchard (citrus)	28	23	5	–	28	–	16 500	5 milk cows
	Milk – orchard (citrus)	34	22	5	–	27	7	15 100	5 milk cows
Coast	Citrus	28	18	10	–	28	–	15 000	–
Ta'anach[2]	Field crops – citrus	50	45	5	–	50	–	12 500	beef 2–3

[1] Northern Negev.
[2] South-east of the Jezre'el Valley.

areas to be equal in economic and social opportunities with the towns. They are an effort to combine the advantages to society of the family farm as a social unit with the necessary production efficiency which agriculture must attain.

In planning farm types, it is important to look into the future so that the physical base which is eventually decided upon will give flexibility to enable the farm to keep up with changing requirements of the market and changes in living standards. There must be room for improvement and intensification through additional investment. Care must be taken to ensure that farm structures do not become out of data because of faults in planning and insufficient attention to long-term trends in agricultural production. One way of assuring the future is to keep adequate land reserves for future distribution. The planning of many of the settlements undertaken in the early years of the State was inadequate from this point of view, and already

TABLE 6-3

Proposed model farms for moshav settlements

	Dairy farm	Citrus farm	Field crop farm	Export farm
Total gross area	31	30	57	52
Buildings and yard	3	2	2	2
Total net cultivated area	28	28	55	50
Utilization of cultivated area				
Citrus groves	6	20	5	10
Field crops	22	8	50	40
Break-down of field crops				
Vegetables and melons	–	6	12	30
Industrial crops	–	2	22	10
Fodder and hay	28	–	4	–
Miscellaneous	–	2	10	–
Break-down of livestock				
Dairy cows	8	–	–	–
Calves (for meat)	6	–	–	–
Laying hens	–	200	300	–
Water				
Total annual water requirements	17 000	17 600	19 300	26 000
Labour				
Total work days	360	335	375	429
Farm balance-sheet (in IL.)				
Total output	25 060	18 500	21 680	19 950
Output of field crops	–	3 100	13 100	14 200
Output of citrus fruit	3 460	11 600	2 880	5 750
Output of livestock	21 600	3 800	5 700	–
Total input and repayment of loans	18 410	12 100	15 060	13 100
Net income	6 650	6 400	6 620	6 850

the farms are becoming small from the point of view of income potential. However, some of the villages had empty farms, either because they were never brought up to their full complement of families, or because some families found themselves unsuited to the new life and left. A question of policy arose here. Should these villages be brought up to their number of families as originally planned, or should the Settlement Department carry on with new settlements? Two forceful arguments were advanced in favour of the second alternative, one of them based on the equal income opportunity principle.

a) That, since these villages had been planned with insufficient attention to the need for long range planning for rising incomes, the unused land within the villages should be kept as a reserve for the future for the families already living and working in the village. Thus, in the case of a village originally planned for 80 families, but on an inadequate base as regards the future, the land and water allocations should be reserved for the 60 or so families at present occupying the village. Where the original allocation was adequate, it was decided to keep the empty houses for the sons of the village.

b) That Israel has both production and marketing potential for export crops, but the conditions for their cultivation are to be found only in the Jordan Valley and in the unsettled areas of the south. In order to satisfy this high value demand new regions would have to be settled as a first priority of the new plan. Therefore, the planning of the new region of B'sor should be undertaken on the basis of the export farm type.

D. MARGINAL AREAS AND THE INCOME PRINCIPLE

For reasons of security, important settlement has been undertaken in the hills, on lands which can only be classed as marginal. The problems associated with the livelihood of farmers on such land are not easy to solve. Much of it cannot be worked until expensive reclamation is carried out, while the small plots, the need for careful soil conservation, the climate and poor soil conditions, very much limit the planner in his search for a farm type which will guarantee the hill farmer equal income opportunity with his counterpart on the plains.

The first hill settlements were based on fruit trees, vegetable growing, tobacco and poultry, with sheep or beef production for those villages where pasture was abundant. There was no fixed 'farm type' although several suggestions were made and implemented at various times. They were all however limited in their ability to provide the farmer with the desired income, because of the limits of the family farm in seasonal labour. The crops specially suited to hill farming combined in one farm unit, have a very unbalanced labour schedule. In the peak months, the schedule is beyond the labour capacity of the family farm, while for a large part of the year, the farm family has little to do. The market crops which can be grown economically in the hills, need specialist knowledge and a certain scale of operation before it is worthwhile to produce them. Farms can therefore be based on only two or perhaps three of these branches, if their size is to be adequate and if the farmer is to acquire the requisite knowledge. Though this combination would provide an income, it cannot be adapted to the labour schedule of the family farm.

The answer of the earlier planners was to diversify the farm pattern by including several small branches – even milk production – which had labour requirements more widely spread out. The labour problem was more or less solved, but diversification meant presenting the farmer with many farm branches, each needing specialized training and each having too small a part in the farm plan to reach an economic scale of operation. Costs were therefore high, and the farmers' labour wasted. The farm plan was ineffective in providing the required income. Special subsidies were instituted to cover the high water costs prevailing in the hill regions, remissions of interest and loan capital were furnished by the settlement authority, and a subsidy given for labour costs in plantation renewal. Quotas for tree plantings were artificially allocated to the hills when the most economic areas for developing the trees were elsewhere. It is clear that such measures imply waste, and cannot in the long run hold their own against economic forces which demand cheap products of high quality, which hill farms, working under their natural difficult conditions, cannot supply unless a large degree of specialization in only one or two crops with real regional advantage is undertaken.

The writer, in his suggestion for planning the problematic hill areas, has attempted to follow the trend of the economic forces and stressed the need for the rethinking of planning concepts in order to develop viable farm units in these important settlement regions, which can support their inhabitants only through a truly commercial agriculture.[35] He advocated specialization in natural hill crops (tobacco, medicinal herbs, vegetables, seed growing and certain fruits suited to hill conditions) on an economic scale enabling light mechanization and long-term investment. This meant choosing for small regions or even each village, a combination of two or three farm-branches best suited to the lands, size of farm unit and climate. Where natural pasture was available, sheep or beef cattle was also introduced. Poultry was seen as a useful supplementary branch. Such a plan had two main difficulties – the unbalanced work schedule which meant shortage of labour in the summer with unemployment in the winter, and a high degree of dependence on a few branches for the entire income.

The first difficulty was resolved by considering the individual units and villages on a regional basis and linking the family farm to regional industries based on agricultural products. The labour requirements of such industry are complementary to those of the farmer, and provide both a ready market for his products and work for the winter months after the harvest is in. The high labour intensity of the summer months make hired labour necessary:

this is obtainable on a casual basis from regional villages and towns which are built around the industries served by the hill farm production.

The second problem – the farmer's complete dependence on one or two crops for his livelihood – can only be overcome by a form of crop insurance for which the higher average income over the years accruing from specialization and economic operation will form a guarantee against crop failures. This is a commercial proposition to the farmer.

As Israel's search for export markets in high quality agricultural products extends, the specialized farms in the hills will find more and more alternatives for its operation: in fields where it enjoys a natural advantage such as in flower bulbs, seeds and other high value goods. At the same time, by linking the farm to the immediate market – agricultural based industry – each can influence and develop the other to maximum efficiency.

The industry-linked farm unit is a new approach to a very severe planning problem. Its strength lies in considering the farmer not as an isolated unit, but as part of a regional complex. In its wider scope, this enables increased specialization, high efficiency, and therefore higher income. This approach is being implemented in the region of Adullam in Israel*, a project still too young for full social and economic evaluation, though its future may well bear out the soundness of the planning concepts on which it is based.

E. THE MOSHAV AS A PRODUCTION UNIT

The main structure and principles of the moshav or smallholder's co-operative village have been described in Chapter 1 and of family farms in moshav villages earlier in this chapter. More must be said about village structure and its connections with nationwide co-operatives and institutions.

The moshav in Israel receives its land from the Jewish National Fund on lease for 49 years; it subleases it to moshav members who, as long as they work and look after the land, have a high degree of security of tenure. The distribution of land within the village gives each family the same resources, in holdings of the same size, and as far as possible of equal quality. Redistribution of land can be made only by decision of the general meeting, but is undertaken only for very sound reasons, such as the need to intensify farming through irrigation or to release land for additional members or the children of the village. Another reason may be the decision to plant an orchard as one plantation within the village, a matter which alters the overall

* A hilly region southeast of Jerusalem.

land use pattern, and means reparcellation of individual plots. Such work is expensive and may lead to frictions in the village. It is better to anticipate the future by sound, long-term planning of land use.

1. The co-operative framework

The moshav is a co-operative of all its members, registered with the Registrar of Co-operative Societies in Israel. On one hand it is a business organization, dealing with outside firms and institutions as a body on behalf of its members, and on the other, an organization with social obligations to its members. The two main functions of the moshav as a business co-operative are buying and selling for farm operations. Raw materials are bought by the moshav as a whole with credit obtained by the moshav. Members have an account with the co-operative, which grants them an open credit to the extent of their needs to carry out their farm plan. The credit is in the form of production materials taken from the co-operative store of the village. It is given on condition that the members market all of their produce through the co-operative which then credits them with the sales proceeds, deducts an amount for outstanding credit and taxes, and pays out the remainder to the farmer. In many moshavim the members hold considerable sums of money invested in the co-operative and receive interest on them. This enables the village to construct buildings needed for its economic or social life, such as new stores, marketing sheds or cultural facilities. The moshav can also, as a co-operative, takes loans for long-term construction and investment which are recovered through the taxes levied on each member as a result of a decision of the general meeting. The general meeting fixes the level of taxes in relation to the financial position of the village, including direct taxes on production. If the village is based on dairy farming, it takes care of the needs of its members for feed supply, mixing, etc., gives credit where necessary for building of barns, etc. and also constructs the dairy with its cooler. The cost is recovered through the price of the feed or deductions according to the amount of milk marketed. The village may buy its own lorry to ease transport problems, or heavy machinery for the use of its members. The investments are according to the farm type of the village and the needs of the individual farms for co-operative structure to aid their production and marketing.

Through this organization of services, the individual is left free to carry on his farm operations with a minimum of trouble and waste of time, and can rely on the co-operative to supply him with cheap goods and services of high quality. Reduced prices are possible because of bulk purchasing and efficient handling, transport and storage. Research undertaken recently by the Ministry of Agriculture, however, shows that there are great differences in the costs of supplying the various functions in different co-operatives.[36] Some use their facilities at full rate and high turnover and therefore costs are low, others lose money through inefficient management and wasteful practices. There are also differences among villages in the range of functions which the co-operative undertakes on behalf of its members. In some, egg or milk collection is from the individual farm; in others, the farmers take their produce separately to the grading shed or dairy. Some villages grow hay collectively and sell it to their members; in others, each farmer grows a small amount of hay on distant plots at high cost in money, time, and waste in inefficient investment on underutilized machinery.

Such paradoxes do exist in the moshav: one section of moshav life is organized on a very high level of co-operation and efficiency, while others, equally adaptable to co-operative practice, are completely outside the co-operative framework. Research on village accounts and much comparative study and consequent evaluation is necessary to find the level of co-operative action within the moshav which is most desirable from a production viewpoint and acceptable from the social viewpoint of the 80–100 farming members.

The primary village co-operative is not the end of co-operation in the moshav. There are two co-operatives of vital importance for its working which are nationwide. The first is *Hamashbir Hamerkazi*, the co-operative wholesale organization of the labour movement in Israel, the members of which are the agricultural settlements – moshavim and kibbutzim – and the urban consumer co-operatives. This organization supplies agricultural requisites and the needs of the consumer stores in the settlements. Between the moshav (or kibbutz) and *Hamashbir*, there is often a regional purchasing agency founded jointly by several villages so that they can benefit from bulk buying and supply and from simpler organization. The village then obtains its credit and makes its purchase through this agency.

The other main co-operative to which most of the moshavim and kibbutzim belong is the central co-operative marketing society of *Tnuva* which collects and markets (sometimes in processed form) agricultural products. *Tnuva* works on a commission basis and deals mostly with fresh agricultural produce – milk, meat, vegetables, etc. Cotton, sugar beet, groundnuts and similar industrial crops are marketed through marketing boards specific to each crop.

2. Social aspects of the co-operative [37]

One of the basic principles of the Israel moshav is mutual aid. The moshav accepts the principle that the community is responsible for the well-being of its members. Providing that each man does his share within the moshav and does not deliberately neglect his land, the moshav guarantees him a minimum degree of security. Although direct person-to-person help is the basis of this mutual aid, it has now been institutionalized in the moshav, and funds are available for needy members. Education for all children of the village is the responsibility of the co-operative, and is not reflected in a tax structure which charges according to the number of children.

With the great increase in the number of moshavim brought about by settlement in recent years, this principle of mutual aid has been extended beyond the single moshav to the moshav movement as a whole. The movement has established a special fund to which older villages contribute the larger share, benefitting mainly younger ones.

Another basic principle of the moshav is self-labour of the farmer and his family. However, for aging or ill members unable to do all the work themselves, the moshav will hire workers, debiting their account.

The co-operative of the moshav is essentially a co-operative of farmers, but the organization of the village needs a considerable staff to run its office, production and social services. Thus the population has a fair proportion of non-farming members who have, however, equal rights and obligations in most spheres, including the right to a small auxiliary farm. Their number in a well developed moshav is about twenty per cent.

3. The village and its centre

Throughout the foregoing discussion, it has been implicit that the village is the primary organizational unit after the farm itself. The design of villages and the number of families in each, are of great importance as efficiency factors in individual farm operations. Careful thought is therefore given to the physical planning of each village within the specific conditions of its location and proposed farm type.

Before satisfactory decisions can be taken about the location of farm units, statistics and general information are required on the following questions:

a) General topographic features;

b) Soil structure and composition; soil colour, texture, sedimentation, depth, moisture, salinity, slope, stoniness, tendency to flatness and degree of water and wind erosion;

c) Climate: rainfall and its distribution, temperatures and temperature ranges, humidity and dew point, evaporation rates and prevailing winds;

d) Water resources;

e) Vegetation and animal life;

f) Existing agriculture in the region;

g) Marketing possibilities and auxiliary services.

The district is first comprehensively photographed from the air, then mapped, and all outstanding topographical features (hills, valleys and wadis, slopes, escarpments, etc.) are marked and noted. A soil survey of the district follows, the results being expressed in formulae derived from a generally accepted code, giving such details partly enumerated in (b) above. Additional data are gathered on the percentage of chalk in the soil, soil aeration, calcium content, mechanical soil structure, field capacity, the relative acidity (pH) and other characteristics of the soil in the area. The assembled information is incorporated in a Soil Classification Map.

The Settlement Authority considers all the relevant information on soil type, water availability and cost, production and marketing possibilities, and decides the farm type for the village accordingly. The physical planning is carried out on the basis of these findings: the farm planner now has the essential information required to site the village, and a map of the location of the farms is made. A village generally compromises 70–80 farms of uniform type, but the exact number is determined when its location is decided upon.

A Land Utilization Map is now drawn up. This is a graphic plan establishing the village on the basis of the farm type and the topographical, climatological and soil data available. The map shows:

1) Lands to be utilized for a) irrigated plots; b) collective grain farming; c) orchards and vineyards.

2) Lands unsuitable for cultivation but suited for grazing or afforestation.

3) The sites of the farm houses, communal buildings and the centre of the village.

4) Layout of the individual farms.

The main principle of land parcellation is equal distribution of land and parcellation according to land resources available. Some of the irrigated lands are near the farm house, some further away. The fields of all the farms are located as nearly equidistant from the farmhouses as possible. Where the farms are of the industrial field crop type, large irrigation blocks are planned to facilitate use of efficient mechanization techniques (e.g., deep ploughing, spraying). Fragmentation is avoided.

A Water Planning Map is then drawn up, showing:

1) The village connections to the main pipe line;

2) Distribution of the main pipes in the village;

3) Distribution of the smaller pipelines in the individual farm units.
Each farm has its own water meter.

The final map is the Architectural Map, which shows the layout of the village, the position of community buildings, individual farm yards, roads, power lines, fences, landscaping (public gardens, parks) and other features.

The standard houses each have 2½ rooms, with a floor space of 45 square metres. Shower equipment is provided indoors, and dry lavatories are built outside the houses. An indoor water closet is installed later. The distance between the houses is about 30 metres.

Community buildings include a kindergarten, a first aid clinic, a village office and meeting hall. Every village has a local grocery, farm store and an arms depot.

On completion of the physical planning, the scheme is approved by a higher authority, composed of various governmental and other institutions dealing with settlement and agricultural planning. These bodies are involved at all stages of the planning and include the Jewish Agency Settlement Department and the Ministry of Agriculture with the Joint Planning Centre the Ministries of Defence, Interior, Health and Labour and the Settlers' Representative Organization.

Planning of the production unit is next determined. In planning for water, it has to be borne in mind that not more than 16–17% of the total annual water allocation may be used in any one month. Consequently, crop rotation must be determined by the Farm Planning Department (of the Agricultural Settlement Department), working in conjunction with the Regional Water Planning Department.

Planning of production is based on the establishment of water and work norms for different crops in the various months of the year. From these figures an annual water and labour schedule can be drawn up, which also serves to check that the rotation is within the limits of water allocations and available labour. An income and expenditure table for the year is also made on the basis of norms of crop yields and prices.

F. THE KIBBUTZ AS A PRODUCTION UNIT

1. General

Forms of co-operation in the kibbutz are more developed than in the moshav, penetrating both economic and social life to a far larger degree. The moshav corresponds directly to the concept of the family farm, both as an economic unit and in the social relations of the family within the community. In the kibbutz, the labour force and the community are made up of the constituent families, but the village rather than the family is the unit of economic and social life. The village takes over all the economic activities on a communal basis so that private enterprise is non-existent. The farm plan is drawn up and implemented by the kibbutz management, and the members work according to directives as in any managed farm.

The distinction between the kibbutz and the large administered 'factory' farm lies in the fact that the members elect the management and that anyone with sufficient ability may be elected to any of the management positions. The members are, of course, owners as well as labourers, so that there is no creation of a gap between a management and a 'working class'.

Land use is not complicated by individual rights and land parcellation. The members are guaranteed their work and livelihood by the community and undertake to fulfill their obligations towards it. Investment and development are undertaken by the kibbutz as a unit which can seek loans for the purpose. Co-operation in economic activities is complete. It includes the full range of planning, investment, production, buying and selling, subject to the central control of the elected management.

In most kibbutzim, there is marked specialization among the members, both men and women. The very modern farming, which is carried out by the majority, demands skilled people in various branches, and therefore positions tend to be either permanent or else held for long periods. Members are sent on courses to study the branch to which they are assigned and are encouraged to develop their ability. On the other hand, some people are needed in work of a non-specialized nature. Efforts are made to give each man or woman a chance to take responsibility for an interesting task, so that no-one will consistently be left doing general work – a factor which inevitably detracts from the overall specialization of labour. There is, therefore, room for individual expression in work, although it is in no way connected with profit.

2. Social aspects of the kibbutz

As in the moshav, co-operation extends to the social life of the community, where it is no less marked than in its economic aspect. The dining hall, children's house and other communal facilities are features as typical of the kibbutz as are its large fields and farm buildings. Social co-operation is remarkably complete, with women labouring their full complement of hours as well as the men. There is a communal purse, individual members do not handle money, unless they purchase on behalf of the kibbutz or spend their small cash allowance or pocket money. The member identifies his future and that of his children with the group which makes many major decisions affecting his life and provides services from raising the children to health. Since he or she has no personal income, choice of expenditure is limited to that which the kibbutz is able or willing to make. Major items, such as radios or room furniture, are bought by the kibbutz and allocated according to the number of years of membership. Clothing and similar personal items are supplied according to a budget for each person within which he has a certain choice, the size of the budget depending on the wealth of the kibbutz. Some kibbutzim are restricted in items which they can supply beyond the bare necessities of life; others can meet nearly all requests, from books to furniture and other small luxuries and even to trips abroad. However, since kibbutz life is so self-contained, each member has a smaller individual need for consumption items than his counterpart in town. His children are completely cared for in well-equipped children's houses; his meals are prepared and served in the common dining room; cultural facilities are common, and his needs in furniture and household equipment for the two rooms he occupies with his wife are therefore modest. A large proportion of his total 'possessions' are in the form of communal property. He does not need a car, for he finds his social life within the community itself or in visits to nearby kibbutzim or towns and in organized trips. Since the kibbutz is of necessity a closed community, anyone not fitting into small community life finds it difficult to be a member, even though he might favour the kibbutz' full measure of co-operative living and identification with the group.

3. The young kibbutz: problems and guidance

The veteran kibbutzim, founded before the establishment of the State, worked out, as we have seen, a mixed farm economy, aiming at self-sufficiency in food supply. The scale of production in all branches increased as the kibbutz developed. After independence, many new kibbutzim were established in settlement areas, which unfortunately maintained agricultural patterns identical with those of the veteran kibbutzim, giving little consideration to changed economic, regional and district conditions, or to new technical and market forces. Rational planning was subordinated to conservative ideas. In the hill regions, for example, with their thousands of dunams of pasture and expensive water, no attention was paid to the sheep branch. In another case, a kibbutz on the coastal plain, with ideal conditions for citrus groves, embarked on vegetable and fodder farming at heavy losses, because of the ideological approach to mixed farming.

There were many such cases, the balanced labour schedule being given as the justification for diversified farming, ruling out alternatives. The Settlement Authority could not accept this argument. It believed that the economic progress of the kibbutz as in the moshav depended on specialization in a few branches of production *suited to the district* and that supplementary branches of agriculture should only be embarked on after the main branches had been established and consolidated.

Because of poor initial planning in the young kibbutzim, both physical and economic, limited investment was spread over many farm branches, none of which were sufficiently capitalized to work efficiently. The young kibbutzim fell into debt; many, having a burden far exceeding annual output, could not pay interest without further loans. Moreover, since the young kibbutzim were developed as isolated units in defence positions with small membership, overhead costs for essential services were high. Because of interest payments and low income due to bad planning, living standards of the settlers were often low. Since not all members were prepared to suffer low standards when the rest of the country was advancing materially, some kibbutzim lost membership. Remedial action was essential through replanning, finding means of lowering the yearly interest payments and the cost of services. The new settlements were not, however, capable of improving their position on their own.

Since most young kibbutzim had been settled on Israel's vulnerable borders, conditions were hard. In many cases economic difficulties arose from the necessary concentration on border security. The Settlement Department therefore regarded it as a duty to consolidate such settlements as soon as possible. In 1957 therefore the policy of the 'guided Kibbutz' was initiated. Although the organization of kibbutzim fearing undue intervention in its affairs, fought the scheme vigorously, the policy proved itself. Of the 100 kibbutzim built since 1948, most became included in the system.

Under the 'guided Kibbutz' scheme, the Settlement Authority undertook responsibility for the economic consolidation of the kibbutzim. It prepared a new development plan in conjunction with each kibbutz and its central organization, advised in planning and administration, and became actually a partner with the kibbutz in financial decision and investment. The system was operated until the kibbutz became consolidated with the necessary manpower and structure for sound management.

The final development programme has been based on specialization tied to natural resources and regional advantages. An annual farm plan is prepared which ensures a minimum standard of living for kibbutz members.

Special care has been taken to limit services to actual needs, increase the number of productive workdays per member and raise efficiency. The plan keeps adequate reserves for depreciation and renewal of equipment. It gives a detailed financial scheme and takes care that investment in production items is made only from settlement budgets or other suitable financial sources. Adequate working capital is found and easy repayment terms are granted in accordance with earning power. An arrangement has been reached with a bank which releases credit facilities according to the seasonal needs of the annual farm plan, and recalls the loans when produce is sold. Depreciation is covered and nonrenewable loans are paid from current income.

The Settlement Department provides a general instructor from a veteran kibbutz to superintend the plan. Each region has a co-ordinator of development for all its kibbutzim. Special planners draw up the agricultural plan for the kibbutz, superintend its execution and advise the management committee on current problems.

The role of kibbutz organizations has by no means diminished. At a time when the kibbutz movement is suffering from a general lack of manpower, when economic conditions are constantly changing and the execution of the 'Guided Policy' is becoming more and more urgent, the resources and co-operation of such important bodies have become vital.

The guided policy in practice

The principles of systematic farm development, with planning closely related to regional features, can be illustrated by the success achieved with a certain kibbutz in the hill region.[38] Before the kibbutz entered the guided framework, it farmed with:

1) A dairy herd consisting of 20 milk cows,
2) 300 ewes,
3) 2000 head of poultry,
4) 100 dunams of fruit orchard,
5) 300 dunams under irrigation (crop rotation consisting of vegetables and fodder crops),
6) 4100 dunams under grain production.

The first major decision of the reconstruction policy was to exploit the abundant grazing lands belonging to the kibbutz. The high price of water pointed to planning for extensive farming. The dairy branch was replaced by a meat herd. The stock of sheep was enlarged and the fruit orchards improved. The new farm pattern had:

1) 100 beef cows and their offspring,

2) 430 ewes and their lambs,

3) 2000 fowls,

4) 100 dunams of fruit orchard,

5) 300 dunams under irrigation (with a crop rotation consisting of industrial crops and potatoes),

6) 4100 dunams under grain production after the first year.

Animal feed was obtained from natural grazing, and the irrigated land was released from fodder growing for cash crops.

The farm showed a profit after the first year of reconstruction, a reserve was built up to cover depreciation costs and non-renewable loans were redeemed. Minor investments were also made.

Further consolidation took place after the second year. The beef herd was enlarged and the number of sheep increased. The poultry branch doubled. Suitable buildings were constructed and special facilities for housing a poultry meat unit were built. Another 50 dunams of apple trees were planted. Development continued in the third year, as shown in Table 6-4.

TABLE 6-4
Development of a hill kibbutz

2 years after the introduction of the guided kibbutz system	3 years after the introduction of the guided kibbutz system
150 beef cows and offspring	195 beef cows and offspring
450 ewes and offspring	700 ewes and offspring
4000 fowls	5000 fowls
100 dunams fruit trees	150 dunams fruit trees
300 dunams under irrigation with a rotation consisting of industrial crops and vegetables	300 dunams under irrigation with a rotation consisting of industrial crops and vegetables
4000 dunams under grain production	4000 dunams under grain production

The above figures show the adaptation of farming to the characteristics of the region. The grazing lands were suitably exploited, beef and mutton production increased through better use of grain stubble and natural pasture, water use decreased through a more suitable choice of crops, poultry farming expanded to provide a necessary cash turnover, and modest investments in the fruit orchards increased that branch to a suitable size. By the end of the second year, the net income per labour day had gone up considerably, while costs per labour day remained the same.

Payment of loan interest decreased in the first year of the guided reconstruction effort as the result of an agreement reached with banks and other creditors under which short term loans were converted to medium or long term. Overhead expenses decreased through better planning of services. A marked improvement was noted in the annual balance sheets.

Most kibbutzim included in the guided reconstruction policy adopted by the Settlement Department have shown results similar to the one cited above. The policy is a successful one, due to the concerted effort of all concerned – the planners and specialist staff, the kibbutz movements, banks and credit organizations, and primarily the members of the kibbutzim who responded to the system in a spirit of active co-operation. The system offers a means of channelling expert planning and management advice to where it is needed.

NOTES

1. BACHMAN, Kennet: 'Changes in Scale in Commercial Farming and their Implications', *Journal of Farm Economics*, XXXIV, No. 2 (May 1952), pp. 157–172.
2. PARSONS, K. H. and OWEN, W. F.: 'Implications of Trends in Farm Size and Organization', *Journal of Farm Economics*, XXXIII, No. 4, Part 2, Proceedings Issue (Nov. 1951), pp. 893–904.
3. [1] SCHICKELE, R.: *Agricultural Policy* (McGraw Hill Book Co., New York 1954).
   [2] SCHICKELE, R.: *Objectives of Land Policy in Land Problems and Policies*, edited by Timmons and Murray (The Iowa State College Press, 1950), pp. 5–29.
4. ACKERMAN, J.: *Family Farm Problems and Policies in Land Problems and Policies*, edited by Timmons and Murray, pp. 205–218.
5. SCHICKELE, R.: *Objectives of Land Policy*, op. cit., pp. 19–25.
6. SCHICKELE, R.: 'Theories Concerning Land Tenure', *Journal of Farm Economics* XXXIV, No. 5 (December 1952), pp. 734–744.
7. ACKERMAN, J.: *Family Farm Problems and Policies*, op. cit., pp. 210–211.
8. BLACK, A. G.: 'Reflections upon Israel's Recent Agricultural Development and its Relationship to the General Development', presented before the Symposium on The Challenge of Development, The Hebrew University, The Kaplan School of Economics and Social Sciences (June 1957), 18 pp. (mimograph).
9. See also:
   WEITZ, R.: 'Family Farms Versus Large-Scale Farms in Rural Development', *Artha Vijnana* 5, No. 3 (Sept. 1963), Gokhale Institute of Politics and Economics, Poona, India.
10. WEITZ, R.: Op. cit. pp. 228–233.
11. See also:
    [1] ESHKOL, L.: 'From the Single to the Comprehensive, *Economic Quarterly* (Tel-Aviv, June 1961), (Hebrew).
    [2] WEITZ, R.: *Agriculture and Settlement*, op. cit., pp. 193–216.
12. See also:
    BEIN, Alex: *The Return to the Soil*, op. cit.
13. Records of the Committee for Arranging a Settlement Key in Different Regions of the Country, The Zionist Organisation (May 1929).
14. Records of the Committee, op. cit.
15. See: VOLCANI, Y. E.: *The Planning of Agriculture*, op. cit.
16. WEITZ, R.: 'Self Examination, *Davar* (daily), 13/4/1953.

17. WEITZ, R.: 'A New Agricultural Plan', *Davar* (daily), 14/5/1953.
18. WEITZ, R.: 'The Field Crop Farm', *Davar* (daily), 21/5/1953.
19. SCHUTZBERG, A.: 'Answer to the Proposal for an Agricultural Plan', *Al-Hamishmar* (Hebrew daily), 29/5/1953.
20. GVATI, H.: 'The Orientation of Our Agriculture', *Davar* (Hebrew daily), 3/6/1953.
21. BLACK, J. D.: *Introduction to Economics of Agriculture* (Macmillan, New York 1953).
22. LOWE, Y.: *Farm Economics* (Am-Oved, Tel-Aviv 1957), pp. 244–255 (in Hebrew).
23. WEITZ, R.: 'Towards Specialized Farming', Extension Authority Department of Publications (1960).
24. VOLCANI, Y. E.: *The Planning of Agriculture in Palestine* (Rehovot 1937), (Hebrew).
25. 'Proposed Irrigated Farm-Types', Joint Agricultural Planning Centre (Tel-Aviv, July 1954), (mimograph in Hebrew).
26. 'Profitability of Moshav Settlements', Joint Agricultural Planning Centre (Tel-Aviv, September 1959), (mimograph in Hebrew).
27. 'Economic Analysis of Established Family Farms, 1953–1958', The Falk Project for Economic Research, Fifth Report 1959 and 1960, pp. 179–180.
28. WEITZ, R.: 'The Field Crop Farm', *Davar* (Hebrew daily), 21/5/1953.
29. For more details see: 'Besor Region, Physical Characteristics and Sketch Plan for Settlement', The Jewish Agency, Settlement Department (Jerusalem, July 1961), p. 51 (mimograph in Hebrew).
30. State of Israel, Extension Authority, 'Research and Experiments in Peanuts, 1960–1961', Publications Division Bull. No. 45, p. 48 (in Hebrew).
31. Extension Authority, 'Summary of Crops for 1960–1961', Extension Bureau for the Negev Region, p. 3 (in Hebrew).
32. ROTEM, J. and FELDMAN, S.: 'Early Experiments in Extermination of Early Blight in Tomatoes at Mivtahim (Autumn, Winter, Spring 1960/61)', Preliminary Report, National and University Institute of Agriculture, No. 364 (Rehovot 1962).
33. KEDAR, N., FALBETZ, D., RETIG, N., and FELDMAN, S.: 'Preliminary Conclusions of Experiments on Export Tomatoes, Winter 1961/62', Internal Publication of the National and University Institute of Agriculture (Rehovot 1962), (in Hebrew).
34. 'Profitability of Moshav Settlements', Joint Agricultural Planning Centre (Tel-Aviv, September 1959), (mimograph in Hebrew).
35. [1] WEITZ, R.: 'Linked Farm, A Proposal for the Planning of the Hill Region', *Economic Quarterly* (Tel-Aviv) No. 16, (1957), (Hebrew).
    [2] WEITZ, R.: 'Agriculture Linked to Industry as an Approach to Regional Planning in Problematic Areas', paper presented to the FAO Study Group on Problems of Individual and Group Settlement in the European Region, second session, held in Wageningen, The Netherlands, August 1958.
36. STERNBERG, A. and DIMAR, N.: 'Economic Analysis of Services in Co-operative Small-holders Settlements', Extension Authority, Pub. Div. Bulletin No. 39 (Tel-Aviv 1961).
37. INFIELD, Hendrik F.: *The Sociological Study of Co-operation – An Outline*, (The Education Department Co-operative Union Ltd. Stanford Hall, Loughborough, England, May 1956).
38. WEITZ, R.: *Towards Specialized Farming*, op. cit., p. 48.

CHAPTER 7

REGIONAL PLANNING

A. THE REGION DEFINED

The plan presented in Chapter 5 for Israeli agriculture for the next ten-year period, dealt with the national level and that of Chapter 6 with the farm unit. The intermediate level, which is of decisive importance in determining the framework of production, must now be considered. This is the regional level, on which there is a wide interchange of economic and social activity. It is extremely difficult to define the region since it is a compound of many concepts, each related to a different purpose. The many attempts at an inclusive definition, mirror the background and bias of the formulators, so that none are entirely adequate.

One of the first serious attempts at definition was made by Sir Patrick Geddes[1] who based his unit on the river valley: he related economic and social life to geographic factors determined by natural boundaries of river, mountain and plain. Other attempts at regional demarcation centred on demographic features of a country's population, such as the areas inhabited by different tribes or traditional land-use patterns. Other definitions have been based on administrative convenience according to population density, revenue needs and structure of the administrative organization, for instance, regions governed by Provincial Officers.

For comprehensive planning purposes a region cannot be defined according to a specific geographic area, or to the exact number of people living in an area but rather as a *functional* unit, the exact size of which is not determined. A region defined by function can theoretically encompass a family, a village, a zone, a country or a group of countries. At present-day levels of knowledge, however, the integration of agriculture, industry and services cannot be realized on the level of the family or village, while a country or group of countries are too large for the understanding of the real day-to-day interaction of the economic, social and other factors. The Zone therefore remains the only unit appropriate for regional development through the comprehensive approach. There is, of course, no absolute size for a region; in practice, however, size is limited to that area in which the development action can be practically realized, which depends very much on the scope of the investment in agriculture, industry and services, the judgement of the planner and the particular conditions of the country.*

An attempt to define the region in terms of size (and function) was made at the Second Rehovot Conference in Comprehensive Rural Planning in the Developing Countries which took place at the Weizmann Institute in Israel in August 1963. In the course of the debate (27.8.63), Prof. Schickele remarked: "The minimum size of a region (is determined by)... the level of aspiration of the people you want to deal with, rather than in any other terms, and large enough to require comprehensive planning; and by comprehensive planning I mean to control or guide the interrelationships between agriculture, industry, and the various service trades.... So, small enough to be handled and large enough to be – you might say – a replica of a microcosm of a social order."[2]

This book deals with the comprehensive planning of rural development. The word 'comprehensive', as defined for this purpose, means the integration of agriculture, industry and services based on the thorough understanding of the interaction between economic, social, institutional, political and environmental factors. Such integration can only be achieved through planning the region.

B. MOTIVES FOR REGIONAL PLANNING

Regional planning is undertaken for a variety of interrelated motives. The common denominator of regional schemes implemented in various areas of the world, is resource development, but secondary motives have always provided the driving force leading to the implementation of these schemes. Without such motives, resource development itself would never be undertaken. The main motives which have led countries to embark on regional development projects are considered below.

* During the preparation of the present book, the concept of regional planning and its institutional framework were re-examined and thoroughly analysed by the author. The results have been presented in the following articles: R. Weitz: 'Regional Development and Programming', *Studies and Monographs* VII, Washington D.C., Pan American Union General Secretariat OAS, 1966; – with Levia Applebaum, 'Administrative and Organizational Problems of Regional Development Planning in Israel', Paper presented at the 1967 Annual Meeting of Directors of Development Training Research Institutes of OECD, Montpellier, France, 1967; 'Analytical and Institutional Approaches of Regional Development Planning', Paper submitted to the Workshop on Regional Development and Regional Development Planning held on the occasion of the 15th Anniversary of the Institute of Social Studies at The Hague, 1967.

1. Pilot regions

Since a country setting out to develop its agriculture and economy as a whole often starts with little experience, it often is advisable to select particular regions, concentrating first efforts on them in order to learn fully how to proceed and what mistakes to avoid; as the author put it in one of his lectures:

"Comprehensive development activities which require a complex professional team on the one hand and the full collaboration of the local population on the other, necessitate concentration in specific regions, both in regard to manpower, as well as investment. Such concentration will enable comprehensive mass activity, which will show results within a reasonable time."[3]

Pilot regions have been used in Israel to prepare the way for other development, experience gained from one region being extensively used in the planning of others. This is illustrated later in the chapter, where the lessons learned from the planning of the Lakhish Region are related to the prospective planning of a new region.

2. Regional development to improve local conditions

The spur to regional planning and resource development may be the pressing need to improve conditions in a particular depressed area. The most outstanding example of a regional scheme which had as its main purpose the improvement of local conditions was the Tennessee Valley Project, initiated in 1933 as part of the 'New Deal'. The valley was, previous to its development, an underdeveloped region with serious resource problems. Its economy was mainly agricultural, not unlike that of the many over-populated areas in other parts of the world. The lack of other activity meant that raw materials were shipped out of the region for processing, while without industry it could not advance. Profits were created elsewhere in the country from regional products which could have been processed on the spot.[4]

The project was in fact the first experiment in regional development. Its most outstanding characteristic was the establishment of a regional authority which 'seeks the simultaneous achievement of a number of diverse objectives by means of a series of multipurpose structures'. The Authority was given the task of developing the resources of the region, harnessing the waters of the river and promoting the establishment of secondary industries to process the raw material products in order to augment the incomes of the inhabitants of the region. The project is just one of many examples which had as its purpose the improvement of local conditions.

3. Comprehensive land settlement schemes to relieve population pressure

The great poldering schemes carried out in Holland have as their main aim the reclaiming of agricultural land from the sea, to enable re-settlement of farmers whose land becomes swallowed up by urban growth or whose holdings are uneconomic. The re-settlement of some farmers leaves more land available for those who remain. Moreover, a policy of intensifying polder settlement through a comprehensive plan of urban development is very important in relieving the population pressure in the western part of the country. The polders afford too few possibilities for a complete solution of these difficulties in Holland, but their contribution is substantial.

During the last fifty years of poldering projects, much has been learned. Lessons gained in the reclamation of each polder have been of great use in ensuring better and more efficient settlement of subsequent polders.[5] For instance in the original concept the polders were to be purely agricultural areas. For various reasons, such low density settlement is uneconomic and inefficient in the provision of social services. The most recent polders are settled according to a comprehensive regional approach in which industry and services are introduced to complement the agricultural basis.

The Zuiderzee regional development project is a good example of integrated economic, social and physical planning to build a modern farm community with auxiliary services.[6] As in most of the other regions mentioned, there is a semi-autonomous regional authority to carry out the work.

4. Economic motives

National policy may at a certain period be dedicated to raising national income through better resource utilization. This has often been the spur to the development of virgin lands. One of the best examples of a *region* which was planned primarily to meet economic targets is the Gezira scheme in the Sudan.[7] The area is a clay plain of five million acres lying between the White and Blue Niles. About one fifth of the area has been included in the Gezira Irrigation scheme[8], with about half a million acres cultivated each year, primarily in cotton.

The physical conditions of this large region are homogeneous, in contrast to most other development projects. The Scheme appears to have been based on functional considerations. The administrative organization of the region is semi-autonomous, with a triple partnership as one of its most important features. The tenants receive 40% of the cotton crop as the first partners, and have full tax-free rights to grow other crops as an incentive to improvement.

The second partner is the Government, which also has a 40% share of the cotton crop. The third partner is the Sudan-Gezira Board, a public body, with the remaining 20%. Today, this region supplies three fifths of the Sudan's total cotton and cotton seed exports by value, and contributes over fifty per cent of the revenue of the Sudan Government.

Another example is the Columbia River Basin in the north-west of the United States. This is an area which has been developed according to carefully selected economic and social criteria. The strong link between agriculture and industry, the growth of new towns and the multi-purpose use of the river waters are indications of the comprehensiveness of the project. There is, in this case, no separate regional authority to develop the basin, but the Bureau of Reclamation has given its offices in Spokane and Ephrata some autonomy in the technical and economic implementation of the plan. One of the main aims has been to increase purchasing power and incomes in the region, and in the country generally.

5. Settlement of immigrants or migrants through resource development

The Instituto Agrario Nacional of Venezuela is engaged in comprehensive settlement schemes which combine immigrant attraction and settlement, with settlement of the native population in areas of resource development. Jungle has been reclaimed and previously empty areas brought into production. At the same time, valuable new human elements have been settled in the country in permanent occupation. The scheme has been successful through a combination of sound technical planning with ample provision for social and economic services; the progress of the area is marked.

The main aims of the settlement schemes are embodied in the Immigration and Land Settlement Act of 1936. The Act specifies that the Venezuelan Government apply itself by every means, direct or indirect, to promoting immigration and land settlement in the Republic, thus endowing the concept of controlled immigration for the attainment of specific ends within a legislative framework. Resource development in Israel, with the settlement of immigrants as its main aim, is discussed later in this chapter.

A common organizational approach is apparent in all the projects discussed above. The specific national motives are implemented through autonomous or semi-autonomous action at a functionally defined regional level. In some cases there is an autonomous regional authority, directly answerable to the National Legislature (TVA). In the Columbia River Basin, the regional authority is an extension of the Bureau of the Interior, whose offices in the region initiate land settlement and encourage regional and resource development through inter-agency co-ordination. Israel's Land Settlement Department carries out settlement in different parts of the country and operates on a regional decentralized basis.[9] There seems to be a strong common element in the approach of multipurpose resource development. TVA includes power, industry, navigation, recreation and flood control. The Yarkon pipeline to the south of Israel boosts agriculture and new town development, and the reservoirs are future recreational areas.

In all the examples quoted, regional planning has proved itself as the means of undertaking comprehensive development. The conclusion to be drawn is that any country embarking on a development programme should from the outset weigh carefully a regional planning approach which includes the principles of comprehensive planning as far as they are understood today.[10] The development effort should be localized through the delineation of regions in which the effort is to be focussed. From these areas, a pilot region should be chosen, so that the process as it applies to local conditions can be evaluated and data obtained for the better planning of other regions.

A comprehensive approach to rural planning must integrate the three main development factors in a way conducive to social and economic progress.

a) Agriculture, which must always be the primary basis of rural life and development;

b) Industry, which must be introduced into rural areas for the following reasons[11]: local raw material, produced by agriculture or by other sources, cannot otherwise be exploited; industrial workers create a consumers' market for farm products; farms need industrial products; surplus labour from the farm can find employment; infrastructure is stimulated far more by industrial growth than by agriculture; industry provides a tax base on which sound services can be built for farm as well as townpeople;

c) Services. Improvement of the economic situation on farms must be linked with a sound community life. Rural people will leave the areas unless they have the same opportunities as townpeople for education, health, recreation, etc. Towns, which act as service centres, provide alternative sources of employment to farming. They help raise the community and living standards of rural people to a level which more closely approximate the national average.

These three factors can each be examined from economic, social, security and environmental points of view, which give added force to the above reasons for their integration in rural planning.

C. ASPECTS OF COMPREHENSIVE PLANNING

1. Economic

The integration of agriculture, industry and services is of great importance in ensuring the success of a regional development scheme. Modern agriculture is advanced in its theory and practice, but the higher the level of agricultural science which is applied, the more is the rural structure dependent on industry and services. Manufactured raw materials, fertilizers, insecticides, feedstuffs, building materials, water piping and machinery – are all bought by farms from external sources. Without them, no modern agriculture is possible; it is only industry and commercial organization which have enabled the great strides in farm practice that have taken place in recent decades. Even though many of these materials can be produced outside rural areas and imported, the commercial channels must be available locally.

At the other end of the farm production process, agricultural produce has to be marketed, and a significant part of it processed in mills, factories, grading sheds and the like. Because of the bulk of most agricultural goods, and the perishability of many, this is best done in the rural areas. Without it, the raw material will generate no secondary and tertiary employment in the region, nor will profits be generated for local inhabitants.

By locating agricultural processing in rural areas, together with services necessary for farmers and their families, a selfgenerating growth process can be initiated. A concentration of industry attracts more industry. From the basic raw material processing operation, a chain of industries can be established: textile manufacture based on the production of raw cotton, for example, or animal feed manufacture from the by-products of oil pressing. If an industrial complex starts to grow within a rural area, greater opportunities are offered to agricultural planners to introduce new crops linked to the industrial complex. Where a flour mill exists, it may be possible to establish a sugar factory, oil pressing plant or cotton industry, according to the agricultural possibilities and the vigour of the planners in pushing both the production of new crops and attracting the corresponding industry.

By adopting a comprehensive approach in the initial planning period, an industrial and service pattern can be worked out to correspond with the agricultural possibilities, a pattern which then becomes part of the general implementation plan for the region. Far wider horizons for economic development are thus opened than if the original planning concentrates only on developing agriculture. A mixed regional economy has far greater possibilities for advancement and success than has one limited to the production of primary products. This is true of regions as it is true of national economies. Diversification leads to horizontal and vertical economies where public services and infrastructure are supported fully by an integrated economy with a high money turnover. The very act of investing in industries and services in a pre-planned regional pattern stimulates employment and economic activity beyond that envisaged in the plan.

2. Social

From the social point of view there is a great advantage in adopting a comprehensive approach to the development of rural areas. Social purpose is created by projects which aim at economic betterment: the integration of industry and services with agriculture provides the framework in which human welfare or social requirements can be satisfied. Nucleated community formation in villages and rural towns gives a basis for social activities and organization within regions, leading to mutual understanding and better living. The physical relationships created through regional planning must cater to social needs: this is the essence of the interaction of social considerations in comprehensive planning.

A physical structure, incorporating a balance between farms and villages, villages, towns and infrastructure, enables valuable social forces to enter a region. Industry brings professional people to the towns, and their demands and abilities give impetus to maintaining a high level of services which, eventually becomes common to everyone in the region. Their way of life has an important 'demonstration' value for the local population, and their wider social consciousness can find expression in the local municipal life. National institutions, by opening branches in the regional town, involve people in the civic framework, and the encouragement they give to local services which is important in raising the level of social life in both town and village. The establishment of a network of communities with many common interests comes about as the result of intensification within the region due to the introduction of industry and services. The wider opportunities created by this balanced development stimulate the mobilization of local initiative.

3. Political and security

A new and imaginative project, such as a comprehensive land settlement scheme, can become a banner for awakening pioneering elements in developing countries and a fresh call to idealistic youth. Regional development provides a challenge and an incentive to practical idealism, leads to visible results, and can therefore be an issue of great political significance in national life.

From the security point of view, settlement and development schemes can be used effectively to close off dangerous borders. However, agricultural settlement alone is not strong enough for this purpose. The co-ordinating point of a region is its town, and the wealth and activity of a developing industrial complex is very important in supporting the defensive position of villages. A healthy, well-educated population which understands the political and security importance of its development efforts is better equipped to provide this support.

A further point, probably valid for many regions which have security importance, is that military expenditure can be important in the commercial life of a region. If the regional authority integrates the development of its services and industry with security requirements, it may be able to get defence expenditure to help its own investments.

4. Physical and environmental

On the simplest level, physical features define the location of a region, but they may also *participate* in its resource pattern. Rivers, mineral deposits or fertile land form an active part of a region's functional life. This conceptual distinction between the active and passive roles of physical features can facilitate the work of the planner. The relative weight of the two roles in the development of the region depends on the part which the economic resources play in the total economic activity to be developed. The delineation of regions must therefore pay attention to the nature of physical features mostly as a result of the second role – namely, the participation of the features in the regional economy.

Environmental considerations in the integration of agriculture, industry and services have two main aspects – developing habitability and developing hospitability.

Hospitability involves making physical conditions conducive to a settled life – encouraging pride in landscape, providing recreation for town and country people alike, making distances between essential service points short and ensuring easy access to them. Architectural plans for towns and villages, incorporating the different functions of each, should ensure harmonious landscape design and physical layout.

Hospitability depends very much on habitability. A region should attract people to come and live there; through careful land use planning and zoning according to type of function, the region can be made pleasant. Cultural and recreational facilities may be essential in the development of a new area if it is to attract and keep the skilled people on whom its hopes rest.

D. THE GROWTH OF THE REGIONAL PLANNING CONCEPT
IN ISRAEL

Prior to independence, primarily because of geo-political factors, no regional planning existed, while regional co-operation was limited in scope. With the Jewish population dispersed in different areas, inter-settlement contacts were loose, the distances involved and security problems preventing day-to-day contacts. The unstable security position made it imperative that each settlement form a closed self-sufficient unit, both in economic and social services. Joint services, therefore, existed only in areas with a dense and contiguous Jewish population, such as parts of the Jezre'el Valley and the coastal Plain. As a rule, however, Jewish settlements were scattered among non-Jewish villages: the cultural differences between Jews and Arabs prevented interaction and reinforced the tendency to self-sufficiency of the Jewish settlements.

Side by side with such looseness of contact and self-sufficiency, however, went the ideology of co-operation as already explained, expressed in purchasing and marketing organizations, and also political. Thus there was a framework on which closer co-ordination could later be built.

During the first years of statehood, there were no marked changes in settlement policy, despite the massive immigration described earlier. The young State had very little infrastructure, agriculture was not fully developed, basic industry almost absent, and the bulk of the population was concentrated in the large towns of Tel-Aviv, Jerusalem and Haifa. The rural Jewish population was concentrated mainly in the central region. The South was almost completely empty, and large unsettled areas in other parts of the country were open to marauders.

During the first stages of mass immigration, the three cities started to grow at an alarming rate. They contained the most favourable conditions for investment, while no organizational machinery was available to settle the rest of the country. The vulnerability of this policy was soon realized. Not only were the cities expanding more rapidly than housing and services, but the difficulty of holding the borders stretching along the edges of empty wastes became apparent.

A policy of population dispersal was undertaken through the establishment of secondary industrial centres such as Ber-sheba, Ashkelon and others. The way in which such centres should be supported and developed was not known; the main reason for their existence was security; the means by which industry could be attracted to them were not worked out. They did not have a firm economic basis for several years, and many of their inhabitants left for the main urban centres. At the same time, agricultural settlement was undertaken in the south-central region, in the hills and in the Jerusalem Corridor, to make those areas secure. Each village was built as a separate unit with no inter-village co-operation and no attempt at integration of industry. Israel is small, and the new small towns built as secondary industrial centres, were usually situated in or near the agricultural settlement areas. But no conscious effort was made to connect the economy of the farms with that of the nearby towns. The town of Affula, for example, in the Jezreel Valley had been established in 1925 as a regional town in an area of intensive agricultural settlement. However, the farms produced milk, other livestock products and citrus, which were marketed directly to the established cities on the coast. Affula, completely by-passed, stagnated for many years.

Within a few years after independence various economic, social and ideological factors prompted both the settlers themselves and the agencies responsible for settlement projects to establish co-operative patterns on a regional scale.[12] The opportunity which unfolded after the establishment of the State for the development of entire districts gave an impetus to the creation of regional frameworks. A regional structure enabled a number of settlements to pool their resources in setting up institutions and enterprises on a much larger scale than would have been possible had each acted separately, reducing operational costs. Their large scope also facilitated the introduction of technological improvements and the employment of high-grade technicians and experts. These joint enterprises included, for instance, the operation of heavy equipment and the sorting of agricultural produce.

In addition, certain activities were developed on a regional scale which the individual settlement would have been unable to afford altogether, indeed, whose very existence only became possible through joint effort. Examples of this kind of enterprise are product processing plants and secondary schools.

Along similar lines originated the idea of establishing central economic and consumer services for a number of settlements, thus making them more accessible to the rural population and reducing maintenance costs.

During the past few years the need for non-agricultural employment has tended to encourage the advancement of regional co-operation both in order to supplement the earnings of the farmer's family in marginal areas and to absorb within the rural area surplus manpower, especially second-generation settlers.

The first group of settlements planned from the start on a regional basis lay in the Bashit area, west of Gedera, in southern Israel. There five settlements now constitute a kind of multi-neighbourhood settlement, each neighbourhood having its own local services, with a joint service centre in the middle. Three of them were established in 1950, the other two in 1953. The Aseret service centre, which also houses the staff employed in services as well as technicians and experts, was set up in 1954.

Immediately afterwards a further cluster of settlements was planned in the Sharon, combining three moshavim round a joint centre – the Kefar Ya'betz settlement complex. Two of these settlements were established in 1950 and the third in 1951, while the centre developed chiefly as a cultural-educational nucleus together with housing for teachers and instructors.

In the following year two additional regions were planned along parallel lines – the settlements of Bakura in the Bet She'an valley, and the 'Sharsheret' settlements. The Bakura complex includes three moshavim with a joint service centre; two of them were established in 1951 and the third in 1952. The 'Sharsheret' settlements were founded between 1951 and 1953 in two groups, one of four settlements and the other of three, each group with a joint service centre. The first group, 'Sharsheret' proper, was planned and established in 1951 and included a number of housing units in the service centre for experts and technicians. The second group, 'Wadi Sha'aria', was established in 1953. Although centrally situated and providing housing facilities for technical and professional staff, the service centre here does not occupy a separate area but is attached to one of the settlements.

The Bakura group was established in an attempt to solve a totally different problem. At that time (1952/53) the planners of the northern district were concerned with designing a large village able to bear the cost of adequate services, while avoiding excessive distance between the settlers' houses and the village centre on the one hand and between the settlers and their holdings on the other. This led to 'core' villages (nucleus villages), made up of neighbourhood units at Bet Yosef-Yardena, Bakura and Yael in Ta'anakh.

The foundation of the multi-neighbourhood moshav Yael in June 1953 formed the first stage in the development of the Ta'anakh region. In fact, Ta'anakh itself is only part of a larger region and was planned in such a way as to be able to be integrated within a more extensive regional structure, consisting largely of veteran settlements. Development of the joint centre of Yael began in 1955. In 1956 another two similar moshavim with three neighbourhood units and a service centre were established.

Planning of the Ta'anakh settlements was based on a pattern of a large multi-neighbourhood settlement, with a neighbourhood centre serving the three neighbourhoods. Each neighbourhood has some 60 farmers; a few craftsmen and professional people, with a minimum of local services. The main services are located in the joint centre catering to the entire large village. The neighbourhoods are adjacent to each other so that the distance from the neighbourhood centre to the joint centre is about three-quarters of a kilometre.

The centres of the multi-neighbourhood moshav settlements provide the principal services which the settler requires, e.g. kindergarten, elementary school, youth club and sports grounds, communal hall, clinic, co-operative store and supply store, storage shed. The centre of the 'Hever' settlement also has a tractor station serving all the settlements of the Ta'anakh region. The joint centre includes part of the housing for service workers. The plan took into account the fact that a higher standard of services was obtainable in the neighbouring town of Afula, which lies in the heart of this region. The planners also allocated land for a regional centre to develop in the course of time, and there has indeed been a recent tendency to establish services of a medium standard in Ta'anakh, a process which is now at its beginning.

The 'Manes' settlements (Menuha, Nahala and Segula) in the South were likewise planned and established in 1953. This group was afterwards integrated in the regional structure of the Lakhish Region.

All these developments were isolated phenomena in the history of new settlement, following the establishment of the State. Only from mid-1954 onwards, with the planning of the Lakhish Region, did regional planning become a generalized system. The planning and settlement of the Lakhish Region thus constitutes a turning point in Israel's planning concepts.

The new planning concept crystallized along with the change from the mixed farm to the specialized as discussed previously in this chapter. The introduction of the field crop farm in 1954 created an opening for the introduction of industry to rural area, the building of towns for industry and services, and for the inception of comprehensive rural planning in Israel.

Once industry and services began operating in the region, possibilities for more industry presented themselves. Because the basic planning was conducive to expansion, the town would suddenly 'get moving' after passing through a brief period of unemployment and slow investment growth. Industries and services expanded beyond anything conceived in the original plan. Industry not based on the agriculture of the region started to locate in the town, and it was seen that comprehensive development was the way to draw industry out of the big towns, dispersing it throughout the country.

The Lakhish area was the first comprehensive regional planning project undertaken in Israel. Thereafter, the system was applied as a matter of course.

In 1958, the Adullam area was planned on comprehensive regional principles.* Four settlements of 50 families were established and in 1961 a fifth settlement was added. Two rural centres were set up, though no regional town in view of the small number of settlements. The centres, however, included industrial projects, offering the farmers additional employment and income, necessary in the hilly marginal region which does not permit attainment of a full income level from agriculture alone.

Aside from the development of new regions on the principles of comprehensive planning, areas previously settled were converted to the regional structure with rural centres appropriately located, and the transfer of services from the villages to the centres. Though this transfer has not been possible under all circumstances, the general tendency has been to implement it wherever conditions have permitted.

A special Committee appointed by the Settlement Department has proposed that in addition to the existing 53 centres in Israel a further 26 be established in different parts of the country.[13]

The idea of regional co-operation has not been the sole prerogative of planners and policy-makers.[14] Among the settlers themselves the recognition dawned of the importance of joint activities on a regional scale. As a result various patterns of regional co-operation have developed in many places, each with its own special features in line with local conditions and the character of the settlers. A clear distinction is noticeable particularly between regional patterns formed primarily by collective kibbutz settlements and those evolved by moshav settlements.

In most cases the regional council is the main instrument for joint action, even if the scope of activities does not necessarily coincide with the council's area of jurisdiction. In fact, a regional council is formally a municipal body, having a similar function to local councils or municipalities, e.g. maintenance of schools and medical institutions. In certain instances, however, the councils have set up separate registered co-operative societies to provide various services to the villages within their jurisdiction: economic services such as slaughter-houses, refrigeration plants, dairies, sorting and packing sheds; consumer services, such as regional laundry or bakery. Frequently members of the Council are also members of the Board of this society, though financially and managerially the co-operative society functions as a separate body.

An example of what is basically inter-moshav co-operation is the regional council of Be'er-Toviyya which functions for 19 settlements: 14 young

---

* Adullam Region is located south-east of Jerusalem.

moshavim, three old-established moshavim, one kibbutz and one moshav shitufi. This regional co-operation, extending to all the 19 settlements in the council, started with the foundation of the township of Qiryat Malakhi in 1950 as part of the national plan for the dispersion of the population. The town was designed as a service centre for the rural area and as living quarters for the service and professional workers employed there. The regional council of Be'er Toviyya, which had jurisdiction also over Qiryat Malakhi, established a number of industrial enterprises and services near the township, but Qiryat Malakhi grew so fast that by 1958 it separated from the regional council and was given independent municipal status.

In addition to the co-operative framework of all the settlements of the council, there also are subsidiary co-operative organizations in the four sub-regions of groups of adjacent settlements. Each such block contains five or six settlements having joint services, e.g. schooling, cultural activities, a tractor station, which are not, however, located all at the same place. Some of the professional people and service workers live in the older rural settlements, while others live outside the area. The regional council is now trying to put up separate living quarters for them in one of the sub-regional centres.

Awareness of the idea of regional co-operation is greater in the collective kibbutz sector than in the moshav sector and the extent of regional activities is more comprehensive. In most cases the chief motive for inter-kibbutz co-operation was first and foremost economic, but in the course of its evolution, ideological motives sometimes also came into play. An interesting example may be found in the regional council of Sha'ar Hanegev, which is considered by many as the most advanced in its level of regional activities.

The Sha'ar Hanegev regional council comprises 11 settlements – ten kibbutzim and one moshav. The main participants in this regional co-operation are the ten kibbutzim. This regional co-operation started in fact as early as 1950, when the settlements of the council collaborated in the establishment of a green-fodder drying plant (which has in the meantime closed down). At a later stage further joint enterprises were set up, mostly on a site especially allotted for the purpose near the township of Sederot. The regional centre has remained without actual residents; the active participants in regional co-operation and their senior staff live in their kibbutzim and other workers live in the town of Sederot.

The regional activities are more extensive than in Be'er Toviyya and are concerned with the following spheres of activity:

a) municipal services

b) rural industry (cotton gin, slaughter-house)

c) heavy equipment (potato lifting, cotton picking, etc.)

d) services (central garage, petrol station, regional laundry)

e) education and culture (e.g. regional elementary and secondary school and amphitheatre).

According to the regulations all the settlements must take part in all these regional enterprises, the establishment of which has been decided by a two-thirds majority. In certain enterprises, e.g. the slaughter-house, settlements not belonging to the regional council also participate.

Patterns of inter-kibbutz co-operation, with certain differences, exist also in other parts of the country, e.g. in Upper Galilee, the Jordan Valley, the Bet-She'an Valley, Gilboa, the western part of Emek Yizre'el (Jezreel Valley).

As Israel gains in experience in the planning and execution of regional co-operation, the concept of regional planning in general continues to de-velop. The villages of the Lakhish Region serve as a good laboratory. The farms in it are prospering and advancing; the handling of the many ethnic groups seems to have worked out; people who were not farmers before they migrated to Israel are able to cope with the most up-to-date of farming organization and methods. Today with a second generation approaching working age, the problem of employment outside the farm becomes acute, for each farm cannot absorb more than one son. A recent sociological survey has shown that as a rule the youth remain in the village rather than migrate to the cities. Hence industry has to be introduced into rural centres serving four or five villages and comprising a sub-region in order to solve employment. The sub-region will then become strengthened, better services introduced, while the economic and social potential of the region will become raised. With industry and services integrated at a lower level than the regional town, the comprehensive approach will have developed to a stage where it is close indeed to the life of the farmer.

Without a policy of population dispersal and the building of compre-hensive regions in Israel, a dangerous situation would have arisen. Economic growth would have become topheavy with concentration in the cities and a wide gap between town life and country life as has happened in so many other countries. The old story of migration from the rural areas to the city would have taken place with people seeking better standards and services than those possible on the farm. Because Israel's problem was of new settle-ment rather than the modernizing of old farming regions, there would have been areas where settlement could not have taken hold, and ethnic groups which could not have been given the chance for integration and modern-ization at a possible pace, and within a group life. With the rural areas on a lower standard than the rest of the country and without the stimulation of industry and service centres, agricultural methods themselves would likely not have advanced as rapidly as they have.

Israel's regional development has been compressed into a very short period of time and is still going on. It could not have come about as rapidly without massive investment and a ready corps of skilled people. However, even for those countries which do not have the same possibilities of finance or skill, Israel's experience is valuable. Whether the investment programme is for ten or for fifty years, and whether development is to be rapid or gradual, success can only come about if the underlying factors which affect planning are properly understood and taken into account.

E. INDUSTRY AND AGRICULTURE IN COMPREHENSIVE PLANNING

With formulation of the policy of population dispersal in Israel, a formula had to be found to guide the dispersal of industry. Services can be planned to follow population concentrations, but the dispersal of the primary source of employment – industry – is a more important and problematical factor.

1. The advantages of integrated industry

Industry has brought prosperity to many nations, but its overall, long-term effect on rural society has often been negative because of its concentration in urban areas. In order to avoid an exodus and the destruction of rural communities, industries must be integrated into the rural economy as we have already suggested. Such an integration is necessary even in western countries where the rural population has declined in proportion to the total population and sometimes in absolute numbers, but is even more so in developing countries where the percentage of rural population is still high. Developing countries are faced with the need to come to terms with modern technology and organization without disintegration of their cultural life and social structure, nor have they resources to handle mass movements of people to the cities. Hence the introduction of industries into rural regions is almost imperative, and can be one impetus to general economic development.

As we have mentioned, industries can provide on-the-spot employment for surplus labour, and can give opportunity for skilled non-agricultural elements, eliminating one of the chief causes for the rural exodus. The type of industry is irrelevant as long as it is adaptable to the general framework of rural life.

The market for agricultural produce which industrial workers in an area create, is particularly conducive for perishable products which give high returns per labour unit but which cannot be shipped to distant markets unless special facilities are available. In developing countries where transport and marketing systems are in early stages of organization, the existence of an enlarged local market encourages the farmer to produce more than he requires for his own use. This is the first step from a subsistence to a market-oriented economy. The added value produced by industrial development will also directly increase the income of farmers and their purchasing power of industrial goods, raising living standards all around and keeping the gap between town and country from developing, or at least getting out of hand.

The advantages of industry in rural areas are not always realized by policymakers in developing countries. Experts in industrial development are often Westerners of urban background, whose approach stems from previous experience in developed countries. They are inclined to calculate the profitability of an industrial plant on the basis of the industrial process alone, without taking into consideration the broader economic picture.

Certain errors of judgement are apt to be committed, the first relating to the size of the plant. In developed countries usually the bigger the plants the more 'economical' they are, but for developing countries the great investments required render such plants unsuitable. The second error relates to location, which for big plants is naturally confined to urban areas. As a result, an important potential lever for rural development becomes lost.

In a certain East Asian country, for example, primitive oil presses were used in rural areas to extract oil from groundnuts and sesame.[15] In view of the inefficiency of these oil presses, it was decided to abandon them and build large modern factories in the big town. It was also decided to expand the the cultivated areas of oil seeds, and thereby increase their output. The outcome was the exact opposite of that expected. Transportation costs grew enormously; instead of the oil being transported, which has a relative high value per unit of weight, the low-value seeds had to be transported to far-off cities. Moreover, no appropriate market could be found for the oil cake, which in the farms had been used for animal feeding. Had the oil industry been developed by means of modern machinery, but in relatively small plants located within the rural areas, all the desired effects could have been achieved. The economic calculation which had been made exclusively for the industrial process, showed quite correctly, that large enterprises were more economical, but it did not give the full picture. This example clearly indicates that an industrialization programme should be evaluated on the overall impact on the rural area rather than on profitability alone.

2. Types of rural industries

From the point of view of rural development, industrial enterprises can be divided into three categories: processing industry, linked industry and auxiliary industry.

i. *Processing industries*

A processing industry located within the producing area eliminates the intermediary stages between farm and factory. Transportation costs are greatly reduced, with the farmer's share in the added value increased. The establishment of such industries necessitates thorough economic examination to determine the optimum size and correct location of each enterprise. Distances from the fields where the raw materials are cultivated have to be taken into account. The siting of the third sugar factory in Israel, which is discussed on pp. 273–275, serves as an example of a desirable planning approach.

ii. *Linked industries*

In marginal areas where the agricultural season is not long enough to secure an adequate living for the farming population, a suitable industry may fill the gap by providing employment in the slack season. Such an industry must fit into the work schedule of the farm and may therefore be termed a 'linked industry'. The introduction of industry enables the farmer to discard those unprofitable branches of agriculture which had served merely to balance his yearly work schedule, using industry instead for the balance.

An example is to be found in tobacco farming in Israel.[15A] The climatic and soil conditions of a certain highland region in northern Israel are favourable for the growth of tobacco, which has a slack season between October and February. As most of this season could be devoted to sorting, grading and fermenting the tobacco leaves, a suitable industrial plant was set up. Today with some additional branches, tobacco farming, and the linked industrial processes, present a balanced and profitable economic system.[16]

A different type of linked industry consists of the manufacture of parts for later assembly in one central enterprise. Thus, in Japan, certain optical and electrical instrument parts are being manufactured in villages and then transferred for assembly to a regional undertaking.[16A]

The advantage of linked industries is that the work can be done in combination and co-ordination with agricultural work by the settler or his family in their free time or slack season and that no rigid schedule is required as in other industrial enterprises.

iii. *Auxiliary industry*

These are industries which may be promoted in rural areas without having any specific connection with local conditions. The purpose of locating such auxiliary industry is to absorb surplus manpower. A great variety of under-takings fall into this category, including diamond cutting, manufacture of jewellery, ceramics, etc. Two enterprises of this kind have recently been established in Israel: one is for the production of fancy goods (ashtrays, candlesticks and the like) and the other produces feed-concentrate containers.

The distinguishing feature of the auxiliary plant, as compared to any ordinary factory located in an urban industrial centre, is that its labour requirements are adjusted to the surplus manpower available in the rural area.

These three categories of industry can and should be integrated into the rural development plan and used as an important and sometimes essential impetus for its realization.

3. Siting industry in rural areas

The first attempts to decentralize industry in Israel were not guided by any real understanding of the main issues involved, since the process was largely experimental; subsequent experience showed up a number of planning mis-takes.

The main unknown factor was the optimum size of a new development town for a rural area. Centres were planned for a population of from 5 to 10000 people, but soon proved that they were not viable at that size. Costs in service provision, road connection and municipal organization were too high. Nor was an area so small able to attract skilled people for the few industries established there. The absence of such a social element and of real urban services was keenly felt. Soon afterwards, because of the slow growth of a number of these centres and of continued immigration which had to be settled, planning aims were raised considerably. It was realized that towns would have to have a minimum of 20–30000 people before their development could get under way. This figure has already been recognized as too low, since once a town begins its development, it tends to grow of its own accord. If the physical planning limits its expansion, stagnation sets in with important momentum lost.

Present policy is to concentrate on a few development towns in order to bring them to initial viability. There is active Government support in the encouragement of initial industry, and thereafter other industry is allowed to find its way to the town with the Government influencing development by its policy of immigrant housing schemes. Centres which are attracting industry are given a greater housing allocation so that labour supply and local demand can be co-ordinated. The whole policy of strengthening the new centres was greatly helped by the boom period through which Israel industry passed for several years, by the great investments made by public and private enterprise in industry, housing and services in the development towns and by the firm demand basis given by continued immigration.

With regard to agricultural industry, experience soon showed that it is best located in the rural towns, as illustrated in the following section.

4. Siting of Israel's third sugar factory

In order to fulfill the requirements of the Seven-Year Plan for Agriculture drawn up in 1953, which called for the introduction of sugar beet to Israel farming, two sugar factories were built, one in central Israel and one in the north. These were meant to receive beet from villages of the south as well as elsewhere in the first stages of irrigated farming. The Yarkon pipeline, completed in 1954, added thousands of dunams of irrigated land, primarily in the south, to the country's total. The farms drawing from it were of the industrial crop type in Lakhish and the Negev regions. The third factory had therefore to be located in the south.

The responsibility for its exact siting devolved on a council of Ministers made up of the Ministries involved in the financing and support of the factory. The Government searched for private enterprise to build the factory, but since it contributed a large share of the investment capital in the form of loan and since it paid the subsidy on sugar beet, it was vitally interested in everything connected with the proposed factory. Since the government had begun building development towns in the south, its choice was limited. The factory had to go to one of these towns and not necessarily to the optimum location. Any grave errors made in siting the towns was naturally reflected in the efficiency of industries located in them.

There were two towns which offered possible sites – Ofakim in the Western Negev, and Kiriat Gat in the Lakhish Region. Of the two, Ofakim was the site favoured by many, because of its more urgent need for industrial plants. The planning staff in Lakhish were convinced that Kiriat Gat was, objectively speaking, the better place of the two. Naturally, a project such as a sugar factory is of great importance to a development town, first because of the direct employment it creates and the stimulating effect this has on trade generally; secondly, because an industry of that size draws other industry in its wake; and thirdly because farm planning can be more intimately linked with the factory if it is close to the farm.

The director of the Lakhish Region initiated a survey[17] of the economics of siting the sugar factory in which the respective advantages of Ofakim and Kiriat Gat were compared. The survey was based on the following criteria:

a) The optimum size of the plant for Israel's conditions.

b) The area of irrigated land needed to supply such a factory assuming sound crop rotation with sugar beet once in a five-year cycle.

c) The average yields of sugar beet in the two regions and the importance of the crop for the representative farm types.

d) The transport costs for sugar beet, lime and fuel to the factory from farms and supply areas. This calculation was made on a kilometre/ton basis from the two towns to each of the villages and to lime and fuel sources.

e) The transport connections of the two towns with national centres.

f) The relative suitability of the two towns to absorb a plant of the size and complexity of the sugar factory.

g) The possibilities of starting secondary industries around the factory from the raw materials.

h) The use of pulp for animal feeding in the two areas around the towns.

i) Possibilities of extending the harvest season by sowing spring beet.

The report concluded that one factory of a capacity of 2000 tons per day at full development would be cheaper than two smaller ones because of the difference in overhead and administrative costs. The depreciation and interest charges on a larger factory of 2500 tons per day were found to be high per ton of processed beet. Such a large factory was not found necessary in any case for the potential sugar beet area.

The area of beet cultivation was estimated at 18 000 dunams in 1959 for villages within a radius of 30 km of Kiriat Gat (including the administered farms and private farms), 9000 dunams for villages and farms within 31–45 km radius, and 9500 dunams for villages and farms within a radius of 46–70 km. The total was thus some 36 500 dunams. This area was the same whether the factory was sited at Ofakim or Kiriat Gat.

The report found that yields in the Negev were higher than in Lakhish on the average of three years of sugar beet growing, but pointed out that the Lakhish settlements were very new and would quickly catch up to the Negev settlements in yields. In this respect there was therefore little advantage to Ofakim as the site for the factory.

The next stage of the examination was a kilometre–ton calculation. The total yields of beet from each of the producing villages in the Negev and in Lakhish were multiplied by the distances from the two towns respectively. This gave the advantage of one centre over the other as far as transport costs were concerned. Kiriat Gat showed an advantage of 400000 km–ton, an estimated saving of $ 6000 per season. A similar saving on lime and fuel transport brought the total estimated saving to $ 12000.

Kiriat Gat was better placed than Ofakim on the national road network and was also on the railway line – important factors in sugar production, not only for the transport of the raw beet, fuel and lime, but also for the transport of sugar beet pulp back to the villages for animal feed. In addition, the attractiveness of Kiriat Gat as a place to live was stressed. Professionals for the other factories in the town had already begun to move in and similar conditions could be guaranteed for the sugar plant-workers. In 1958, Ofakim had not yet reached such a stage.

Finally, the report stressed that the Negev settlements could grow early potatoes and export crops, while Lakhish was more dependent on sugar beet as a main crop in the rotation. The settlements around Kiriat Gat would therefore be more reliable suppliers of sugar beet for the new factory, which, it was suggested, should have an initial capacity of 1500 tons of beet per day with an ultimate capacity of 2000 tons.

This report played a decisive part in the decision of the Ministers to locate the factory in Kiriat Gat. The plant commenced operations in the 1961 season.

5. Industries for the rural centres

We have seen that in the regional co-operative system of Sha'ar Hanegev, 10 kibbutzim and one moshav participate, with one other kibbutz participating partially. The total regional population is 3200 people, of whom 1800 are full village members. The total population of each of the settlements is less than 200.

The small population per village means that there is no possibility of each one providing its own complement of services – single village schools are of low standard and high cost, and there is insufficient revenue to cover other municipal services such as social welfare, health, etc.

In the drive to gain specialization of the country's agriculture through allocating quotas and restricting the spread of particular farm branches to certain areas, Sha'ar Hanegev is very much limited in the output of animal products and fruit, and has had to concentrate on industrial crops.

In view of this position, the villages decided on the following five points of action[18]:

1) *Establishment of industry* based on agriculture, which would sell higher value products than raw material. So far a green-fodder drying plant, a cotton gin and a poultry slaughter house are working. An abattoir for sheep and cattle is being planned. The local enterprises directly benefit the region itself.

2) *Joint investments for production:* The small villages have little possibility of buying heavy equipment on the scale needed for the production of the whole range of crops they grow. The regional co-operative buys, maintains and uses this machinery on behalf of the members. The machines work over 2500 dunams of sugar beet, 6000 dunams of cotton, manure spreading over 8000 dunams, liquid fertilizing over 80000 dunams, potato lifting, sowing alfalfa and grading of apples. The co-operative also maintains: a garage, a refrigeration plant, a sorting shed for potatoes, a citrus packing house and a packing house for other fruit.

3) *Services:* Joint services reduce production costs directly or indirectly. They serve individual settlements and common activities. A large general shed has been built to include a spare parts store for the region; a fuel station and weigh bridge are in use, and an office in Tel Aviv has been opened. Marketing of poultry and dry alfalfa are cared for by the organization. Dairy cooler services, electricity supply, roads and offices are provided in each village. A restaurant and a new regional laundry are other activities of the co-operative.

4) *Regional education* includes a number of related activities:

a) A regional school for all classes which has successfully integrated the small classes existing previously in the villages.

b) A regional secondary school and trade school in co-operation with the township of Shderot. The teachers and the services (laboratory, library, sports, workshops) are on a high level.

c) An evening institute for adults. 15 groups meet once a week under the instruction of teachers from the University. This opportunity was not available before co-operation started.

d) Functions for all age groups and particularly for youth. These are very important in countering the isolation which settlers sometimes feel.

e) A regional amphitheatre which brings the best plays and shows to the Negev. Performances are open to the general public and are attended by people from many settlements not included in the co-operative.

f) Municipal services – the most striking of these are the regional clinics and swimming pool.

The functioning of the regional services is governed by several basic principles, among them that all the villages must share in *all* of the activities, that all members of the villages who work in regional activities receive the same wage, and that all profits are divided according to the amount of produce supplied by the villages or by the use they make of the services.

Participation of the kibbutzim in the regional co-operation framework has within the villages and the kibbutz movement raised questions on the eventual effect of such regional organization on the individual kibbutz. The regional enterprises need highly skilled managerial manpower not subject to yearly changes at election time. There is fear that a 'management class' will be produced, with drastic effects on the kibbutz structure and ideology. This and other questions raised by regional organization are the subject of earnest debate within the kibbutz movements, which cannot overlook the necessity of co-operation between member villages if they are to prosper in the mechanized agriculture of today.[19]

The type of industry and services set up in Sha'ar Hanegev are typical of those planned for other rural areas in Israel, including the rural centres of Lakhish. They are of two main types – service activities for production and social needs, and industry which processes produce of the farmers. It is clear that such industry is insufficient for many of the rural areas, particularly where the increase of mechanization has released women from agricultural work, and where, as in Lakhish and some other areas, the youth want to remain at home but cannot *all* become farmers. In these cases, industries must be found which on the one hand use female labour and on the other provide work and careers for youth. Agricultural industry is not in all cases the answer, nor must it be. If labour and capital are available, enterprises which can function in a small village environment, without the complicated services which a large town provides, are suitable for a rural or regional centre. The kibbutzim, which for several years have been intensifying and diversifying their employment possibilities through the introduction of industry, provide a precedent. In the bigger kibbutzim, many of their industries have no connection whatever with agriculture: They range from furniture manufacture to art work, from light metal industry to plastic factories. The original drive to industrialize the kibbutzim arose from the need to provide work during slack winter months. In the second stage, industry began functioning throughout the year, while the need for it was intensified by the presence of older members unsuited to farming and of others who wanted a kibbutz way of life but not farming; by the increase in mechanized farming and by the drop in profitability of some farm branches. Often an industry was established because the diverse kibbutz membership included an expert in a particular trade or occupation. The industry was built around him.

In order to enable moshavim and regional groups of settlements to initiate industry, the Jewish Agency and the Government established a company called 'Rural Industries, Ltd.'. This company planned regional services for a number of areas, of the type described above for Sha'ar Hanegev, as the first stage. For some areas, small industry will be set up which can utilize rural labour, such as, for instance, jewellery and art work in rural centres of the Jerusalem Corridor or ceramics in places where the necessary raw material is available. At a later stage, if the early experiments are successful, then industry may be introduced into individual villages, and a start made in cottage industry.

The size of a viable rural centre has yet to be determined, and can only be known as centres are strengthened by industry and their actual progress noted. No certain method of determining the minimum or maximum size of a rural centre has yet been found.

F. REGIONAL PLANNING IN LAKHISH

The Lakhish Region was the first area intensively settled on the basis of full comprehensive planning from its inception. The area has had a long history from Biblical times when its strategic importance made it a centre, as well as its fertility. Situated in a south-central plain of present-day Israel, bounded by the Mediterranean Sea on the west and the Judean mountains on the east, it was on the ancient trade routes of Egypt and Mesopotamia. Guarding the foothills to Jerusalem, its settlers, both in the past and now, had the task of holding an area vital to national defence. The entire area is covered with the remains of settlement from periods of the Canaanites, the conquests of Joshua and the Philistines. Under Rehoboam, Solomon's son, a chain of fortresses was built up to Jerusalem, which became subject to Assyrian and Babylonian invasion and conquest. After Nebuchadnezzar, the Maccabeans and the Romans fought on Lakhish battle grounds, and later the early Christians moved in. In the seventh century C.E. armies from Arabia swept the area, leaving it waste and desolate, except for a few extensive villages. The area later became one of the main battle grounds in the War of Independence of modern Israel.

The experience of that war proved conclusively that such a vital area could only be held safely if it were closely settled. The first villages were built as defensive outposts on the borders, mainly kibbutzim settled by Israeli pioneers who lived in constant danger of attacks by infiltrators from over the border. Moshavim of new immigrants were built along the coastal area to link settlements of the Gaza area with more northerly population centres. The central Lakhish Area remained empty except for two kibbutzim established before the State and a few moshavim established on a dry farming basis in 1950. Because the wide empty spaces left a free passage-way for infiltrators between Gaza and Jordan, it was soon decided that they had to be settled. The great immigration of these years made it imperative as well to extend the area of settlement southwards, giving newcomers security and employment in agricultural production. (Israel at that time lacked a base for industrial expansion on a scale sufficient to absorb the inflow of groups of Jews from all over the world.) These were the reasons which spurred the then Prime Minister, David Ben-Gurion, and the Head of the Land Settlement Department, Levi Eshkol, to proclaim the importance of regional settlement in the Lakhish Area in 1954, giving it a priority which enabled the Land Settlement Department to carry out the implementation of an existing overall plan in a relatively short time.

1. General description of the area

The Lakhish region as defined by the settlement plan contains 900000 dunams. Approximately 460000 dunams are arable, of which 250000 are suitable for intensive irrigation. The area is divided into three distinct geographical sections:

1) The coastal strip in the vicinity of Ashkelon, which is a continuation of the citrus belt. Here the soil is light and underground water resources can be tapped by wells. The climate and soil conditions make the area suitable for most types of intensive irrigated agriculture.

2) East of the coastal strip, the soil is heavier and richer, most of it suitable for irrigation, but underground water in effective quantity is lacking. Soil is deep, rainfall low and unreliable. Grain farming is possible, though at risk of drought, so that the limits for settlement on such a basis are low. Erosion is marked in places, although the deep soil means that reclamation is possible.

3) Further east, approaching the Judean foothills and the Jordan border, the landscape is eroded and hilly and the areas suitable for irrigation limited. Grain farming is possible, although the rainfall is unreliable. Grain stubble, on the other hand, can be a useful supplement in feeding pasture animals.

The settlements which had been established in the region some years before the start of the scheme furnished much useful data on rainfall, soil conditions, etc.

By 1965, seventeen years after the first settlement and ten years after the start of the regional scheme in the central area, Lakhish had become a flourishing agricultural region. Its production formed a considerable part of the total national agricultural output, and the planning ideas worked out there greatly enriched the understanding of regional development in general.

2. Lakhish in 1965

Table 7-1 shows the number and types of settlement in Lakhish in 1965.

TABLE 7-1
Farm patterns in Lakhish

Moshavim	32
Kibbutzim and Moshavim Shitufi'im	16
Institutions and administered farms	9
Total	57

The final plan for the settlements of the region includes 4780 family units. At present some 3200 families have been settled.

There are three rural centres, Nehora, Merkaz Shapira and Even Shmu'el. Several other centres are in various stages of planning in areas built before the full Lakhish Regional Plan was conceived and put into operation.

The farm types are distributed in the region as in Table 7-2.

TABLE 7-2
Farm types in Lakhish

Farm type	Number of villages
Dairy	18
Field crops	23
Orchard	3
Extensive pasture	4
Institutional and administered farms	9

The total irrigated area was in 1965 96 000 dunams of which 22 400 are under fruit, mainly citrus. The irrigated field crops occupied 73 600 dunams, divided as follows [20]:

Industrial crops

Cotton	4400 dunams	
Sugar beet	8900 dunams	
Groundnuts	700 dunams	14000 dunams
Vegetables and Potatoes		26000 dunams
Fodder		10200 dunams
Hay		3400 dunams
Irrigated Grains (Cereals)		5000 dunams
Miscellaneous		11200 dunams
Total		73600 dunams

Dry farming occupied 148000 dunams, divided as follows:

Wheat and Barley	90000 dunams
Sorghum	10500 dunams
Legumes	7500 dunams
Hay and Silage	20000 dunams
Green Manure	800 dunams
Fallow	18500 dunams
Miscellaneous	700 dunams

The marketing of milk amounted to 13000 tons, of meat to 450 tons; 22 million eggs and 1800 tons of poultry meat were sold by the villages.

Yields have been continually rising since the inception of settlement. A survey carried out in 1963[21] showed that tomato yields rose during five years (1955–1961) from three to five-and-a-half tons per dunam; cucumber from 1.5 to 2.5 tons per dunam (some farmers reached yields of 3–4 tons); sugar beet from 3 to 5 tons (a number of settlements reached an average of 7 tons per dunam with 17% sugar); groundnuts from 206 kg per dunam to 300–330 kg, with some villages reaching an average of 400 kg; and cotton yields from 200 kg per dunam to 300–320 kg.

Water use in the region in 1965 was approximately 42 million cubic metres, all of it applied by sprinkler irrigation. Of this amount, 39 million m^3 were from the Yarkon pipeline, and 3 million m^3 from private wells in the villages near the coast.

The present water consumption is 60% of the final planned allocation.

3. The planning concepts

Before and during the survey phase, the overall plan was being worked out. Several of the basic concepts which determined the particular form of the plan in Lakhish (discussed earlier in the book) brought home to the planners that regional planning was the only method which could develop an integrated settlement scheme.

The concepts were as follows:

1) The whole region was national land, available for close settlement, without limitations of location by interposing unfriendly areas, as was the case before Statehood.

2) It had been decided to undertake specialized farming in Israel.

3) The breakdown of the mixed farming philosophy led to the breakdown of the idea that each village should be a closed socio-economic unit. This opened the way to regional co-operation.

4) The main economic base of the new region was industrial crops, which demand a large scale of operation, heavy mechanization and special marketing organization, in the framework of smallholder co-operation – the moshavim.

5) It was necessary to base the settlement pattern mainly on the moshav to meet the social demands of the new immigrants.

6) It was necessary to settle each village with one ethnic group only in order to avoid inter-group friction and clash of culture.

7) It was therefore necessary to provide a settlement structure which would enable the gradual integration of the several groups, avoiding a danger of isolationism in the villages.

8) The equal income opportunity principle, which has as its corollary the moral (and practical) need to provide education services of at least urban standard, had to be implemented. Otherwise, the system would not be just, nor would the settlers stay on the farms.

9) It was necessary to provide a suitable community framework for instructors, administrators and service personnel, which would encourage them to live in the region.

10) It was necessary to process the agricultural products locally and thus keep the wealth originating from the industrial crops within the regional complex.

To satisfy these needs and concepts, the regional pattern already briefly described was evolved at three levels, each corresponding to a different function within the overall scheme and variously interrelated. Each of the three levels, the village, the rural centre and the regional town, to a large extent, has a life of its own: each unit is a community with its own structure and life, and each engages in trade and activity with the others.

The villages, mainly moshavim, are the primary units of the production process, based on industrial crops and vegetables and planned physically for the efficient production of these crops within the moshav structure. The housing plan and physical structure of the village are designed to aid the production process and the advancement of community life. The village has the minimum complement of service buildings – office, small clinic with a nurse, synagogue, kindergarten, club room and small co-operative store (grocery).

The eighty families comprising the village are not enough to form a self-sufficient community for certain services and activities both in the social and economic fields. The village has to be included in a wider framework which can provide it with the required services at a reasonable cost and with a high standard.

The function of the secondary unit – the rural centre – is to provide a site for the community and production facilities for which the village is too small. Four or five of the primary units co-operate in setting up the rural centre at a location convenient to all. It offers a meeting-place for different village groups, and a community for the administrators and service personnel working in the centre and in the villages.

The regional town, in the original settlement plan, has two main functions, apart from community services for the town population itself: the processing of industrial crops grown on the farms and the provision of services on a larger scale, greater than which four or five villages can provide at the rural centre.

This system of villages and rural centres is termed 'The Composite Rural Structure'. The system and the town, which serves four of these structures, are now described in more detail.

G. THE REGIONAL TOWN

A vital part of the Lakhish regional plan is the urban centre called Kiriat Gat. Its function, as we have noted, was twofold – to be the site of the industrial processing of the cotton, sugar beet, groundnuts and vegetables produced locally, and to provide the urban services in commerce, administration, education, health, recreation and government which were beyond the scope of the rural centre. By means of this organic connection between the town and the surrounding agriculture, two main objectives would be achieved: a) the population could be dispersed in the rural areas in the specialized farming system based on industrial crops; b) industrial expansion could be achieved outside the large towns. Without farm production, the town would have no basis for its industry, while without its industry, the farms would have no market for their industrial crops.

1. The development of Kiriat Gat

The plan called for an urban centre of some 5000–8000 people[22] engaged in production and services for the town and the region. The town would complement the rural centre in supplementing the social needs of the farm population, and the balance which is so essential between town and country would be artificially created in a very short time without having to wait for the completion of a protracted natural process. The plan, through the physical relationship it created between village and town, automatically provided the balance.

The town began functioning in December 1955 when the first immigrants arrived and were settled in huts and soon afterwards in houses. Immigration pressure was so great, however, that some immigrants had to be placed in temporary hutments where some of them stayed for two or three years. Those who had not moved out using their own resources, were moved into flats built especially for the purpose. Other small housing was built for new immigrants as a temporary stage until they could save enough to move into a larger flat. Conditions for home purchase were very easy, both for families moving to a better style of house, and Israelis coming into the town to staff the offices and the technical posts in the factories which the new immigrants could not fill.

The majority of the population in the first two years came from Morocco and North Africa, mostly unskilled people who had to be taught a trade. From these countries came as well people skilled in office work, banking and administration, who quickly learned the language and settled down in productive employment. Immigration from eastern Europe brought skilled artisans, business people and a few technical and professional workers, together with many old and unskilled settlers, who found conditions hard. A scheme to attract immigrants with small capital from England and the Western countries was widely publicized but brought few families who actually made the town their home. Farmers who could not settle in the villages for various reasons were given houses in Kiriat Gat. The town soon had a population composed of people from some thirty different countries, ranging from India and Burma to South America and South Africa, all the countries of the Mediterranean, Europe, Asia and a large number of Israelis. Skilled people leaving kibbutzim found an opportunity in Kiriat Gat for good cheap housing, and many found work connected with agriculture. In the first years, this pioneer element was important in getting the town off to a good start.

This population, from so many different lands and cultures, had to build the town and community from scratch. The Ministries of Labour and its Housing Division, the Jewish Agency through its Settlement and Absorption Departments, the Ministries of Development, Trade and Industry, Welfare and others, co-operated to plan and establish the town through an inter-ministerial committee and a local administration. At first, the head of the Settlement Department Regional Offices also headed the town adminis-tration; he was later replaced by the elected mayor of the newly formed municipality. The municipality then took over the running of the town in close co-operation with the Ministries, which undertook various investment programmes. Great emphasis had to be placed on social and community work in order to unite the various groups in building the town, and avoid antagonism which could all too easily have arisen under the raw conditions of a new town of so many elements.

In accordance with the plan, the first housing units were built with a fair-sized plot of land to serve their inhabitants as a secondary source of income. But the rapid influx of immigrants into the town in the first few years completely changed the character of the housing. Instead of a target of 5000 inhabitants within five years, possibly reaching 8000 later on, the larger figure was reached already in 1960. Two- and three-story apartment buildings replaced the rural atmosphere and the town landscape changed immediately. A policy of putting immigrants directly into their permanent homes was adopted as far as possible. As the number of Israelis increased and gardens were developed around the flats, semi-detached houses and in public places, the town became greener and more attractive. The commercial centre with some 57 shops and community and office facilities opened and the raw edges became somewhat smoothed off. In 1962, the population was 12000; in 1965, 17000. Growing emphasis was placed on landscaping and public facilities, because the town was trying to retain as many of the arriving immigrants as possible and also to attract veterans. The success of the town, the rapid growth of the surrounding villages and the continuing immigration led to a revision of the original plan of 8000 population to a forecast of 80000. The commercial centre and community facilities were to be extended accordingly, and the architectural plan was being brought into line with the new situation.

2. The position in 1965

The town's progress report for 1965 stated that the town's population rose from 2750 families in May 1962 to 4300 families in 1965 with a total popu-lation of nearly 20000. New housing units were in preparation for new immigrants or veterans over the coming years. The school programme was correspondingly being enlarged. The number of primary schools in 1965 reached seven; two more were under construction. The number of pupils in the seven schools was around 3800. There were two secondary schools with some 380 pupils. Similarly, clinics, public playgrounds, parks, green spaces, trees and other facilities were being expanded. An internal road system was being completed and extended, and the commercial centre was enlarged to include 50 shops; the construction of 70 more shops was completed in 1965 as well as a large public hall with bank and office buildings; an additional cinema, a swimming pool and various public buildings. Sports facilities were planned for the town.

A Development Company was set up in 1962 to encourage industries of many types to come to the town. The Company acts as a lever in the town's development in place of Government help which, now that Kiriat Gat is on its feet, is going to younger or weaker towns. The company has brought an electronics factory to the town, has established together with a Development Company of some of the villages a plant for conserves, has participated in the construction of a cone-making plant for the textile industry, and is in contact with other firms which express interest in the town.

The textile chain has been expanding in the number and type of firms. A large wool combine, to employ eventually a thousand workers, has started work; it now employs 500 workers. Small factories, employing thirty or so workers, have been set up in different kinds of activity; the number of workshops for carpentry, metal work, garage repair, laundrying, etc. has increased. A plant for the grading and packing of flower bulbs, a new and successful branch on the farms, has been operating since 1961. Potato grading for the region has been firmly based in the town in conjunction with the cold storage plant. It is expected that the number of similar service industries based on the agriculture of the region, will be increased as the farms increase and diversify their output in field and vegetable crops.

The report shows that the town's budget grew from IL. 639000 in 1959/60 to IL. 729000 in 1960/61, to IL. 1028000 in 1961/62, to IL. 1240000 in 1962/63, IL. 1800000 in 1963/64 and IL. 4000000 in 1965/66.

Total investments in 1965/66 reached IL. 27.2 million out of which IL. 13 million was in housing.

3. The growth of employment

At first, the rapid increase in the population outstripped the investment programme and a large part of the labour force was dependent on relief works for minimum subsistence. Forestry, road construction and various other public works projects were carried out in the surrounding area and the town itself. Many of those in relief work were old or unfit people who nevertheless had families to support, from whom only a minimum of work was asked, but the problem of providing steady work for the able-bodied and capable among the workers was severe. The unemployment was exaggerated by the very newness of the town; industry did not want to venture into the struggling young area, but preferred to wait for someone else to make a start in the provision of roads, services, homes and a community for the skilled and professional staff, without which an industry cannot work. The lack of developed artisan shops with a known record was also felt. Firms weighed the alternative of staying where they were or moving to the new zone, where all was uncertain. The efficiency of the local administration, the adequacy of the services and the ability of the labour force were all in doubt.

The Government, through its development budget, encouraged private investors to set up firms in Kiriat Gat. It participated in the original investment by up to 70% for an approved project. The response at first was not large. A cotton gin began work in the town, followed by a cotton fine-spinning mill. Together they began the textile chain of industries. A celluloid tape factory, an ice factory and cold storage plant, a transport company, a diamond polishing plant and a few workshops were established.

In 1959 an economic survey was carried out to provide background material for the architect commissioned to plan a housing suburb.

The economic survey gave the details in Table 7-3 of the employment position in the town.[23]

The decision to locate the third sugar factory in Kiriat Gat gave a great boost to the confidence of the town and encouraged other investors, who realized that such a large industry would bring further development in its wake. The decision was taken in 1958 and was significant not only because it meant that the town would get an important and badly needed source of employment, but also as the commencement of a new chain of industry and food processing. The survey included this in its analysis of the possible industrial future of the town, and predicted that at the stage when the town would have a population of 12000, the labour force would stand at approximately 3240 of whom 2500 could be employed in industries based on the agricultural produce of the region and in services for the town and the surrounding rural population. Some of these industries were discussed in the survey. It was pointed out that the other 740 jobs would have to be provided by industry not based on the agricultural potential of the region. It was

TABLE 7-3

Work position in Kiriat Gat in February 1959 by firms and professional structure (not including relief works)

Place of work		Man-aging staff	Techni-cal staff	Admini-strative staff	Super-vising staff	Semi-skilled	Un-skilled	Total	% Administrators of total labour force
Spinning mill		10	13	14	20	263	40	360	16
Gin		1	3	2			1	7	
Weaving		1	2	1		20	2	26	16
Diamonds		1		1	1	20	2	15	20
Tape		1	2	1	1	10	1	16	31
Bakery		1				10		11	27
Ice					2			2	100
Immigrant furniture		1		2	2	15	10	30	17
Workshops			26			4	15	45	58
Settlement department offices			19	18		9	4	50	
Town council		5	2	12	5	5		29	
Schools		3	46	3			10	62	
Police		1			1	30	1	33	
Offices and institutions				20				20	
Building		3	8	5	12	50	60	138	20
Commerce	90							90	
Various	105							105	
	195	28	121	79	44	426	146	1039	

recommended that a development body, composed of citizens of the town and the Town Council, be set up to seek out industry and bring it to the town, offer it all possible help in overcoming the difficulties of establishment and also back it financially. The preparation of a long-term development plan was recommended to serve as a base for the activities of the development body, and as mentioned, a Development Company was set up in 1962.

During 1959, the pace of employment provision in relation to the population growth increased. Immigrants were still arriving in numbers, and some of them began to enter trades and to follow their old skills. The temporary shopping centre moved to the commercial centre and the municipal offices and banks took over permanent buildings. Other offices opened, and the administration and commerce of the town became firmer and more reliable.

The number of private workshops increased as the building programme, including the sugar factory, gave an air of promise to small enterprise. Firms started to express interest in the town as a possible location, since the objective conditions there and the increasing difficulty of obtaining sites and labour in the big towns were giving a powerful incentive to the national policy of dispersion of industry.

In 1960, with the further advances of more workshops, three weaving factories, construction of the sugar factory and the housing programme, employment was created. The figures for December 1960 showed 600 working in industry, 140 in agriculture in the settlements and institutional farms around the town, 150 in services, 150 in business, 200 in building and 110 in other work inside and outside the town. There was a total of 1500 *permanent* work places for 2500 work seekers from 2475 families. The position was as follows:

Permanent work	1500
Relief work	450
Old and handicapped workers	150
Army	150
Welfare	150
Training	30
Others	170
	2600

The planning position at that date, however, showed a very different picture. The expansion of industry and services corresponding to the investment plans of the firms coming to Kiriat Gat meant that a shortage of labour was forecast for the time when plans would materialise. The balance, as estimated in December 1960, was:

Work places under construction		1070
Expected additions to the labour force	550	
Additions to labour force through special effort in recruitment	250	
	800	800
Estimated future shortage*		270

* Note that only 800 of the 2600 labour force detailed in the previous table were available for recruitment (relief work, army, training and others).

In a five-year period distinguished by a grave employment problem, the town's development made an all-out drive to receive more immigrants and housing units, and to attract more veteran Israelis. The policy at first was to take people to the towns and then worry about providing work conditions. The boom in development was, however, influenced by the concentration of population. Industries sought a cheap location, government help in investment and the abundant labour supply which government helped to train. This, in Kiriat Gat, led to the situation of the predicted labour shortage described above.

The sugar factory began operations in April 1961, and by the end of that year there were 2450 permanent workers of 2850 in the work force; the beginning of 1964 saw the following division of labour:

Industry	1100
Agriculture	250
Workshops	400
Services	550
Trade	250
Building	250
Seasonal work	400
Others	150
Sub-total	3350
Relief work, old and handicapped workers, others, Army, Welfare, training, etc.	400
Total Labour Force	3750

At the end of 1965, the total labour force reached 4900 out of which 1970 were employed in industry.

Once an industrial nucleus had been formed, the town grew of itself, so that government participation in investment and financing of approved projects could be severely cut and transferred to other centres further south, though there continued to be heavy government investment in housing and related services. Such development, in no small measure, was due to the inclusion of the town in the Lakhish regional plan. Not only sugar beet, but new crops introduced on the farms offered industrial possibilities, particularly in vegetable processing, the grading and packing of bulbs, potatoes and other produce for the home and export markets. The growing cotton in the villages gave the start to the textile chain through ginning, spinning, weaving and eventually to garment manufacture. Agricultural industry could continue to grow with other industries grouped around it. By including a good-sized town in the region, industrialization could begin.

H. THE COMPOSITE STRUCTURE IN PRACTICE

1. The Rural Centre and its Villages

In order to better understand the regional programme, it will be useful to focus on one rural centre and its villages – Nehorah, one of the two working rural centres in the Lakhish Region. It is situated on a hill overlooking the seven villages, five of which make full use of its facilities, and two only partial use, such as of the school. Of the five villages fully attached to Nehorah, the closest is half a kilometre away, and the furthest approximately three kilometres. Each of the five villages was planned as a community of eighty families and to be culturally homogeneous. At Noga, the settlers are from Kurdistan; those at Otzem come from the Atlas Mountain area of Morocco; Zohar is composed of immigrants from Tunisia; the Shahar settlers are from various parts of North Africa; and those at Nir Hen are mainly veteran Israelis. The degree of homogeneity varies: the settlers at Otzem are from the same community in Morocco, while those at Shahar come from different towns and regions. These variations in composition are important factors in determining a village's development, as we have seen from the chapter on the 'human factor'. Each village includes a local synagogue, grocery store, nursery school and infirmary. These services must be in each village, since shopping at a regional store is too cumbersome (although it might be cheaper), since separate synagogues are necessary for each ethnic group, and since a local infirmary is needed to screen patients and deal with less serious cases on the spot.

From an agricultural point of view the crop programme in each village is roughly similar: it is based on intensive irrigated cultivation. Industrial crops, such as cotton, sugar beet and groundnuts, as well as vegetables, are the basis of the crop programme.

The groups of settlers who came to the Nehorah complex were ill-prepared to become farmers. With the exception of the Kurdistani settlers, some of whom had formerly been shepherds, none of the immigrants had had any farming experience. The settlers were first employed as farm labourers, under a management, in order to avoid placing them immediately within the independent farming system, in which they were likely to fail. The land of each village was rented to a contractor, who guaranteed to employ the settlers in agricultural work. In addition, the villagers were allocated a three

or four dunam plot behind their homes to raise vegetables for home consumption. A team of Israeli instructors was sent to each village to be responsible for teaching farming skills and managing the village's financial
system. This administered farm system was a temporary arrangement: as the
villagers gained skill and confidence, the land was transferred to them. On
becoming independent, each settler received two irrigated plots of land, one
of about eight dunams and the other of about sixteen dunams. Poultry, beef
or sheep and citrus branches were added to the farms. The transition from
an administered system to independent farming varied between the five
villages: in three villages the settlers asked that the land be apportioned after
two years of farm experience, while in the other two villages the administered
system lasted for four years. Thus it was only a relatively brief period of time
before the villagers became independent, modern producers.

Similarly, each community developed as a separate sociological unit: just
as the villages became agriculturally independent according to their own
pace of growth, they also developed their own institutions in different ways.
Their rate of acculturation measured in terms of the adoption of new
practices, or the assumption of 'Israeli' ideals has varied. For example, the
assumption of new co-operative practices has been more easily attained at
Otzem, where the settlers are from rural Moroccan areas, than at Noga,
where they are from Kurdistan. The villages have maintained their distinctive
cultural traditions; although modified to accommodate to the new conditions
in technology and organization, habits and ceremonies characteristic of each
community have remained dominant. The villages have also developed indigenous organizational leadership: in each community, elected officials participate in planning and directing village affairs. Although village instructors
are still assigned to each community, responsibility has increasingly been
assumed by the villagers themselves.

The rural centre of Nehora contains both technical facilities and homes
for the staff who operate them. Along one edge of the centre, a regional
school, clinic and cultural centre were built. The school, which serves the
surrounding villages, has an enrolment of over five hundred pupils, from
grade one to eight; evening classes for teen-agers and adults are also conducted at the school. The educational complex includes a library, gymnasium
and dining hall. A cultural centre seating six hundred persons was built
nearby: films are shown each week, and the hall has accommodation for
plays and concerts. The regional clinic is staffed by a resident doctor and
trained nurses; it contains modern technical equipment. Other medical
specialists – pediatricians, psychiatrists, etc. – regularly see patients at the

clinic. A regional tractor station with full maintenance facilities and a vegetable grading shed were built in another part of the centre. The modern tractors and combines which serve the area are located and serviced at the tractor station; vegetables from all the surrounding villages are brought to the grading shed where they are graded, packed and shipped to market. Other services in the Centre are: a Bank, regional Council offices, a supply store and two mechanical workshops.

Homes for twenty-five families were built at Nehorah; each home includes two rooms and a kitchen, and the plots have ample room for gardens and fruit trees. The homes are occupied by the resident specialists – the doctor, nurses, school, principal, teachers, mechanics, tractorists and other technicians. Most of the families living at the centre are veteran Israelis; although many originally came as volunteer participants in the new project, Nehorah has become the permanent residence of several of them.

While the villages have grown as independent units, the ties to the rural centre bring people from each village into regular contact with one another. The children from all surrounding villages and the centre itself attend school at Nehorah. The school is the centre for other activities as well: parents' committees meet with teachers, and holidays are occasions for general school-centred celebrations. Excursions, summer-time activities and special-interest groups and evening classes are organized by the school staff, bringing together the settlers from the different villages. Large in size, and with a permanent staff, the school is an integral part of the entire development programme. The clinic likewise functions as a point of contact as do cultural events, such as plays, films and lectures. During the summer of 1959, Nehorah was the site for a regional 'county fair', in which settlers from all the villages participated; there were various activities, including ploughing and other contests.

2. The composite structure's economy

While it is too soon for full economic evaluation of the composite village structure based around Nehorah, some preliminary conclusions can be drawn. On a village level, the predominantly successful adjustment to farming is noticeable. Progressive agricultural extension methods have produced satisfactory results; the settlers take a keen interest in new crops and techniques, and some have already invested savings in agricultural equipment.

The rural centre has given technicians and agricultural extension workers their own community near their work. This was of great importance for attracting a permanent staff of trained persons, who could live and serve the new community without a feeling of intellectual isolation. The centre is a natural focus for group extension activities with facilities that could not have been provided for isolated villages.

Table 7-4 indicates the success of extension as manifested in increased agricultural production. The yields reflect only those of the villages established as part of the composite structure.

TABLE 7-4

Comparison of average annual yields (1956–1965)[24]
Yield in tons per dunam (0.427 acre)

Crop	1956	1965
Tomatoes	3	5.5
Cucumbers	1.5	2.5
Potatoes	1.5	2.5
Sugar beet	3 (15.5% sugar)	5 (17.5% sugar)
Cotton	0.2	0.33
Groundnuts	0.2	0.29

Table 7-5 shows the rising of farm incomes and the reduction of supplementary income through relief work in the five clustered villages.

TABLE 7-5

Estimated composition of average family farm income[25] (1955/56–1964/65) in IL.

	1955/56	1956/57	1957/58	1958/59	1959/60	1960/61	1964/65
Average farm income	600	900	1700	2400	3000	3200	5100
Administered farm wages and relief works	1000	1500	1000	500	100	–	–
Total average annual net income*	1600	2400	2700	2900	3100	3200	5100

* This income is after the deduction of direct expenditure only; indirect expenses like depreciation, interest on investment and village taxes were not deducted from total value of output.

The concentration of services in the rural centre is already showing technical and cost advantages. Agricultural equipment, for example, serving 400 rather than 80 families, provides a larger range of modern equipment than any one family or village could maintain, and at a lower cost.

Since, however, one tractor station is insufficient to attract and pay high quality management or even provide a long enough season to employ high cost equipment, a company has been formed to organize decentralized tractor services for several regions. Equipment can, if necessary, be transferred for short periods from one region to another. Such interregional relationships show definite cost advantages, while the large scale of this mechanization body, enables it to experiment in new techniques to fit the special requirements of its small farm customers. Amongst other operations, sugar beet harvesting, vineyard cultivation and citrus spraying have benefitted. The company has become a major supplier of cultivation services, with a steady price fixing policy based on real costs which effectively lower the price of cultivation and smoothes out excessive seasonal prices. The competition between the company and private contractors ensures healthy costing.

Centralizing the grading and packing facilities within the rural centre has improved the quality of produce marketed and has lowered costs considerably. These are a result of the much larger turnover within the central station as compared to one in a single village, more suitable investments in layout, buildings and machinery, and the utilization of trained staff working over an extended season. A comparison of grading costs made in 1961 between a single grading station serving new villages, showed a difference of IL. 3 per ton to the advantage of the latter.[26]

The single pick-up point enables efficient investment to lower loading costs, while the large output means a high percentage of full lorries to market with consequent transport saving. Since the station is not in the village and is professionally managed, grading is free from the pressure of farmers to accept sub-grade produce. This has meant better wholesale market prices for the produce of central grading stations.

The need to maintain both social and economic services on a high level involves considerable overhead expenditure. Such overhead, when spread amongst farm families, has a direct influence on the level of the farm income. The problem therefore is how to organize rural life in such a way that the service costs are not an impossible burden on the farm family. In 1957, the average level of taxes of a number of older villages was examined.[27] The average municipal and social tax level in the sample villages examined was IL. 1336 per farm unit annually. The average municipal and social tax level of the new villages served by a rural centre in the Lakhish Region in 1960 was only IL. 338 per farm family annually.[28] Table 7-6 gives a detailed breakdown of the different costs per family.

This table shows only the order of difference since it is not based on an adequate sample, nor on similar farm types, and can therefore point only to preliminary conclusions. The older village pays more into national insurance and to the sick fund, partly due to the higher gross farm income level. It also enjoys a higher standard of services. However, if the appropriate

TABLE 7-6
Comparison of taxes (1960 prices)

Old-established village		New composite village	
	IL.		IL.
		Village taxes directly collected and a	
Village taxes	480	2% commission on market returns	158
Other services	314		
Sick fund	415		108
National insurance	127		48
		Participation in school feeding	24
Total	1336	Total	338

equalization in gross farm income and in service standards is made, family taxes for covering costs in such a new settlement remain very much lower than in the older village. An estimate was made that the tax level would reach IL. 610, excluding property and tenure taxes. The difference can be illustrated still more clearly by adding depreciation costs, interests on circulating capital, property taxes, etc. Table 7-7 shows an annual amount of IL. 1326 more per family in the older village.[29]

TABLE 7-7
Comparison of depreciation, taxes and interest (1960 prices)

Cost item	Old-established village	New village
	IL.	IL.
Depreciation	1000	700
Municipal and social taxes	1336	610
Property and tenure taxes	110	110
Interest on circulating capital	400	100
Total	2846	1520

Regarding depreciation costs, this table shows a IL. 300 difference per family in favour of the new composite village. The difference of IL. 300 per family in interest charges comes from the differences in total annual village expenses: IL. 30000 for the older village as against IL. 11000 for the new. While the villages are not fully comparable, since different farm types are involved, the order of the figures points to significant savings in service costs.

The cost advantage of service concentration is a factor that is encouraging kibbutzim to form either their own centres or to participate with moshavim in the establishment of rural centres.

From a social viewpoint, the existence of a rural centre has lessened the settlers' sense of isolation. The rural centre and its activities have strengthened their participation in the entire settlement project. It has become the meeting place for the leaders of the different villages, who tend to discuss common economic and organizational matters, a fact rapidly leading towards district local government. A similar process can be observed regarding religious matters. In one of the centres, religious seminars are regularly held, with the participation of the religious leaders of the different communities.

The regional school functions as a centre breaking down ethnicity through its various activities. Although cultural fusion between the groups of the immigrant generation proceeds slowly, there is evidence that it is happening in the second generation. It is too early to draw conclusions, but the evidence does suggest that school play-groups are not based on ethnic factors, and that bonds of common Israeli attitudes do join the young.

As with centralizing of equipment and other agricultural services, experience indicates that centralizing cultural and social services is far superior to the earlier system of village autonomy. The regional school provides better facilities, at a lower price, than local schools: not only are the classes larger and the equipment more adequate, the teaching staff has become permanent and conducts school-wide, co-ordinated programmes. The improvement in the health service is striking: where the doctor used to visit a village twice weekly, he is now continuously available. Functioning as a permanent staff, the medical team has developed a sustained family and community health programme.

Composite planning and village cluster development have stabilized rural life for these communities. Only 17% of the families settled in the composite villages have left agriculture, of whom 7% left in the initial settlement period before the farm homesteads were built and the regional infrastructure developed. This percentage of transfer is much lower than the national average for settlement, which is about 31%.[30] Part of the success in integrating newcomers can be attributed to the method by which families were introduced to agriculture. The initial administrative farm system reduced early fears and insecurity, and offered settlers an opportunity to learn the intricacies of intensive irrigation farming without the expensive economic and social consequences of trial and error. In this way the crisis of adaptation was reduced. However, there is no doubt that the composite structure as a whole, played its part in the integration.

3. The development of an individual village

To illustrate the growth of the villages more clearly, the economic development of one in particular will be described.

The village of Noga was founded in 1955 as one of the villages of the Nehora regional complex. It was planned as a field-crop farm with eighty farming units on a total area of 4090 dunams. The structure of each unit for the final stage of development was planned as follows:

	Irrigated Dunams
Plot near the house	2.5
Main plot for irrigated field crops	22.5
Citrus	3
Dry farming	15
Total	43.0

The water allocation for the final stage was to be 15100 m³ per unit and 1 208 000 m³ for the village. The maximum amount which the pipe network was to carry in any twenty-four hours was 7000 m³.

Noga formed part of the administrative farm system previously described. The newcomers came from transit camps and temporary quarters into which they had moved on their arrival in Israel from Persia and Kurdistan. At Noga, they lived first in prefabricated quarters and worked for the building company on house construction, then worked for a company as agricultural labourers on the lands which were to become their own.

1957 was the first year of independent farming, and the regional farm planning section followed up the year's activities with an economic survey, quoted below[31]:

Population

53 families arrived at Noga in 1955 and another 14 in 1956. In 1957/58, the 67 families had a total population of 362 (81 men, 90 women and 191 children). During the years, six families left the village, to be replaced by sons returning from the army. Of the 67 families in 1957/58, six did not work their farms.

In the autumn of 1957, the 61 farmers worked 629 dunams of irrigated land. The total crop area in that season was 306 dunams. In the spring, each farmer was allocated more irrigated land, to bring the total up to 1259 dunams; 706 dunams of crops were grown in the spring–summer season. A noticeable feature of that year was the large number of settlers who worked in partnership with other settlers in the village. Most of the partnerships broke up in succeeding years.

Table 7-8 shows the area of each crop planned for the village by the farm planning section, and the area actually sown by the settlers in 1957/58.

TABLE 7-8

Planned and actual areas sown in Noga (1957/58)

Crop	Autumn/Winter Sown	Spring/Summer Planned	Spring/Summer Sown	Total Sown
Potatoes	156	130	142	298
Tomatoes	78	65	84	162
Cucumber		65	49	49
Squash			40	40
Beans		65	44	44
Green onions	79			79
Dry onions			37	37
Miscellaneous	15	65	49	64
Total vegetables	328		445	773
Cotton		260	124	124
Groundnuts		130	124	124
Total industrial crops		390	248	248
Fodder	—		16	16
Total crops	328		709	1037

The table indicates the emphasis on vegetables rather than industrial crops. This was because the farms were very small and vegetables used the labour of the family intensively to give a fair income. Some farmers grew vegetables over and above those entered in the plan. These were vegetables which were unrestricted in quota, or which were not subject to a minimum price.

Average yields are given below:

Cotton	230	kg/dunam
Groundnuts	284	kg/dunam
Spring potatoes	2	tons/dunam
Autumn potatoes	0.8	tons/dunam
Tomatoes	5	tons/dunam
Cucumbers	2	tons/dunam

Some of the settlers sold their produce outside the regular organized marketing framework. Because they received credit from the Settlement Authority, almost all production expenses are known. An estimate was made of the produce marketed privately. Incomes for the village as a whole, and as an average per member, are given in Table 7-9.

TABLE 7-9

Output, expenditure and income for the village (1957/58 current prices)

Organized Marketing			Expenditure	
Crop	Quantity	Output		
	kg	IL.		IL.
Autumn potatoes	123399	19928	Credit used for production	37094
Autumn tomatoes	106933	16834	Water cost	33394
Green onions	189025	12545	Taxes on production	5078
Miscellaneous		4168		
Total from autumn crops		53475	Total expenses	75566
Spring potatoes	276952	34610		
Spring/summer vegetables		56321		
Cotton	28540	27908		
Groundnuts	13066	7142		
Total from Spring/summer crops		125981		
Total		179456		

1. Farm income	IL.	Per member (61 members) IL.
Total income from organized marketing	179456	2940
Total expenditure	75566	1239
Income	103890	1700
Estimated addition through private marketing	24010	400
Yearly total output from farms	203466	3335
Total income from farms	127900	2100
2. Outside Employment	24018	400
Total Net Income	151918	2500

Out of the 61 working farms, four had a net income, including income from outside work, of less than IL. 1000; 21 reached incomes between IL. 1000 and IL. 2000; 36 reached incomes of over IL. 2000. Of this latter group, eight farmers earned more than IL. 3000, and four more than IL. 4000.

If employment outside the farm is not counted, then 25 farmers earned more than IL. 2000, while 5 suffered losses.

In 1958/59 more sons took up farming in the village and there was a considerable increase in crop area. The area under each crop, the yields and turnovers are shown in Table 7-10.

TABLE 7-10

Output, expenditure and income (1958/59)

Crop	Area (dunams)	Yield (kg/dunam)	Total turnover IL.
Cotton	106	303	35000
Groundnuts	356	300	63704
Sugar beet	105	3500	19053
Autumn potatoes	146	1000	35853
Spring potatoes	138	2000	33486
Vegetables (autumn)	128		38902
Vegetables (spring)	598		135138
Total	1577		360697

Expenditure	IL.
Water (805000 m³)	44275
Materials	45578
Tractors, etc.	17000
Other costs	32000
Total	138853
Outside work	17785

Net income	Village	Per farm (69 farms)
From Agriculture	221844	3215
From Outside Work	17785	257
Total Net Income	239629	3472

1958/59 thus showed a very satisfactory increase in yields, turnover and farm income. The introduction of sugar beet proved fairly successful, although yields were rather low. The efficiency of water use improved. Taxes for the year, including National Insurance, sick fund and village taxes were IL. 300 per member. (Income tax is not paid in the villages in the first years.)

By 1959/60, the number of families had reached 77, of whom 74 were working. The irrigated area of the village reached 2436 dunams, of which 276 dunams were under citrus planted 1956/57. In addition, the village had 1331 dunams of grain land, 173 dunams non-agricultural land, including roads, etc., and 150 dunams for farm yards and houses making up the total area of 4090 dunams. This area remained stable.

The village plan, drawn up for the year by the Farm Planning Section, totalled 1080 crop dunams; actual plantings reached 1962 dunams. The members planted more cotton and miscellaneous crops than were allowed for in the plan. On the other hand, the area under groundnuts, sugar beet and some vegetables was less than planned.

In 1959/60 poultry was introduced to the villages. Each farmer was allocated 300 day-old chicks on the settlement budget. This branch proved successful. Attempts in earlier years to introduce sheep had not been as successful, with the exception of a few members who built up fairly large flocks.

In 1960/61, the turnover of the village passed the one million pound mark, and the village took on the appearance of prosperity. In the year 1961/62 irrigated crop areas as shown in Table 7-11 were sown. [32]

TABLE 7-11

Irrigated areas (1961/62)

Season	Crop	Area
Autumn	Potatoes	552
	Vegetables	839
	Sugar beet	59
	Other crops	50
		1500
Spring/Summer	Potatoes	117
	Vegetables	362
	Cotton	140
	Groundnuts	167
		786
Year's Total		2286

In later years the total cultivated areas remained stable, in 1964/5 the irrigated field crops reached 2295 dunams, the non-irrigated area was the same as in 1960 – 1300 dunam.

In 1964/5, crops as shown in Table 7-12 were sown. [33]

TABLE 7-12

Total crop area (1964/65)

Irrigated	(Dunams)
Vegetables	873
Autumn potatoes	502
Spring potatoes	105
Sugar beet	380
Groundnuts	100
Irrigated grain	320
Other crops	15
Total irrigated field crops	2295
Citrus orchard	285

Non-irrigated crops	Dunams
Winter grain	630
Fallow	670
Total	1300
Total cultivated lands	3880

In addition one million eggs were marketed and the sheep herd numbered 500 head of ewes.

Noga has to rely considerably on hired labour, particularly in certain seasons of the year. This enables the turnover to increase, but reduces the percentage return per crop dunam.

4. The other villages of the sub-region

The other five villages which, together with Noga make up the sub-region around the rural centre of Nehorah, did not develop quite as rapidly or reach the same yields and intensity of farming as Noga. However, their progress has been on similar lines over the years. The irrigated land area has gradually increased in all of them as in Noga, and the crop patterns have been more or less the same. This is true as well of the two other villages which use Nehorah only partially.

The development of the five primary villages in the Nehorah complex is fairly typical of that of all field crop farms in Lakhish generally, although their average output is higher.

5. Future development

The major part of physical planning in Lakhish has been concluded. The villages are built and functioning. Each year the crop area increases, yields go up and the farmers become more capable of directing their own farms. More and more, the villages are taking their affairs into their own hands. Through elected committees and representatives they are releasing themselves from dependence on the administrative services of the administered village system. Agricultural instruction is changing its emphasis, and the villages are fully integrated into the extension service of the Ministry of Agriculture. Incomes are rising; the settlers are rapidly approaching the stage of full village independence when they will sign the contract with the Settlement Authority and start to repay their loans.

It is necessary to consider what the task of the composite structure will be when the guiding hand and purse of the Settlement Department are less evident and the area runs itself on local authority lines: what will be the nature of future endeavour?

There should be two main lines to the continuation of the work: one furthering the development of the physical and organizational frame of village and rural centre in order to keep pace with changes in agricultural practice and the needs of the market. The other line should be the fostering of community development on both village and regional level.

In both these lines – in the improving physical facilities and in the strengthening community ties – it is the settlers themselves who will take far more initiative than in the first stage of settlement. The Settlement Authorities still give active monetary and advisory support to the development of the villages and the rural centre, but much future growth will undoubtedly be through the work of the local councils.

When one considers further improvement in the physical facilities of the sub-region, it has to be remembered that they were laid out with specific functions in mind, which today are being carried out, but which must expand in the future. Small agricultural villages were built, looking towards a rural centre for their social and economic services. The aim was mainly to gain cheaper and better services and provide a means of bringing culturally distinct groups together and eventually achieving integration. If the future of the region is to be assured, the centre must not stagnate at the level determined for it in the original plan. As the plan unfolds, new functions become revealed and should be followed up, so that the structure can meet new demands, the most important of which is the absorption of surplus manpower from agriculture.

The original plan did not include an extended economic role for the rural centres. The tractor station, repair services and centralized agricultural supply stores were the first conceptions. Later, the difficulties of marketing vegetables of several different sorts, in varying amounts and from many small farmers, led to the building of a grading station at the centre connected to an ambitious plan of pre-packing and processing, some at the centre itself, some in the regional town. At first, only the grading stations were operated. Because the rest of the programme was not fulfilled, they were not a real economic attraction to the settler to bring his produce from the village to the centre. Nevertheless, the higher standard of grading fetched a higher price for the produce, and the saving on transport to central markets enabled some compensation to be paid to the settler for the greater distance to the central station. Plans are now in process to extend the grading operation and introduce vegetable processing, cold storage and possibly packing in the centres.

At the same time, small industrial activity connected with agriculture, which will strengthen the farms and increase the wealth of the region, is being sought. Since the farms of Lakhish are based on industrial crops, and the livestock carrying capacity is limited to a small number of sheep, the farmer, whose income is rising each year, has very little opportunity for investment within his farm. On the other hand, the things he produces undergo processing in factories in the rural centre or in the town. If the farmers can be convinced of the soundness of investment in these enterprises, then the rising amount of money available for investment which cannot find an outlet on the farms, can be gathered and used to develop industry, raise the demand locally for the farm products, and give a strong stimulus to the introduction of new crops with industrial potential. It will also provide the settlers with a secondary possession to divide among those of their children who will not inherit the farm.

As for the strengthening of community ties of the sub-region around the rural centre, a stage has been reached in which there is need of professional social scientists, both to determine the possibilities of social ties, and to activate them.

Many villages have successfully built up a flourishing community life and today run first-class co-operatives or collectives with a high standard of service and a satisfying community life. This applies to villages settled by immigrants of all the ethnic groups which have come to Israel. But Lakhish represents something different, which has yet to be fully exploited – a fuller life through regional organization. The possibilities have to be explored, discussed and developed before the true measure of success of the Lakhish experiment with its several different ethnic groups 'and common service centres can be evaluated. Today the villages are functioning as agricultural units and the centre has been successful as a provider of services at a level which no other organization would have made possible. The main test of the scheme is still in the future – whether this particular structure will develop into a region with a fuller life than that possible in the traditional self-contained village. If, for example, its youth which cannot be absorbed on the farms, remains, making its home there and engaging in other activity besides farming, one of the tests of the future will have been met.

There is no doubt but that the approach of Community Development, the applied social science, can be helpful in strengthening the social ties of the sub-region, even though until now it has been concerned primarily with activating the small group on the village level. Like the sub-region itself, Community Development is new in Israel and has not yet been fully exploited. Nor are there as yet enough trained professionals for the work.

I. REGIONAL ADMINISTRATION

The administrative structure of the development region, with its various cells and co-ordination in the planning of settlements in the region, are to a large extent dependent on the human factor. This factor, with all it involves – such as regional co-operation and co-ordination, has a decisive effect on the administrative structure, which is delineated, in its form, by the Settlement Authority. Discussion of the administrative structure is therefore preceded by consideration of the human factor in its administrative context.

1. The human factor in the region

i. *The human basis of co-operation*

The planning and implementation of development plans, without resorting to totalitarian methods, is achieved on the one hand through the discreet actions and interventions of Government and on the other through the formation of community and co-operative structures, through persuasion rather than imposition. Community and co-operative structures come into existence through volunteer action stimulated by outside pressures which can be of an economic, social or physical nature.

A community cannot be established through formal planning or organization alone. The social fabric of a community is mainly determined by direct personal relationship and mutual understanding. Therefore, the term 'community' cannot be applied to groups of people whose only common link is the supply of economic needs in a routine impersonal manner.

When a community leads people to extend their mutual co-operation beyond the limits of spontaneous and informal actions, the range of community spirit and action grows into a more formal co-operative structure. If, for example, a group of farm people meets economic deficiencies through the pooling of efforts for common services, working in intimate personal co-operation, the formation of a rural community is achieved.

One result of such community formation is the rise of accepted, influential leadership. Such leaders develop from the community, and are characterized by strong personalities which retain original ethnic elements but can cope with adjustment to the new conditions.

The size of a co-operative community for agriculture will vary according to needs. If the pattern of farming is very diverse, the functions of a formal co-operative will be many; such a multi-purpose co-operative can serve a small number of members efficiently provided that the total annual turnover is high. To ensure a high turnover, the loyalty of members must be secured; mutual co-operation is required and, therefore, the community must have homogeneous features such as ethnic ties, attracting one member to work with another.

In Israel, homogenous patterns of community settlement were adopted, as we have seen in the chapter on the human resource, both to reduce the crisis of adaptation in the first years of settlement, and to provide minimal conditions for the transitional growth of village communities to co-operative villages. The co-operative structure of villages became stronger as farmers became more involved in a diverse range of farm activities, particularly related to animal husbandry. This increased the dependence of the farmer on his co-operative for basic economic services and, therefore, as the financial turnover of farmers increased from year to year, the co-operative structure and village management gained strength.

On the other hand, with the specialization in agriculture, and particularly with the reduction of profit margins in farming, it became necessary to set up co-operative wholesale organizations which could, through large-scale purchasing and marketing operations, reduce costs for agricultural requisites and market margins. National producer and consumer co-operatives were thus developed and the village became a primary co-operative dealing with producer and consumer distribution and associated with vertical and national co-operative structures.

The requirements for specialization not only have forced the link-ups with national co-operatives, but have reduced the viability of the primary co-operative within the village. Because the production of industrial crops depends on processing facilities that are not associated with high transport costs, yet are on a scale larger than viable in the village, it becomes necessary to locate processing and storage facilities in the region of specialization, as in a rural centre.

The functional delineation of regions for specialized agriculture can be determined at the national level. The physical planning of the region, whereby conditions for production are provided, can also be imposed from a national level. However, the actual development within a region cannot be successfully implemented unless conditions are provided for the farmers and settlers of that region to participate actively in decision making. The problem is now how to mobilize farmers of specialized agriculture, to participate fully in the growth of the region if, on the one hand, the functions of the village co-operative have been reduced, and on the other hand, vertical co-operative structures are in existence, which can be incompatible with the region's development. There seems to be only one solution: horizontal regional co-operation and co-ordination.

ii. *Regional co-operation*

Regional co-operation between farmers and regional authorities must aim at a positive process for achieving balance and fullness of development, whereby the individual and the area as a region can benefit. The first prerequisite for such a region should thus be its independence from political boundaries. Secondly, the economic activities of the region can be used as the base for regional co-operation: for example, a region producing industrial crops acquires credit for its farmers by linking the marketing process of the product to one market board or agent, and a regional credit organization can be established with the same crop used as security for credit monies advanced. In this respect, it is relevant to point out that administrative arrangements with banks in the region can be established to reduce the costs of credit services to the minimum. By such regional credit services, farmers can learn the value of inter-dependence, while the presence of credit organizations in the region can prevent their mismanagement.

Over the last eight or nine years, such regional credit organizations have been established, and this in turn has encouraged banks to open branches in the region, in the service both of the farmer and the regional credit organizations.

Another important sphere for regional co-operation is in the fields of marketing and processing, since the smaller primary co-operative of the village cannot handle the produce of its farmers efficiently. A regional co-operative for marketing can handle produce through an adequate scale of operation. Even in cases where national co-operative marketing agencies exist in the bigger cities, farmers can benefit from regional marketing co-operatives which enable them to decide on the disposal of their products before its transport out of the region. By pooling resources, storage and refrigeration, facilities can be established so that products can be withheld from the market if prevailing prices in the main centres are unsatisfactory. The Dutch auction system, as it exists in various regions of Holland, is an example of this pattern of regional co-operation. Farmers are shareholders of the auctions, and through their own elected management can decide on the disposal of their products.

As has already been indicated, a regional marketing structure has begun to develop in Israel. There are regional grading stations which reduce costs and improve efficiency of packing and transport. A good example is in the handling of potatoes for which a regional marketing and cold-storage system have been established. There is room for further investment in regional grading stations, and for a permanent staff maintained throughout the year.

That which makes farmers willing in Israel and elsewhere for regional co-operation, is the economic saving which is gained through large-scale and more efficient undertakings.

iii. *Regional co-ordination*

The regional authority as an institutional framework has the function of a catalyst. The farmers who are engaged in their individual farming operations often require initial support for establishing regional functions. Success has even been encountered in the case of the regional authority investing in services and, thereafter, selling those assets back to farmers. This process is now taking place with regard to cotton gins in Israel.

In most cases the regional authority attempts to co-ordinate resources and operations for the region as a whole. For instance, in the Lakhish region it was felt that spraying by helicopter would enable the production of cotton on scattered plots. The regional authority then obtained the agreement of all cotton growers in the region to sign on for spraying by helicopter at fixed prices. Another case of such regional co-ordination was when the regional authority became instrumental in facilitating the purchase of residual sugar pulp for the region from the factory processing sugar beet in the same area. The regional authority is often the body that levers investments for the development of the region.

Regional government need be no more permanent than the issue it seeks to solve; its boundaries may change as conditions change; its functions may change as the times demand. Experience in Israel has shown that the regional settlement authority originally concentrated on the basic needs of settlement in a functionally defined region. However, as villages and farms consolidated, the regional authority could decrease its administrative staff, but had in many cases to increase its professional staff, who function today as a co-ordinating unit, bringing the results of research and know-how to the notice of the farmer and encouraging him to make the final decisions. A good example of this process is the introduction of new crops and varieties. The development of a new crop now takes place more gradually than through direct interventions formerly. Innovations now take root through communication of information, financial arrangements encouraging a particular process and through administrative co-ordination – a slower process but one utilizing the willingness of the farmer, and hence sounder.

Regional co-ordination, and through it, co-operation, is a means by which the natural leadership within smaller individual communities can be brought to the fore to serve a wider regional community. The problem is one of stimulating a regional loyalty which can be directed towards co-operative efforts for the region's economy and community. This can be brought about through a careful plan of action, in which a public relations campaign in the community plays no small part. The process can be hastened by a discreet regional investment programme, both for social and economic needs.

Although regional co-operative undertakings have been started, regional co-operation in its widest sense has probably not been fully achieved anywhere. It would seem that such achievement must start with the physical planning of the region, whereby town and country are closely related. Only then can a framework be evolved for close interaction between urban and rural communities.

2. The departments

The following sections examine the administrative side of Israel's experience in regional planning.

The Settlement Department of the Jewish Agency has been responsible for planning and carrying out all rural settlement in the country, working through a decentralized structure (for details see Chapter 8). When the Lakhish Region was delineated in 1954, the Department set up regional offices at first in Ashkelon, and later in the new regional town of Kiriat Gat. Since its establishment and up to 1961 the region was divided into two districts, each of which enjoyed a fair amount of administrative autonomy within the general regional organization. Planning and policy-making were done on the regional level with district participation. Implementation was left to the districts with regional supervision. The region had five sections in its first years: administration, farm planning, irrigation, technical, finance. Extension was included under Administration. Each of the sections received technical direction from the corresponding national section of the Settlement Department. The administration was responsible to the central office of the Settlement Department, farm planning to the Joint Planning Centre; and the others to their respective counterparts. The local sections were on the one hand the representatives of the national branches at the regional level; on the other hand, their work was co-ordinated through the regional office and its Director. Main directives came through the vertical channel from national branch to regional section; most of the day-to-day work was carried out at the regional level by the sections working in close horizontal co-operation under the director. Co-ordination was achieved by regular meetings of the regional directorate which consisted of the director, his deputy, the two district directors and the heads of the sections. This was the policy-making body in which work priorities were laid out and translated into a programme of action. Once the policy was approved, the section heads were responsible to the directorate for its implementation. If one of the sections disagreed with the part laid down for it by the directorate, the dispute was decided by the administration and the respective branch at the national levels – though this never happened in practice.

The vertical organization was decisive, but representatives in the field had sufficient autonomy to co-ordinate their work programmes according to the needs of the region as decided by its directorate. Budgets, plans and proposals had to be approved by the national branch before a section could undertake implementation, but once approval was obtained, the work was carried out with little reference to the national authorities. A detailed example of how the process functioned is given later.

Each of the five sections was as follows:

i. *Administration*

The director of the region headed the administration and had on his immediate staff a deputy director, the two heads of the districts, an assistant for clerical work and personnel, an assistant for credit and marketing, and secretarial staff. He was responsible for co-ordination, budgeting and the regional and district administration. For the first three years he was also responsible for the administration of the new town of Kiriat Gat.

The deputy director (an agronomist) supervised planning and implementation as it was jointly carried out by the various sections, supervised the day-to-day co-ordination and timing of the activities of the several sections and was directly responsible for the extension network of regional and village instructors through the district directors.

The district heads were responsible for implementation in their districts and for the administration of their villages. There was a district officer for each six or eight villages, who co-ordinated the staff work in the villages. The staff consisted of a general instructor, an agricultural instructor and a home economics instructress. Each district had a unit of each section attached to it for direct implementation work.

ii. *Farm planning section*

The planning section was headed by a chief planner; it had two units:
  a) Survey, parcellation and mapping.
  b) Farm planning and management.
One planner worked with each district and one with the kibbutzim of both districts. Two agricultural economists for economic research survey and marketing problems (in so far as they were connected with farm planning) were also on the staff.

iii. *Water section*

The section was headed by a chief engineer supervising two units, one for planning the irrigation networks from the main take-off points to the villages and within them, and one for costing and preparing tenders, supervising the contractors and carrying out maintenance. Each district had a sub-unit of the water section working with it.

iv. *Technical department*

The Department was headed by a chief engineer supervising two departments:
  a) Architecture and village planning.
  b) Costing and preparation of tenders for construction of houses, public buildings and roads within the villages.

v. *Finance and budgeting section*

The finance section was headed by an accountant responsible for three units: salaries, investment and credit.
  All sections, except for finance, therefore, consisted of a unit for planning and one for implementation under the same head. One of the strong points of the Lakhish Region (although it is by no means unique in this respect) was this tie-up between planning and direct implementation within the context of full inter-sectional co-ordination and co-operation.
  This administrative framework carried on its duties from the beginning of the plan until 1961 when it was changed, as described later.
  An example of the way in which the sections worked together in the regions and districts in planning a new village was as follows:

3. Co-ordination in planning a new village

The following is a summary of procedure from the point of view of unit co-ordination.

a) The farm planner seeks the site of the proposed village, the total area and land classes, which are based on a prototype determined at the regional level and approved or modified by the Joint Planning Centre and the Settlement Authority Directorate.

b) Approval of the site is given by an independent commission of the Ministry of Health, Interior, Defence, Settlement Authority, Archaeology, Transport, Agriculture.

c) A soil survey is made for further detail. The farm planning section transfers the detail to maps of 1:5000 scale.

d) A land utilization map is then drawn up according to soil type and farm requirements. The mapping stage of parcellation starts. The land utilization map is used in the first years to plan the activities of the administered farm.

e) The mapping of the water pipe network starts, with the parcellation map as the basis. Here, the take-off points from the main pipe line and the full internal pipe network are shown.

f) Contracts and tenders are prepared for deep ploughing and land levelling by the farm planning section and the finance section.

g) Contracts and tenders are prepared for the water network by the water and finance sections. The contracts specify a time limit.

The farm planning and the water departments supervise the work according to the contracts which each signs.

h) Finally, the architectural plan is completed by the Technical Sub-Department, a branch of the Jewish Agency, affiliated to the Settlement Department, which has its staff in the regional offices together with the Settlement Department staff. They work at this stage mainly with the farm planning and water departments and the district offices.

The co-ordinated plan and implementation programme are approved by a committee composed of the various sections involved and the district officer, under the chairmanship of the deputy director. He is then responsible for initiating and carrying through the implementation of the plan.

J. STAGES OF PLANNING AND IMPLEMENTATION ON THE REGIONAL LEVEL

Once the administrative organization had been set up, the actual settlement work could begin. A number of fairly distinct stages defined the period during which the budget of the Settlement Authority was used to build the villages, and the Government was starting the regional town. The work has not yet finished; the Authority is still active in the region, but its functions are slowly taking on new aspects as has been described. A considerable part of the budget has been allocated; the villages are built and working; soon they will be independent. The process of bringing them to this stage involved a many-sided effort by all of the sections and by their corresponding national bodies.

A description of each stage of development presents some of the intricacies of co-ordination:

1. Stage one: the basis for land allocation

i. *Survey and plan*

The area of Lakhish was not well known as an agricultural region before the present settlement was undertaken. The few settlements existing there prior to the new colonization scheme had collected some information over a number of years about soil conditions and climatic characteristics including rainfall, but much remained unknown. Irrigated farming had not been practised before, while industrial crop farming was new to Israel as a whole and had to be learned from the beginning. The particular regional plan evolved was an experiment, calling for new departure in physical planning of villages and region. New social problems had to be tackled in working out the relationship of the settlers to the rural centre and regional town. The town itself was new in conception, and the way in which it would progress could only be tentatively assessed.

The National Soil Survey gave useful information about soil characteristics and types. This information was supplemented by an areal survey which gave the basis for comprehensive mapping of the area. A pasture survey was undertaken in the eastern part to assess the carrying capacity and possibility of economic improvement. Land use maps were prepared, and the siting of the villages determined.

The siting of the villages in Lakhish was decided upon by a committee composed of town planner, architect, water engineer, agricultural planner and security officer, working both in the office and in the field. The internal plans of the villages themselves were laid down by the town planner and architect. The agricultural planner checked that the planned plots met certain agricultural requirements regarding slope and other factors. Huts were put up on the sites of the houses and in several cases the immigrants were brought straight from the ship to the village. In others they were moved over from temporary transit camps. Two villages were settled by Israeli urban families through the 'Town to the Village' movement.

At the same time, the authority started to build up a staff of general and agricultural village instructors. These men were recruited mainly from older established villages; many came in response to a call for volunteers to help in the new settlement area for a limited period. Some left their homes and went to live either in the rural centres which started to go up at the same time as the villages or to the town. Those from older villages kept on with their own farms with the help of their families, staying in the villages or centres only during the working week. Regional instructors, administrators and technicians were recruited and provided with housing in the town of Ashkelon or in Kiriat Gat itself. A garage to provide transport facilities was added to the regional office. The team started work, augmenting its numbers as the need arose and as the scope of the activities broadened.

ii. *Transition to independence*

During stage one, the preliminary parcellation and land use maps within each village were worked out. The land was leased under contract to a private company or to companies of the *Moshav* Movement which initiated the administered farm system. The settlers, on completion of the building programme, started to work as labourers on the administered farm, while the detailed planning of individual plot parcellation and the irrigation network for the first plots were yet being completed. The first stage of irrigation covered large blocks without the individual plot networks. The main pipes were laid, and the administered farm was intensified through irrigation and industrial crop farming. This meant more labour for the settlers as well as their first practical experience in intensive agriculture.

The crop plan of the farm was worked out by the farm planning section and the contractor, since the administration had to ensure provision of the maximum number of work days and also that the land would not be misused under the system. The Jewish National Fund started a large afforestation programme on Class D lands, lands neither suitable for cultivation or pasturage, which provided work days.

Meanwhile, community work started in the villages, organized by the regional administration through the districts and district officers, who supervised the general instructors in the villages. Home economic and social instruction were also given by a permanent instructress in each village. This work had two immediate aims: to organize the villagers as a group and get them to realize the advantages of co-operation, and gradually to bring them to take on their own independent farms. The Authority was not willing to do any individual parcellation until a sizeable group of settlers in any one village demanded their plots at the same time.

This policy was very important. It meant that group responsibility was exercised and that the newcomers received their land as a right, not as a favour thrust on to them by a process of imposition. The break with the country of origin and settlement under difficult conditions made great demands on the people. The people had to learn Hebrew and acclimatize to the ways of living and working in Israel; at the same time, a refugee mentality had to be avoided. Hence the transition period was necessary, during which a livelihood without responsibility for a farm or, collectively, for the village, was possible. Time had to be allowed for the immigrants to be brought to the understanding that the village and farms were to be theirs, and that once they accepted their land, they would be responsible for working them and for running village institutions connected with co-operative marketing and with general affairs.

With the forming of groups willing, indeed demanding, the release of land from the contractor for individual farming, the first stage had come to an end. The villages did not reach this independence all at the same time, for some villages took up to as many as four years before they were willing to start individual farming. These latter had had severe social problems; in one village in particular, only concentrated and devoted work in community development to remove antagonism prevented the village from breaking up entirely. A suggestion had been made to retain this problem village as a worker's village, but the alternative of resolving the social problems was adopted. The result was, that, today, the village is on a par with most of the others in the region.[34]

Other features of this first stage were the start of a system of data collection covering population, production and inventory; and the start of a register of budgetary allocations to the settlers. As the budget was expanded, the settlers received their house and equipment, for which they signed receipts. Individual files were kept. The settlers were forbidden to sell or in any way transfer, any of the property they received on the budget.

Book-keeping in the villages started, but it was only concerned with items such as village taxes, sick fund and the payment of wages from the administered farm and public institutions employing the settlers.

2. Stage two: independent farming

i.

As groups started to come into the regional offices and demand their land, the second stage commenced. As the irrigation pipes were laid, and the land deep-ploughed (an expense borne by the investment budget), the administration prepared its credit schemes and also intensified its efforts in the villages towards forming committees and co-operatives, or at least co-operative inclination and understandings. Apart from the intrinsic merits of such attitudes in the villages, they were essential to the working of the credit scheme which was the lever to the whole process of introducing the settler to the *business* of independent farming. Obviously, the settlers had merely done agricultural work; they had made no independent decisions, nor laid out money in sowing and tending their crops. Independent farming meant a completely different sequence of operations and one which needed the most careful explanatory work, encouragement, and faith to see through. The way in which the settlers were treated was vital at this point, for on it depended their whole outlook on future farming operations and the Authority. The group attitude was of extreme importance because without it, there would have been no organized marketing or farm planning; these two operations are the key in getting a credit system on its way, and with it the entire settlement machinery.

The procedure was as follows. The farm planning section prepared a rotational plan for the five or ten dunams which the farmer had at this stage. The emphasis was on vegetable production for the following reasons: the plentiful labour supply per farm for working such a small piece of land, the fact that the farmer had to get a large part of his income from it and that the Authority was interested in his attaining quick results from his enterprise. In the first year or two, practically no industrial crops, apart from groundnuts, were grown outside the administered farm, which continued to operate on the reduced area of land at its disposal. The farm plan, which was fairly standard for each member in a village, was built on crops which would give the farmer a minimum income for this second stage, to be supplemented by other sources guaranteed by the Settlement Authority. The quotas of vegetables received for the whole region were divided among the villages and the individuals on allocation. Most vegetables had a guaranteed minimum price.

The village as a whole had to undertake to market the produce co-operatively, and on this assurance, credit was given according to the needs of the farm plan for the different crops and the months of the season. The planners established norms of labour days necessary for each crop, water needs and expenditure and income for each month of the growing season, and therefore a full farm plan could be drawn up. On the basis of this approved farm plan the administration drew up lists of credit requirements which were paid out at two-monthly intervals to the settlers in the villages by the finance section, and signed for as each payment was received. The money was given in the form of credit slips cashable in the local store for fertilizers, seeds and other raw materials, for tractors and for water bills which were paid for by the village as a group through the secretary who collected the slips from the farmers according to the amount of water registered on their meters.

The slips were not cashable in the grocery store of the village – the farmers had a cash income from outside work which they used until the income from the crops came in. In difficult cases, an advance on the crop planted and its maturing was given, but the lack of credit for living expenses gave a strong incentive to the farmers to start planting and earn from their farms. Any other method would have encouraged them to neglect their land so that they would quickly have got into debt. An independent attitude would have been harder to foster.

ii. *Marketing problems*

The village marketed its produce co-operatively through its own grading station to *T'nuva*, the national marketing co-operative which includes most of the agricultural villages in Israel. *T'nuva* sent a cheque for the produce to the Settlement Authority, which according to the amount of credit out-standing for each farmer, as was listed by the village book-keeper, deducted up to half of the return. Should the crop have been unsuccessful, less than half could have been deducted, on the recommendation of the district officer.

Thus, by tieing credit to farm planning and to organized marketing of cash crops, the money was there as and when needed, in a convenient form, while its return to the Authority was assured. Should any settler use the credit for crops other than those for which it was given him, or if he marketed to private merchants outside the organized scheme, the Authority could, and often did, refuse him credit.

Merchants often paid higher prices for some vegetables, but they only took those in short supply on the market. The settlers sent the other vegetables to *T'nuva* at the grading station and received the guaranteed price. This meant an unfair burden on *T'nuva*, since it operated the price funds mentioned in an earlier chapter. Unless it collected the high price crops, it could not subsidize the low prices which some vegetables may bring in certain months and years. On this point, there has been considerable conflict of interest between *T'nuva* and the villages, which has persisted until recently. The settler is very much tempted to market to a merchant who comes to his field and buys an ungraded crop at a higher price than the station offers. The merchant's costs are low, and the settler does not have to pay credit debts or village taxes. He pockets the cash. In some cases, *T'nuva* closed the grading stations and refused to take the produce of a village if a large proportion of the members sold through other channels. Then, an obvious thing happened. The merchants suddenly had no interest even in high price crops; they said the quality was no good, and that prices had fallen. Though they were willing to take the most profitable crops at a price which was well below the price in the station, they completely refused to take the crops which on the market were depressed in price. Therefore, since the station had been closed, the crops stayed in the field, the settlers sometimes suffering heavy losses. It was not long before they requested that marketing through *T'nuva* be renewed.

Another major problem in organized marketing through *T'nuva* was that a village sent all its produce, graded and packed, to the main collecting centres. There it was checked for grade and quality by means of a sample. If the check revealed that the produce was below standard, the whole consignment was returned to the village. The village could not however re-divide the consignment among the owners, since it was impossible to mark each sack or box exactly. Thus, all the settlers lost, whether their produce was good or not, because someone had sent a bad lot which was chosen for the sample. *T'nuva* also paid for the produce a month or two after it had received the accounts from the market sale. The new villages had no reserves to last out this period, and the delay discouraged them from producing vegetables. Their grading was below standard, the organization with empty boxes and sacks was poor, and the villages lost much money through secondary marketing arrangements which were badly organized.

The Settlement Authority saw this as a crucial problem, affecting the success of the whole settlement effort, and realized that *marketing is an integral part of the production process*. It therefore approached *T'nuva* with a proposal that a subsidiary company be formed for the new villages which would meet their particular needs. The Authority undertook to bear 25% of the possible commercial loss of the company. The company was founded and worked on a different basis to the parent company. It took over the grading in the village sheds, and, according to the ruling minimum price, gave the settler a receipt stating the quantity, grades and written price. The company organized the despatch and allocation of empty boxes, and paid the village the base price within two weeks. If the crop sold for more than the minimum price, the difference was paid at a later date. The produce was dispatched in bulk according to its grade. The money came in one cheque to the Authority which deducted the credit according to an individual list submitted by the village where the accounts were checked against a copy of the receipts. The members were paid the difference by the village after deduction of village and personal taxes.

With the start of the new company, an experiment in centralized grading was started at a rural centre for three villages. Each member took his produce individually to the grading station, from where the produce was sent to central markets for the three villages together. Each member got a receipt as before, and according to this, was paid and his credit deducted. The system saved considerably on transport costs to the market; through more efficient grading, a better price was obtained in many cases. The organization of empty boxes was also simplified. The problem was the 3 km distance the settlers had to go with their mule carts as against 1 km to a village station. A small subsidy per ton of vegetables was paid from the savings the company made on costs by operating only one centre, to compensate the farmer for the time taken.

The responsibility of the village for operating a grading station was thus removed. This is a function carried out by co-operatives in the rest of the moshavim.

The progress of the villages was anything but uniform. In some, strong leadership appeared, often centred around one family. They sometimes subdued the other unorganized members in village affairs. In others, two strong family groups appeared, splitting the village, making progress towards a co-operative doubly difficult. Some which showed no strong family groups, developed one or two natural leaders who found fairly general support in forming committees or running village affairs. Sometimes, these natural leaders were absent, or not strong, with the result that a new committee was chosen every three months, with a consequent lack of continuity. The responsibility of the administration in guiding the villages through this difficult period of first establishment and village organization was very heavy. It rested mainly on the general instructor and the district officer. The turnover in general instructors tended to be high, for the work was arduous; the economic position of the settlers was far from secure, particularly as they were taking the risks of individual enterprise for the first time. The people often tended to be very aggressive, particularly if things went wrong.

Meanwhile the tractor station and the grading station, both situated in the rural centre, started to play a larger part in the settlers' life. The regional school was consolidated and the beginning of cultural activities became evident. The Regional Council became a factor in the municipal life of the fast developing region.

3. Stage three: secondary plots

During the time that the farmers were working on the one small plot of land near the house, preparations were being made for the second allocation of land. The areas to be allocated (on which the contractors' lease was not renewed) were surveyed and the parcellation and water plans made. Tenders were issued for the necessary work, and the pipe networks were laid. The large blocks served by the main pipes were divided into plots of twenty dunams or so, according to the lay of the land and the size of plot near the house. Secondary irrigation pipes were installed, and all made ready for the farmers. The farm planning section prepared plans for the villages, based on sample plans for individual farms. Industrial crops were given precedence over vegetables, since in the large area neither the labour capacity of the farmer, nor the limited quotas for vegetables, would allow main emphasis to be laid on their production.

The plots were, for one or two years, shared by two farmers, each working on his own half. In a few of the villages, cotton and sugar beet were planted on whole blocks by the settlers in co-operation with a large measure of help, organization and persuasion from the Authority. The irrigation system made this co-operative effort possible. Main operations were carried out over the area as a whole, and the members were debited for the cost according to their share in the block. Irrigation, weeding and picking were done by each settler separately, and payment was according to the yield on each man's plot. A main reason for the failure of the system to take hold was that the irrigation programme of the separate settlers could not be co-ordinated with a spraying programme to cover the whole area, since the tractor with the spray equipment would get bogged down in the wet parts. After this trial, several partnerships did form among members who took advantage of the particular physical planning scheme to work their lands as one unit. Generally most of these partnerships broke up after one or two years, although some, usually with a strong family basis, continued and prospered.

Marketing of the cotton, sugar beet and groundnuts was done by the members through the village which, together with the region, decided on times of harvest for each member. The region worked together with the gin and the sugar factories in regulating the amount of cotton or sugar beet harvested according to the needs and capacity of the factories. At this early stage, there was enough factory capacity to take the region's products without serious hold-ups, even though the Kiriat Gat sugar factory was not yet working. Economic research carried out in some of the villages after the first year of industrial crop farming showed that these crops were profitable, although in general, incomes in the villages from agriculture were still low. Relief work programmes such as reforestation were cut to a minimum to give the settlers additional incentive to work their land and live by it. Moreover, the labour exchange in the towns discouraged the villagers' demand for urban work, and therefore they had no alternative.

With the increase in work and incomes in the villages, some of the social problems disappeared. The villages became hardworking communities; it is in many ways remarkable just how soon they did settle to the arduous work involved in their new life. Efforts to organize the villages and seek out stable leaders for the posts of secretary and treasurer, water inspector, tractor organizer and other seasonal tasks continued, and the general instructors started to hand over the administration of these and other jobs to elected officers. Perhaps surprisingly for the social structure involved, these posts usually went to the younger members, although the real authority in communal matters rested with the elders in most cases.

In the town, the sugar factory started to go up, several firms started building, and others tentatively enquired about opportunities there. The population continued to rise, and there were the beginnings of an urban atmosphere.

4. Stage four: towards full resource allocation

By the fourth year, the villages were already well established and producing. As far as the budget and planning allowed, the rest of the secondary plots were laid out and handed over to farmers in those villages which were ready to receive them. Each man took on a full plot of twenty dunams in addition to the one near the house. Poultry or sheep were introduced to some villages. The area of all crops grown increased, although the land in some places was found too heavy for economic production of groundnuts. Economic research showed that incomes and yields were generally going up, although in some villages danger signals started to appear. Due to over-exploitation in certain vegetables and neglect of a sound rotation, some land became unfit for certain crops and had to be rested from them.

The social patterns in the villages became more apparent; in most cases they were not oriented towards a fully organized, registered co-operative. The financial turnover of the members and the villages increased considerably. A definite pressure of savings was felt, and the need for an outlet discussed. Since the field crop farm does not give much possibility for a farmer to invest, the planners started to think of regional organization for investment in industries at the village centre or the town, connected to agriculture. Some farmers objected that they had not been given the chance to invest in the sugar factory. Others started talking of the need to build cold storage in the rural centre. Generally, marketing and investment problems became important points of discussion in the offices and among some of the more advanced farmers.

Progress was also felt in the extension service, which started to refine its instruction in more advanced points of picking, irrigation, field work and planning with the farmer. The national organization of extension was changed at this time through the establishment of an Extension Authority under the Ministry of Agriculture, to include both new and old villages throughout the country. Offices were opened in each region and the regional specialists came under their jurisdiction. The village instructors continued to work for the Settlement Authority and were aided by the specialists. The Central Extension Authority supplied national specialists and extension material for the use of the regional instructors.

During this stage, the pressure on the regional office generally began to slacken in certain fields, notably technical, irrigation and surveying. Most of the main physical planning and implementation had been done, and there was no need to keep full staffs in the sections. The administration also found its work lessening in intensity as the villages started to work as communities and to manage themselves. The region passed into a stage of consolidation, in which the newly-made farms and villages had to become fully productive, efficient and self-sustaining units within a village framework. Accordingly, the independence of the region was reduced: the two sub-districts were fused to become a single district within the Negev (Southern) Region of the Settlement Authority. The staff was reduced from 240 in 1956 to 154 in 1961, with no subsequent increase; physical planning, surveying and similar services were obtained when needed from the main regional offices.

The District Director was made a member of the Directorate of the local branch of the Extension Authority, and the Director of the local extension service was made a member of the District Directorate. In this way, full co-ordination was achieved between the two bodies. Five district officers shared the administration of the district under the director and his assistant, and the new organization started work on the problems of the rapidly developing settlement area as a closely knit team.

5. Stage five: consolidation (co-operation reconsidered)

The present stage has the formation of strong, well organized and efficiently producing villages as its main aim. They have to be brought to full independence within a regional municipal framework, with an organizational structure suitable for credit facilities, marketing, investment and services. In the *moshavim* of Israel in general, all of these functions are provided through the co-operative which belongs to a regional council supplying municipal services and sometimes economic organization.

In the new villages of Lakhish, co-operatives have been slow in forming. This was for several reasons, of which the most important was probably the unfamiliarity of the settlers with co-operative organization and the way it binds members in so many spheres of life. Other main reasons were connected with the structure of field crop farming and the cash turnover of the village. Crops such as cotton or vegetables do not offer the co-operative as many functions to perform as a system based on animal husbandry where there is a continual process of buying and selling and a real need for daily services of feed mixing and distribution, milk cooling, egg collection and marketing. The field crop farms have busy seasons for preparation and sowing, and others for harvesting. Both operations do not particularly need a village co-operative for their successful functioning. In a certain sense, the converse is also true: the scale of production and marketing in these crops is too large to be contained in the limited framework of one village, and only regional organization can effectively provide the necessary facilities such as tractor station, processing and grading, as we have seen.

Because the villages are not strong co-operatives, they have, with the active promotion and backing of the Settlement Authority formed Regional Purchasing Agencies working with twenty or so villages which supply credit to farmers individually and purchase on behalf of the stores in the villages. The credit, one of the main functions of a co-operative, thus by-passes the village. There are no indications that the village wants to be responsible collectively for the credit needs of its members. Credit is returned to the purchasing agency through deductions from market returns, on the basis of lists submitted by the village. The village management does not handle and administer large sums of money, and formally there is therefore little need for its members to form an organized, registered co-operative. There are no main economic or social pressures which demand formal co-operation in the running of a village.

Co-operation of course exists in all the villages on other levels: meetings are held, officers are elected, book-keeping services are paid for. Certain necessary operations are organized within the village on the basis of full co-operation between members, and the village is formally registered as a co-operative.

In this context, sub-region and region must be considered again. Farmers, with the help of the Settlement Authority, will have to form Regional Growers Associations, or similar representative bodies, to which each farmer can belong individually. The bodies must be responsible for credit provision, marketing organization and improvement, channelling of investment to village and rural centres, and the other functions which the field-crop farms are unable to undertake separately because of their social patterns and small size. Special organizations for special functions may be necessary. A start has been made in the Regional Purchasing Agencies which seem to be working satisfactorily. The idea of regional co-operation to replace some of the functions of the village co-operatives is still too new and untried in Israel to be fully evaluated. Several years must elapse before its suitability can be assessed.

K. LESSONS FOR FUTURE SETTLEMENT PLANNING

Each wave of settlement in Israel has been dealt with on the basis of lessons learned from previous waves. This is an historical process stretching over the last fifty to seventy years. Agricultural development and the growth of settlement have so strained the country's water resources that only a few more regions can be contemplated at present for settlement. One of these is the B'sor Region, west of Ber-sheba, which is to be based on the export farm type. It will be instrumental in achieving the targets set in the Development Plan, and will be even more significant in successive decades. The majority of settlements will probably be on the moshav pattern, and the settlers will be the present generation of Israeli Youth. They will be mainly drawn from the already developed agricultural areas in the north and centre of the country. Therefore, the pressing problem of cultural integration, which was such a prominent feature of the Lakhish project, will be absent.

When the Lakhish project was initiated, it provoked an active discussion in the country because of the revolutionary nature of the physical layout and crop production patterns proposed, in relation to existing settlement ideology. Now, many of these 'revolutionary' ideas of 1954, such as inter-village co-operation based on the composite rural structure, the separation of farm lands from the home, and a farm system not based on animal husbandry, are fully accepted. Some have even been applied in the older settlements of the country. The application of these principles to the planning of the B'sor Region is fully accepted. The B'sor Region will, however, undoubtedly introduce many changes of its own, some of them perhaps just as revolutionary in their own way as were those introduced to Lakhish.

Co-operation between villages is now accepted and this will no doubt be a cornerstone of the B'sor planning as well. Marketing organizations will probably work on a regional basis, representing the farmers of the region. Probably the organization for marketing groundnuts for example, will include all the growers in the region. It is hoped that it will not be organized on a narrow village basis with direct village ties to national groundnut marketing organizations, but with each member marketing through the regional company. The region will therefore be based on horizontal co-operation between farmers rather than vertical integration in national marketing frameworks of villages, which is the case with the traditional moshavim. Vertical organization strengthens the village and ignores the region; horizontal co-ordination takes functions out of the village and makes the region the organizational framework with its wide possibilities of large scale production. It is thought that for the type of production to be undertaken in B'sor and the larger village which will probably be built there, a regional organization will meet the needs of the members in a fuller and more efficient way than could traditional, all-inclusive village co-operatives.

NOTES

1. GEDDES, Sir Patrick: *Cities in Evolution* (Williams & Nargate, London 1949).
2. WEITZ, R. (Ed.): 'Rural Planning in Developing Countries', Report on the Second Rehovoth Conference, Israel, August 1963 (Routledge and Kegan Paul, London 1965), p. 101.
3. WEITZ, R.: Lecture presented by the author at the International Conference on Fundamental Problems of Agrarian Structure and Reform in Developing Countries; Berlin Tagel – Deutsche Stiftung für Entwicklungsländer; May 12–16, 1962.
4. See also:
   [1] LILIENTHAL, David E.: *TVA: Democracy on the March*, 20th Anniversary Edition (Harper & Bros., New York 1953), pp. 294.
   [2] PRESTON, J. Hubbard: *Origins of the TVA* (Vanderbilt University Press, Nashville 1961).
5. CONSTANDSE, A. K.: 'The Changing Settlement Pattern in the Netherlands' in *Changing Patterns of Rural Organisation*, Papers and Discussions of the Second Congress of European Society of Rural Sociology (Oslo 1961), pp. 290–301.
6. LINDENBERGH, A. G.: 'The Enclosure of the Zuiderzee and the Reclamation of the Polders in the IJssel-Lake As seen from the Standpoint of National Economy', Zuiderzee Polders Authority (Zwolle, The Netherlands 1958).
7. [1] 'The Gezira Scheme', Great Britain, Central Office of Information (June 1950).
   [2] GAITSKELL, A.: 'The Sudan Gezira Scheme' *African Affairs* **51** (1952), pp. 306–313.
8. LAMBERT, A. R.: 'The Sudan Gezira, II: The Irrigation Scheme', *Geogr. Magazine* **9** (1939), pp. 127–144.
9. WEITZ, R.: 'Rural Development Through Regional Planning in Israel', *Journal of Farm Economics* **47**, No. 3 (August 1965), pp. 634–651; reference on p. 651.
10. WEITZ, R.: 'Rural Planning in Developing Countries' op. cit., Part one, pp. 54–56.

11. See also:
   [1] HALPERIN, H.: *Agrindus, Integration of Agriculture and Industries* (Routledge and Kegan Paul, London 1963), pp. 214.
   [2] WEITZ, R. with the assistance of ROKACH, A.: 'Agriculture and Rural Development in Israel: Projection and Planning', The National and University Institute of Agriculture, Bull. No. 68 (Rehovot, Israel, February 1963), pp. 120–125.
12. See: 'Regional Cooperation in Israel', The National and University Institute of Agriculture, Settlement Study Centre, Rehovot, pp. 7–10.
13. 'Planning of Rural Centres in Israel', Agricultural and Settlement Planning Centre (Tel-Aviv 1964), (mimograph in Hebrew).
14. 'Regional Cooperation in Israel', op. cit., pp. 15–17.
15. Personal observation of the author on a Far-East Tour, 1961.
15A. See also:
   [1] WEITZ, R.: *Agriculture and Settlement* (Am-Oved, Tel-Aviv 1958), pp. 273–280 (in Hebrew).
   [2] WEITZ, R.: 'Linked Farm, a Proposal for the Planning of the Hill Region, *Economic Quarterly* 16, (1957), Tel-Aviv (Hebrew).
   [3] WEITZ, R.: 'Agricultural Linked to Industry as an Approach to Regional Planning in Problematic Areas', Paper presented to the FAO Study Group on Problems of Individual and Group Settlement in the European Region, Second Session (Wageningen, The Netherlands, August 1958).
16. See also: WEITZ, R.: 'Agriculture and Rural Development in Israel: Projection and Planning', *op. cit.*
16A. Personal observation of the author on a Far-East Tour, 1961.
17. ARGOV, L. and ABT, V.: 'The Siting of the Third Sugar Factory in Israel', Lakhish Planning Section, Settlement Department, The Jewish Agency (Kiriat Gat 1958), (mimograph in Hebrew).
18. [1] Agricultural and Settlement Planning Centre, 'Regional Enterprises in Israel' (Tel-Aviv 1965), (mimograph in Hebrew).
   [2] WEILER, R. and APPLEBAUM, L.: 'Preliminary Comparative Analysis of Regional Cooperation Patterns in 12 Rural Regions', in Regional Cooperation in Israel. op. cit. pp. 120–121.
   [3] HALPERIN, H.: *Agrindus, Integration of Agriculture and Industries, op. cit.*, pp. 130–141.
19. 'Regional Enterprise and Regional Cooperation', Union of Kibbutzim and Kvutzot (Tel-Aviv, April 1962), (Hebrew).
20. Agricultural Plan 1965–1966; Jewish Agency, Settlement Department, Negev Region (1965), (mimograph in Hebrew).
21. The survey was carried out by the Planning Section of the Lakhish Regional Office, of the Settlement Department (not published).
22. Report of the Committee for Examining the Establishment of an Urban Centre in Lakhish Region; Jerusalem, 30.12.1954 (not published).
23. BLACK, M. and ABT, Y.: Economic Survey of Kiriat Gat; prepared for architect A. Glikson, coordinator of the research for the Ministry of Housing; (Tel-Aviv, March 1959), (mimograph in Hebrew).
24. Figures submitted by the Lakhish Regional Planning Office, 1965.
25. Figures submitted by the Lakhish Regional Office, 1960 and 1965.
26. Based on calculations made by the Lakhish Regional Office in 1961.
27. LOWE, Y.: 'Kibbutz and Moshav in Israel', paper presented to the FAO Study Group on Problems of Individual and Group Settlement in the European Region (Wageningen, The Netherlands 1958).
28. Figures submitted by Mr. Y. Abt of the Lakhish Regional Planning Office. The figures

are not based on adequate samples nor on similar farm-types, and should therefore point only to preliminary conclusions.

29. WEITZ, R.: 'Land and Water Allocation', *Economic Quarterly* (Tel-Aviv) No. 28, (October 1960), (in Hebrew).

30. Figures based on a survey carried out by the Land Settlement Department of the Jewish Agency in 1961 (not published).

31. BLACK, M.: 'Economic Survey of Noga village', Lakhish Farm Planning Section (mimograph in Hebrew).

32. 'Annual Farm Plan 1961/62', Lakhish Farm Planning Section (Kiriat Gat 1962), (mimograph in Hebrew).

33. 'Annual Farm Plan 1965/66', Negev Farm Planning Section (Ber-sheba 1965), (mimograph in Hebrew).

34. PARDESS, J., 'Community Centre, Even Shmuel', Ministry of Welfare, Government of Israel (Jerusalem, February 1961), (mimograph in Hebrew).

CHAPTER 8

IMPLEMENTATION OF AGRICULTURAL DEVELOPMENT

Implementation of agricultural planning, the actual changing of blue-prints into farm communities and produce, hangs on both the institutional framework or development authority, and on the farmers who raise the crops and grow the livestock. The institutional framework must have either budgetary support for its projects, or else authority to see that other governmental bodies carry them through, such as the laying of water systems, or roads, or the setting up of schools. It must also have a close co-ordination with the planners, or a 'feed-back' mechanism which evaluates the interaction between the plan and the various factors of development, particularly those which cannot be assessed beforehand. Planning and implementation must be closely interwoven, with the plan continuously modified to fit real conditions. The institutional framework must be such that this dynamic interaction can proceed smoothly.

As to the farmer who must carry out the micro-plan, or the growing of crops and livestock according to the farm-plan, in a democracy he is a free agent who cannot be coerced. In a democracy he responds to government appeals as he sees fit, running his farm so as to give himself the best income, labour and capital inputs. He is protected against coercion by the check in the democratic process – that is by his voting rights and representative political institutions. This means that government planners do not control agricultural resources and must carry out their plans with full regard for the farmer's right to act as he pleases within the framework of the law. The development authority therefore has to work through incentives, encouraging the farmers to follow the lines of the agricultural plan. It must also be prepared to underwrite guarantees and the promises it makes to farmers, both from a moral point of view and because, in most fully democratic societies, the farm vote is a strong force.

It is not only that the farmer must be given the proper encouragement to grow crops according to the farm plan, but, as we have discussed in Chapter 4, his success in the actual production of the crops is dependent on his ability and willingness to farm, which are highly complex factors. The cultural background of the people in his village, the village framework in which his farm is set, as well as his relationship to the development authority, are all part of the complex. For this reason, therefore, agricultural research, the passing on of the results of research through extension work, and sociological insights, are important facets to the implementation of any plan.

The following chapter presents three sets of 'activities' which have been necessary for implementation of agricultural development planning in Israel: the institutional framework of the authorities responsible for seeing that the plan is carried out; the economic and fiscal incentives by which the farmers are induced to implement the micro-plan, and the research, extension and sociological activities.

A. THE INSTITUTIONAL FRAMEWORK IN ISRAEL

1. The Settlement Department of the Jewish Agency

i. *As a separate authority*

The organization responsible for settlement and development in Israel has been the Settlement Department of the Jewish Agency. Because of historical accident, it has been a body separate from the conventional governmental framework. Before the establishment of the state, two central agencies dealt with agriculture: the Agriculture and Fishery Department of the British Mandatory Government, and the Settlement Department of the Jewish Agency. Their functions were often parallel, for both engaged in agricultural training, in research and in extension services. Since Jewish settlement was handled by the Agency, the work of the Mandatory Government was primarily for the benefit of Arab farmers.[1]

The duality of institutions continued in a modified way after the State of Israel came into being. The mandatory Agriculture and Fishery Department became the Ministry of Agriculture and certain functions of the Settlement Department were transferred to it. The Ministry took over agricultural research and extension services for the older Jewish settlements as well as for Arab villages, while the Settlement Department concentrated on establishing and maintaining new settlements. It was only in 1966 that the Ministry of Agriculture took over extension services to the new settlements which had, by the time, become more or less consolidated.

Although the separation of the Settlement Department into a special authority was because of unique circumstances such a separation proved fortunate for only such a body could co-ordinate all the different facets of development. Actually, development of areas elsewhere has been planned and implemented by similar separate authorities, such as the TVA in the United States.

ii. *As a regional authority*

Just as separation of the Settlement Department was due to historical circumstances, so was its division into regional offices which came about through experience in settling so many people so quickly. It was through the trial and error of those dynamic years of mass settlement that led to the concept of comprehensive regional planning, which was first put into practice in the Lakhish region in 1954. Planning became the three-tiered system discussed in previous chapters – macro-planning, micro-planning with co-ordination carried out on the regional level.

From Israel's experience, a regional authority, separate yet co-ordinated with other government institutions, has come to be considered by the author the most effective organizational instrument for the carrying out of development.

Three conditions of an institutional nature are important for development and can be answered by a regional authority. They are:

1) Continuous co-ordination on a regular basis among the various economic and service sectors within the development area;

2) An active and continuous interrelation between planning and implementation, that is between those involved in planning and those in charge of carrying out the plan in the various sectors;

3) Flexibility in the development process itself, which is to say, the possibility must be created to transfer resources from one objective to another in the light of changes during the process of implementation.

The conventional (governmental) structure in most countries and also in Israel is conspicuously a sectoral one. The government consists of ministries that deal with separate economic or social sectors, and every ministry has its own hierarchy of directors and administrators, with a similar pattern for all ministries.

Inter-office co-ordination in matters of fundamental – as well as routine – significance is generally maintained at the ministerial level, at cabinet sessions or in ministerial committees, or at the level of directors-general. In any case matters requiring co-ordination at lower levels are also dealt with at the main offices in the capital, notwithstanding that each ministry has representatives in different areas. Official contacts between these local representatives are relatively rare, since they are not authorized to negotiate in inter-office matters, and also because of the fact that the level of responsibilities varies for the local officials in different ministries. Decisions taken within each ministry separately involve individual sectors, and these bear no necessary correlation to decisions arrived at in other ministries. On the other hand, decisions made at cabinet meetings or in a ministerial committee are at quite a different level generally than those pertaining to the individual ministries. On the current routine level, very often that most important to the process of development, the machinery of government operates in a purely sectoral manner.

Lack of co-ordination within the government structure is particularly apparent when planning or implementation of comprehensive projects is attempted. Actually the conventional structure of government serves as an obstacle to development, especially because the communication between the local officials of ministries concerned with the project are rather weak and not obligatory. Often offices of the different representatives within the development area are in different towns or in different sections of the same town. But even if there is a proximity of offices the communication between officials at the regional level generally does not bring any concrete results, due to the fact that particularly those officials who are in direct contact with reality have no adequate powers of authority, while those who make the decisions on inter-sectoral matters sit in the capital. The fact that every problem requiring inter-sectoral co-ordination is dealt with in the capital results in loss of time; moreover, the capital being remote from the development region both in terms of distance and acquaintance with local problems, the decisions arrived at are often cut off from reality and irrelevant to the problem involved. Things look different on-the-spot than they do in the capital.

Another limitation imposed on comprehensive development planning arises from the difference in priorities assigned to a certain problem by the various ministries. Every ministry determines its scale of priorities in accordance with its needs, and in consequence distributes its budget and manpower among the different matters with which it is dealing. Thus, quite often it happens that the various components of a comprehensive development project are given differing priorities in every ministry involved. And the result is that it becomes quite impossible to carry out a comprehensive development plan on target, particularly if it consists of many projects.

Where the separate-sector activities produce the greatest distortions is in the domain of physical planning. It is known for instance that where the economic, social and recreational services are concentrated in one physical centre this results in greater use than when each service is separately located.

Nevertheless, in most countries each ministry prepares its own plan of locating services. For instance, the Ministry of Education may set up educational programmes which bear no relation to the welfare services of the Ministry of Welfare, the health services of the Ministry of Health, the agricultural programme and services of the Ministry of Agriculture, etc. With no co-ordination among the authorities and without overall planning, the result is that the foci of various services are dispersed over the whole rural area.

Since the government structure is sectoral and it is impossible in practical terms to adjust it to comprehensive development activities it becomes necessary to combine – or at the very least to co-ordinate – government services within a framework of a regional authority. In order to overcome the limitations imposed by sectoral government structure it is essential that authority be granted powers sufficient for co-ordinating the activities of the various government offices in the region.

Four general principles guiding the pattern and structure of regional authority are as follows:

a) The regional authority must be able to undertake the leadership in the administrative and professional functions of the development process.

b) The regional authority should be granted powers of authority to act in the spheres of both planning and implementation. Its terms of reference should be well defined as regards both the central government and the local authorities.

c) The regional authority should be provided with sufficient means and manpower to enable it to operate the mechanism of feed-back between planning and implementation, and the intersectoral co-ordination upon which the whole concept of comprehensive development is based. With these powers at hand, the authority will be able to draw upon public and private resources for maximum acceleration of the regional development process.

d) The regional authority should be able to translate the development plan into concrete projects within the scope of the targets and time limits of that plan. The regional authority should be the instrument of implementation of the government, in the same way that regional planning is an instrument of national planning.

iii. *Functions of the Settlement Department*

The functions of the Settlement Department of the Jewish Agency are as follows:

1) The establishment and care of rural settlements until they reach a fair degree of autonomy.

2) The creation of a stable rural community. Most of the immigrants who settled on the land or were directed to agriculture had had no farming background, and even those who did were not familiar with the branches of agriculture current in Israel, with the modern technological methods applied and with the co-operative farm structure which is the norm in the country. To fuse these new immigrant settlers into economically and socially well-integrated rural communities was one of the central tasks of the Department.

3) Physical, architectural, agricultural and economic planning of rural settlements in line with changing conditions.

4) Implementation of planning and development by providing suitable guidance to the settlers – technical and social – with the aid of a special team of instructors.

5) Organization and orientation of settlers in regard to village structure, management of farm co-operatives, loans and credits, marketing, etc.

6) Provision of supplies, structures and means of production for the establishment of viable farms which bring in an adequate income to the farmer.

7) Development of irrigated fields and of fruit trees and citrus plantations and their cultivation until the settlers are able to take care of them.

8) Supporting the settlers during the initial stages until they become socially and economically independent.

9) Providing jobs for members of the second farm generation in the village proper or within the rural area.

10) Consolidation and development of the rural community by the planning and establishment of rural centres, various joint farm and consumer services, rural industries, etc.

11) Conducting social research to study the specific problems of creating a stable rural community and provide specific solutions.

iv. *Structure of the Settlement Department*

The activities of the Department are based on planning and implementation units operating jointly and as close as possible to the district in question. These units are subordinated to the Board of Directors but are given sufficient authority to make independent decisions when necessary. They are directed and administered from the centre through two channels: administrative regulations and professional instructions. Each unit forms a complete entity capable of solving most day-to-day problems arising in the spheres of settlement planning and implementation in its area.

Up to 1950 the Settlement Department was completely centralized in its operations; but after a comprehensive survey of its functions the peripheral system was adopted in a tri-level structure. At the upper level there is the Board of Directors and major divisions, at the middle one there are the Regions and below that 'the sub-regions or districts'. Recently there is a tendency to cancel the lowest level.

The Head of the Settlement Department is a member of the executive of the Jewish Agency. The Board of Directors consists of a Director-General, four deputies and three additional members. Each member of the Board is in charge of a specific sphere of activity in which he has a large measure of independence. The Board determines the general policy and also deals with major issues brought to its attention by the members.

Divisions: Directly subordinate to the executive is an administrative unit dealing with all personal matters. In addition there are various specialized divisions and sections which enable the executive to work out general planning directives, allocate the budget, control implementation of its programmes, recruit new settlers, etc. These units are:

1) The engineering branch – water and building divisions;
2) The citrus division;
3) The plantations division;
4) Contracts and loan-security division;
5) The rural industry division;
6) The population section (integration, social cases, religious affairs, etc.);
7) Statistics section;
8) Social Survey and Research Unit (appointed by the council for social affairs for matters of common interest with the Hebrew University);
9) The Unit for the Care of the Second Farm Generation.

The Agricultural Planning Centre: The Agricultural Planning Centre is a joint institution of the Ministry of Agriculture and the Settlement Department of the Jewish Agency. At present it comprises four separate divisions, each with a well-defined function.

1) *Comprehensive Planning Division.* This division deals with the national planning of settlement and agriculture, that is, with macro-planning. It works out long-range development plans, usually for five-year periods as well as for current annual production programmes. It sets the production targets to be attained in the various branches, draws up forecasts of the distribution of output between the domestic and the foreign markets, and determines general farm development and land settlement policy.

2) *The Micro-planning Division*. This division determines the branch and farm output distribution shown in the long-range and current development plans for the different areas and farms. It also deals with detailed planning of different farm types, evaluates the effect of various modifications in existing farm types and proposes new types as required. It works in close collaboration with the planning sections of the Development Department, which operate on a regional basis (see below for the functions of the Settlement Department) and with the planning officers of the Ministry of Agriculture who also work on a regional basis and are attached to the regional offices.

3) *The Economic Division* (Survey and Consultation). This division examines the profitability of the various farm branches from the economic point of view, lays down planning norms for the different branches, draws up demand forecasts, estimates projected output and input prices, etc.

4) *The Rural Regional Planning Division* deals with all aspects of regional rural development – planning of rural service centres, of regional rural population, of regional agricultural undertakings, etc.

v. *Regions*

The regional units of the Settlement Department are: 1) Northern Region; 2) Central Region; 3) Negev–Lakhish Region; 4) Jerusalem Region; 5) Galilee. In addition there is the Division for Middle Class Settlers, a special unit not confined to any particular area, dealing with settlers who have certain means of their own.

The region represents the second operative level of the department, dealing on the average with one hundred settlements. The regional office carries out development activities in accordance with the general policies laid down by the Board of Directors. Among its responsibilities are the preparation and implementation of the annual budget for the settlements in its area of jurisdiction, the provision of extension services to the settlers and supervision of the activities of the various professional sections (e.g. irrigation, construction, farm planning, etc.).

The regional directors have the authority to act independently in development planning and implementation in accordance with the predetermined plans and within the limits of predetermined budgets.

Districts: The districts are directly subordinate to the regions, as their operational units which do the current everyday work in the settlements. Recently the tendency has been to, as far as possible, do away with the districts and operate directly through the Regional units. Accordingly, the number of districts of the Central Region has been cut down and the same is about to be done in the Negev. The two Mountain Regions, Jerusalem and the Galilee, had not been divided into districts in the first place.

vi. *Regional functioning of the Settlement Department*

In Israel, the regional authority (or regional office) of the Settlement Department went through many stages. At the beginning, there was only one central planning bureau that worked with almost no direct relations with those in charge of implementation in the regional offices. Later on it was decided to combine planning and implementation, and planning bureaus were established in each region. These bureaus were responsible for planning in the region and worked in close relations with the staff in charge of implementation on all levels, while guidance and advice were provided by the central planning body (then the Agricultural Planning Centre). This centre co-ordinated the plans of the various regions. At this stage, the director of the regional office was attributed the responsibility for planning in addition to the implementation powers he had before. Two deputies to the director were appointed for planning and for implementation separately. Actually this structure has not been uniformly established in all regions, a fact that indicates the difficulties involved in maintaining the principle of a close operative relationship between planning and implementation.

The above described structure of the regional authority resulted in a situation where the connection between the planning and the implementation on the village level depended on personal contacts between the persons in charge of each aspect. To eliminate this situation it was suggested to attribute both planning functions and implementing powers to one person, a co-ordinator in charge of several villages. The proposal was actually tested in two regions, with almost no success due to the different abilities required for each function. The tasks of the planner require academic education, while a co-ordinator must have first and foremost an organizational and practical experience. In many cases the existing co-ordinator could not assume planning tasks due to insufficient knowledge, while planners refused to change their professional status and to undertake administrative functions. In order to explain this reaction it is worthwhile to clarify the functions and domains of activity of the planner and the co-ordinator, whose responsibilities are limited to a group of several settlements within the region.

The planner, who is in charge of agricultural planning, is connected in his work with the individual settlements on the one hand and the national bodies for agricultural planning on the other. He represents the settlements in the Agricultural Planning Centre and in the various production boards, where the production quotas for the various agricultural branches are determined. The planner receives from the farmers, or their representatives, their requests for production quotas, examines them and then negotiates with the central institutions.

The planner receives general directives from the Planning Centre in accordance with the development policies, for instance, he is informed on the crops that should be given preference. He also receives basic planning data, profitability calculations for agricultural crops, etc. These directives are applied to the current planning of the settlements. The planner also suggests means for their implementation such as extension programmes, credit for the growing of preferred crops, etc.

Current planning of individual settlements involves the examination of efficiency and profitability of various branches, recommendations for changing branches or modifications in their volume of production as well as supervision and follow-up of implementation. Since the scope of tasks assigned to the planner is rather wide, he usually selects only a sample of farms in each settlement, representing different levels of development. From the well-developed and average farms he derives production norms and indices of production efficiency. These are compared with the national indices provided by the national planning centre, with regional indices and with indices obtained in underdeveloped farms in the villages, thus enabling the planner to determine development trends and to point out demerits that should be taken care of. Except for the planning and follow-up of the farm activities the planner is engaged also in related problems that affect the success of the individual farm unit, such as purchasing, book-keeping, the village co-operative, etc.

The co-ordinator is responsible for implementing the approved plans. He follows each farmer separately and is responsible for the complete and precise implementation of the plan. He is also in charge of the implementation of the annual budget. Since his work is very intensive and requires personal acquaintance with each of the settlers, the co-ordinator assumes responsibility for 3 to 5 villages only, as against 10 to 20 settlements that are dealt with by one planner. That is one of the main difficulties involved in the proposal to combine the tasks of both planner and co-ordinator in one person.

In view of the above, a reorganization of the regional office was carried out, permitting a better operation of the feed-back mechanism in the field. The co-ordinators became directly subordinated to the regional director (or his deputy). Consequently, the planners are subordinated to the regional director through the head of the regional planning bureau who is in charge of planning in the region. The regional director is, therefore, in charge of both planning and implementation, and it is he who operates the feed-back mechanism between them.

The development team: The integrated approach to development requires the establishment of an inter-disciplinary and inter-sectoral team within the authority's framework. The members of the inter-disciplinary team must have the knowledge and training necessary for planning the various aspects of the development project. It is not simple to get all these experts together to work as a team, since every expert is interested mostly in the problems that fall within his sphere of competence. A uniform outlook is rarely found among different experts, due mainly to lack of a common understanding among the various professions. The only way seems to be to concentrate on the point of view of the 'object', that is, of the local population for whom the development project is planned. Even so, a common outlook and similar working methods are essential and not easily obtained.

In Israel an attempt was made to establish inter-disciplinary teams in two spheres of activity related to rural development – research and rural regional planning. An inter-disciplinary research team is working at the Settlement Study Centre in Rehovoth. It was formed on a personal basis rather than on an organizational structure. Experience shows that not every expert, even high-level specialists, is capable of perceiving development problems beyond his specific field of interest. The work carried out jointly by the team resulted in the elaboration of an inter-disciplinary approach to research in many subjects concerning regional development, such as town-country relationship, the structure of production services in the rural area, social problems related to the establishment of rural centres, etc. In many cases results obtained by the team's work are not immediately applicable in the field, but there is an indirect effect which is already felt especially in regional planning concepts.

In practical planning work some important steps were also taken in the direction of comprehensive team work for regional planning. The Settlement Department had initiated the establishment of an inter-disciplinary team that prepared a comprehensive development plan for the Galilee. The team consisted of fourteen experts, representing the various professions and institutions involved in the development of the area. They all worked under the direction of the head of the team. Among the professions represented were: agriculture, economics, industrial planning, infrastructure planning (electricity, water, roads), architecture (urban planning) tourism and sociology. In addition representatives of certain ministries and agencies participated in the planning as consultants, e.g. the ministries of Health and Education, local and regional municipal councils, etc. Recently, two additional inter-disciplinary planning teams were established, one for the Yizre'el Valley, the other for the Jerusalem region.

In view of the experience accumulated in settlement projects in Israel, some changes were recently introduced into the structure of the regional offices of the Settlement Department. Special planning teams were established in each planning bureau, consisting of a comprehensive planner responsible for the aspects of agriculture, agricultural industry and neutral industry, a social planner dealing with demographic and social problems of the development area and a physical planner who is in charge of the physical aspects of the plan. Such a limited team is capable of guiding and supervising the current operations within the area in question.

The organizational hierarchy: Apart from the planning of rural areas there are still no clear-cut frameworks for comprehensive regional planning in Israel. Nevertheless, there are in existence three inter-disciplinary teams as follows:

1) *Inter-sectoral planning team in the Economic Planning Authority.* This team determines the development policy with a view to an inter-sectoral balance, and guides other planning teams on the regional level. It is responsible for planning only.

2) *Inter-disciplinary team within the regional authority.* This team is responsible for both planning and implementation. The planning done by this team is more detailed than the team at the national level and it produces not only directives, but rather projects for implementation. Such a team must be of an inter-disciplinary nature, so as to make it capable of dealing with as many problems as possible within the development area.

3) *Inter-sectoral supreme committee,* in which various ministries as well as political organizations are represented and which is responsible for the development policy. The committee provides public control on the planning and implementing activities carried out by the administrative corps.

In addition to the above there exist 'ad hoc' planning teams, such as the team for the planning of the Galilee, mentioned above.

Rural regional planning is concentrated in the Agricultural Planning Centre which co-ordinates rural development plans in Israel. Specifically this is the task of the division for rural regional planning of the Settlement

Department which forms an integral part of the Planning Centre. This division consists of three units:

1) *Section for inter-rural co-operation.* This section employs planners who are dealing with problems of inter-village co-operation, i.e. regional co-operation, regional organizations, settlement and employment of the younger generation, social problems related to the development of settlements, etc. Four experts perform these tasks:

a) *Planner of regional projects*, who co-ordinates the work of regional planning teams in every region, directs their work in accordance with the national development policy and also acts as a co-ordinator between various institutions and the planners in the development areas.

b) Personal and public services planner, whose task is to determine criteria and indices for the planning of personal and public services, including problems of location and scope of services.

c) Statistician, who deals mainly with demographic questions, employment and problems of the younger generation.

d) Social planner, who is concerned with the sociological aspect of the comprehensive planning.

2) *Section for agricultural and non-agricultural enterprises:* This section recommends and approves the establishment of agricultural and non-agricultural enterprises. The staff consists of a production engineer and several economists who specialize in problems of location and volume of production of industrial plants. The distribution of tasks among the staff is functional, namely each expert deals with the planning of a certain type of enterprises, and is therefore able to study both national aims and local requirements before he decides on the establishment of the plant. The principal function of the planners involves economic studies of the planned enterprises, while the head of the section deals with general problems of a comprehensive nature, such as the location of plants and their relation to the population in the area.

3) *Mapping section:* This section handles all mapping and other graphical expressions of the plans prepared by the two previous sections.

Units of the authority and its manpower requirements: The structure of the regional unit of the Settlement Department will serve here to illustrate the internal structure of a regional development authority.

The structure of the regional office and its functions.

An average region includes 100 to 120 settlements or 7000 to 9000 farm families. Its boundaries are determined mostly by the natural conditions of the area.

The region is headed by a director whose sole responsibility is to the directorate of the Settlement Department. He is assisted by several deputies and assistants, each in charge of a certain sphere of activity: planning, implementation, social problems, administration and organization, etc.

The regional office consists of several sections, one for each of the following activities: planning, water, construction, citrus or plantations (in case such branches exist in the region's settlements). Some of these sections are extensions of the respective divisions at the centre, though administratively they are subordinated to the regional director. Thus, the water section is an extension of the water division, the citrus section is an extension of the citrus division, etc.

For the purpose of implementation the regions were previously divided into several districts, each headed by a director. The co-ordinators mentioned before were attached to the districts. Recently the district level was cancelled and the co-ordinators are now responsible directly to the regional director.

The structure of the regional office and its powers of authority grant it a considerable measure of independence in the implementation of development projects within its area of jurisdiction. Its functions are as follows: Preparation of the annual budget for the settlements in the area and its implementation; economic and physical planning of the settlements and co-ordination with outside institutions.

The budget is prepared in accordance with the general directives of the directorate of the Settlement Department. The detailed budget of each settlement is submitted to the directorate for final approval. Usually, however, the directorate does not enter into the details of the budget of individual settlements, and is concerned mainly with general development trends within the region. The regional office has therefore a considerable independence in the determination of the development programme for each settlement.

The actual implementation of the budget and supervision of activities is handled by the co-ordinators, who work in co-operation with the water, construction and citrus sections. Each of these sections is responsible for implementation in its sphere of activity. The water section plans water projects and irrigation networks for the settlements in co-ordination with other agencies as required, for instance with the citrus section when the irrigation of citrus groves is considered. After the approval of the plan, it is handed over to contractors, selected by means of a tender. The water section supervises the work and is also responsible for routine maintenance of water networks, for which it employs a special maintenance technician.

The construction section is in charge of planning the facilities included in the budget. The implementation is handed over to private contractors, and the section supervises the work.

The citrus section, in citrus growing areas, and the plantation section where other plantations prevail, are responsible for both planning and planting of orchards as well as taking care of young orchards until they become fruit-bearing. At this stage the orchards are turned over to the farmers.

Economic and physical planning are carried out in the planning bureau of each region. The bureau receives general planning directives from the Agriculture and Settlement Planning and Development Centre. The staff includes both comprehensive regional planners and physical planners, engaged in several functions: economic farm planning, physical planning, mapping and surveying, soil tests and statistics. Economic planning involves the preparing of long-term and short-term plans for individual settlements. The former are prepared in accordance with the natural conditions prevailing in the area and the farm-type selected for the settlement. The latter are prepared in accordance with national production plans determined by the Agricultural Planning Centre. The physical planning involves soil survey, designation of crops to be grown on every plot and parcellation. In fulfilling its functions the physical planning section is assisted by the sections of soil tests, mapping and surveying.

Co-ordination with institutions in the region: The regional office maintains continuous relations with other institutions related to the development of the area. For instance, the Ministries of Health and Education are contacted in matters of health and education and the Ministry of Agriculture in matters of extension and agricultural development; settlers' organizations are consulted in social matters, etc. The work programme of the staff is determined by the regional director in co-ordination with sectional directors within the office on the one hand, and the directorate of the Department on the other hand.

The continuous contact between the regional director and sectional directors, and among the sectional directors themselves, guarantees a maximum co-ordination between planning and implementation on all levels, and enables the planner to introduce the changes required as a result of implementation.

The location of the regional authority and administrative measures for attracting manpower to the area. The problem of location is of special significance in the case of new settlement areas or underdeveloped areas. It is not enough to determine the location of the regional authority by administrative decision; suitable means must be applied in order to attract manpower to the place to work and live there. This is essential since the establishment of close relations between the development team (professionals, instructors, etc.) and the local population is a prerequisite for the success of any project. It is well-known from experience, that quite often people who have attained a certain level of education and a standard of living are not ready to move into a rural area or a distant town, where their social needs can hardly be met. This phenomenon is particularly apparent where uninhabited areas are concerned, but also in inhabited areas, where the more highly educated, especially among the younger generation, find it difficult to adjust to life in their home village. But it is particularly this highly educated stratum that is capable of pushing the development process forward.

All the above considerations were taken into account when the Lakhish region was planned in 1954. The first step taken by the Settlement Department was to establish the office of the regional authority in the middle of the region, where there were only fields with neither buildings nor roads, and no infrastructure network. The usual reaction was that no professionals were likely to accept the offer to work in the area, but experience proved that it was easier than expected. First, the new place offered a challenge, and second, there were special administrative arrangements, namely higher professional status, higher wages, etc., that attracted the newcomers.

The problem of how to attract manpower to newly developed areas is encountered in many places in the world. The subject was raised in the present writer's report on a visit to Turkey, where three rules for the eliminating of such difficulties were suggested:

Personnel in regional offices should enjoy better conditions than those with similar rank and function stationed in the Head Office. They should be granted more pay, better housing, higher status, and wider opportunities for professional training.

Personnel out in the field should be encouraged to travel around and maintain direct contact with the people for whom the development is aimed so as to live with the region and feel its pulse, and to be able to give these feelings expression in their development plans. To facilitate this the relationship between the salary and the per diem paid for expenses should be reconsidered.

A system of service in rotation should be introduced between one region and another, and between the Head Office and its branches. This will help give people engaged in regional planning wider horizons and a deeper understanding of their subject, while fostering a healthy esprit de corps among members of the Department.

The authority and the local population: The relations between the authority and the population in the region is of importance on all levels, from personal contacts to relations with the leadership. In a free and democratic regime such relations are particularly significant since the success of the authority is dependent upon the good will and active co-operation of the people.

In the history of the settlement in Israel there are several cases in which no co-operation was established with a result that the settlement project either failed or did not meet the targets. Following is one example to illustrate the point.

On planning one of the northern regions of the country it was decided to establish there a field crop farm-type. Means of production, land and water quotas, field parcellation and village location were all planned in accordance with the farm-type, which requires bigger plots that of necessity cannot be attached to the farm house. This is in contrast to dairy farms where the smaller plots are adjacent to the farmer's house. Extension officers were attached to the villages in order to assist with implementation. In one case the officer was a farmer from a veteran settlement where dairy farms prevailed. Using his own experience as a guideline, this officer advised the new settlers to introduce dairy cows as a principal branch into their farms, notwithstanding the fact that the farms were planned as field crop farms and without consulting the regional and district directors. The settlers were inclined to follow the advice of the instructor rather than that of the directors, since the former was better known to them and consequently had more influence. They therefore demanded the introduction of dairy cows into their farms even though the structure of the farms and the facilities provided were not suitable. The rate of progress slowed down, with some settlers purchasing cows privately and some maintaining a field crop farm. In this case there was a lack of co-operation between the authority's officers on the one hand and the instructor and settlers on the other hand, with the result that the predetermined targets could not be met. The relations with the settlers have to be established with great care, especially when settlers are of different ethnic origins each requiring a different approach. For that reason it is impossible to elaborate universal rules, and the approach should be suited to each case separately.

The main contact between the authority and the settlers is created in daily work. The local officer of the authority is considered by the settler as a representative of all institutions related to the development project. He is responsible for the implementation of the budget, controls the activities of the rural association, participates in meetings of the village boards, etc. The mechanism of the Settlement Department is designed especially for the absorption of settlers in villages, through the village instructor and the coordinator who is in charge of several villages. In addition to agricultural extension the instructor also deals with all the routine problems of the farm management and submits the information to the co-ordinator who negotiates with external agencies as necessary.

The instructor is usually responsible for one village only, or two at the most, and performs his tasks by daily visits to the farms. The efficiency of his work is determined by the extent of confidence the farmers have in him. The more they believe in him the more they request his assistance. The main subjects dealt with by the instructor are the routine management of the farm, acquisition of equipment and the addition of facilities, determination of crops to be grown, application of more efficient methods of cultivation, etc. He also advises the farmer in case of difficulties, such as diseases, pests, and suggests control methods. The instructor's knowledge which is transferred to the settler is partially self-obtained and partially provided by professional instructors of the regional extension service. Another responsibility of the instructor is to assist the backward farms through individual care even though the owners often do not approach him for help.

It is not unusual to find cases where the farmers lack confidence in the instructor. The reasons are many and different from one case to another. For instance, in villages inhabited by different ethnic groups it is enough for the instructor to establish close relations with one group only for the rest of the settlers to stop co-operating. The only solution in that case is to replace the instructor. In some villages the instructor was selected from among the members. This has proved to be successful in certain cases and a failure in others. It seems that whenever the population in the village consists of several ethnic groups, the instructor should not be a member of the village, so that he could stay above internal intrigues and gain the confidence of all settlers. A member of the village can be chosen as an instructor only if he is a born leader and popular in his village. In such a case he might gain more confidence than any other instructor from outside.

There are usually two instructors on the village level, one deals with professional matters (mostly in agriculture) and the other in social matters. The task of the social instructor is to help the settlers to establish a stable rural community, capable of surviving temporary difficulties without desertion of the village. This is done through the creation of a social infra-structure on which rural co-operation, mutual aid, etc. can be based later on. The whole process must be carried out gradually to enable the settlers to absorb new ideas at their own pace.

The agricultural instructor is responsible for the transmission of know-how to the farmer, in form of practical on-the-spot guidance. He follows the farmer all along the season and advises him in daily problems. Most of the instructors are graduates of agricultural secondary schools, or experienced farmers from veteran settlements, or persons that were trained on agricultural farms. The most important rule that the instructor should follow is to adjust the methods he applies to the social and cultural background of the settlers. It has been found from experience of the Settlement Department that the methods used determine to a large extent the rate and quality of farmers' training. For instance it was found that for settlers of traditional origin and those who did not get much education, demonstration is the most effective method. The instructor selects a few farms in which he gives detailed in-structions on all details to be carried out by the farmers. The success of such farms provides a clear-cut proof that the adherence to instructions leads to the desirable results. It is therefore advisable that the instructor become familiar with the mentality of the population in his village, so as to be able to determine which is the most suitable and desirable method to be applied.

With the development of other means of communication between the settler and professional centre, such as the radio, professional literature, journals and even visits to other farms, the farmer has become more and more independent, and the place of the instructor has been taken by the co-ordinator who maintains the contacts with the relevant agencies (for budgets, construction and other purposes) and the regional professional instructors who visit the farm on request and assist in the solution of problems that arise from time to time, lecture on new developments and transmit information from one farm to another. In most cases the regional instructor is a university graduate, specialized in a certain branch. There are, therefore, several instructors, each of whom is an expert in a certain branch or a group of branches.

It is important to establish personal contact between the authority's repre-sentatives and the population also for the purpose of information. Settlers who are acquainted with the objectives and targets of the development plan will more readily join in its implementation, assuming the role of active partners. This feeling of partnership is a prerequisite for the success of development plans, since it is an expression of the confidence and mutual understanding that prevail between authority and settler.

On the leadership level of contact with the people there is one essential requirement concerning the regional director. He must be a person with the capacity of leadership in respect to both his staff and the local population in order to be able to perform his task. One of the major roles of the development authority is to encourage this capacity of its workers. This may be achieved for instance by using leadership abilities as a criterion for promotion.

On the other hand, the authority should encourage the development of leadership among the population, and maintain contacts with natural leaders. These leaders may be extremely useful in helping to implement development plans. A case in point is the development of local leadership in one of the regions of Israel. This is an extensive region, with many relatively new settlements that were established after the declaration of the State. Climate and soil conditions are particularly difficult and the region has always lagged behind other regions in the country. A few years after establishment, the settlers started to organize for the purpose of obtaining better consumer services. This organization was headed by a local settler who was supported and encouraged by the regional settling authority. With this support the organization has assumed many economic functions and handled all purchasing and some of the marketing activities of the settlers. This organization was used as a tool for carrying out many development operations and with a large measure of success due to the fact that the settlers considered it as their own representation. In course of time the organization developed, and a special development company was established, the functions of which are to implement development plans in the area, including the establishment of regional centres and productive plants. Such purchasing organizations and development companies have become more and more popular, and today most of the settlements and regions in Israel belong to similar organizations.

As a conclusion it must be stressed that the functions of the authority in regional leadership are hard to define since that depends largely on the character of the director. When the director is a born leader, the people follow him and thus contribute to the success of the development project, but he becomes a leader only when he gains the confidence of the people, so that they turn to him whenever a problem which requires his intervention with outside institutions arises, and also accept his advice and follow his instructions.

2. The Ministry of Agriculture

In Israel as in other countries, the main function of the Ministry of Agriculture is to provide regular services to the farmer, including extension, animal and plant protection, farm legislation, market control, preservation of farm stability (by means of subsidies, etc.), and support for the development of certain branches and new methods.

When the Ministry was first established in 1948, it was organized into a number of more or less independent regional units each providing a number of services but without any authoritative regional administration. The administrative authority remained vested in the competent central national departments. The regional units thus were unable to respond to the needs and demands of the farmers even within their own regions. The farmers consequently preferred to apply directly to the central departments. When this was realized, the regional units were liquidated and a centralized functional system was adopted, located mainly in Jerusalem and Tel-Aviv. This system was kept up until 1963 when the Ministry went back to the regional plan. Since then, the Ministry consists of a central administration, divisions, departments, and special services (e.g. the veterinary service or the agricultural engineering and productivity institute) and eight regional offices constituted by geographic region.

Administration: Administration consists of the director general and his three deputies in charge of different spheres of operation; the Ministry's legal adviser and advisers in certain specific areas, e.g. on the development of family farms, on production and marketing councils, etc. In addition, the accounts department, the stores and supplies department and the personnel office are directly subordinated to the central administration.

Extension Service: The Extension Service caters to all farmers in Israel now that most settlements established since 1948 have made sufficient progress. Only villages recently settled still have rural instructors provided by the Settlement Department of the Jewish Agency.

The Extension Service comprises several central branch departments and eleven regional extension bureaus. Each of the central departments is responsible for one or more farm branches: animal husbandry, fruit-growing, field crops, farm mechanization, farm economics, home economics and the like. They supervise the work of the instructors and extension workers from the agronomic point of view. Administratively these workers are connected with the various regional bureaus. Farming information and know-how in the different branches is concentrated in the hands of the central departments and passed on by them to the extension workers at the regional level.

Each of the eleven extension bureaus disperses extension workers in the regions. In point of number of persons employed, the Extension Service is the largest unit of the Ministry of Agriculture.

Water Commission: The Water Commission lays down the farm water price policy (on the basis of economic analysis), operates a hydrological service, and deals with matters of drainage and the efficient and economic use of water. It works in close collaboration with the Water Planning Company of Israel, a public corporation in charge of planning the development of water production.

Plant Protection Division: The Plant Protection Division controls and supervises the use of insecticides, fungicides and pesticides, grain storage, the export of agricultural produce, as well as seeds, plantings and tree nurseries; it maintains plant quarantine and control. In the past it also provided extension services in pest and disease extermination but this function has now been delegated to the Extension Service.

Agricultural Development Department: The Agricultural Development Department is in charge of the Ministry's development budget. It accordingly draws up the draft development budget, implements approved allocations and ratifies the grant of loans to various farms. It deals with soil conservation and the assessment and evaluation of damage caused by development works. It also fulfills a co-ordinating function with banks which grant farm development loans.

Regional offices: The eight regional offices carry out various operations on the instruction of the various technical departments and units. Administratively their staff belongs to a particular regional office, but functionally it is attached and subordinated to the technical units. The workers of the regional offices carry out the following functions: agricultural planning and development; soil, water and drainage problems; production engineering; plant protection; veterinary services and extension work.

Of the eleven extension bureaus, eight operate in conjunction with one of the regional offices, while three run their own office, and engage solely in extension work.

Further technical units of the Ministry are the Veterinary Service and the Veterinary Institute, the Farm Engineering and Productivity Institute, the Foreign Trade Division, the Subsidy and Price Department, the Fishery Division and the Division for Agricultural Co-operation in Developing Countries.

In addition there is the Agricultural Planning Centre which is jointly run by the Agricultural Ministry and the Settlement Department of the Jewish Agency as mentioned above.

3. Other agricultural institutions

i. *Agricultural Research Station*

There is one central Agricultural Research Station in the centre of the country, with subsidiary Stations in the different regions. Though affiliated to the Ministry of Agriculture, the Agricultural Research Station enjoys a wide degree of autonomy in its work. Its operations are co-ordinated with the Faculty of Agriculture of the Hebrew University which also engages in agricultural research by means of a roof organization common to both – the National and University Institute of Agriculture.

The Agricultural Research Station is the centre for the country's entire agricultural research. The projects are determined in consultation with the different branch experts of the Ministry of Agriculture, the Settlement Department and the Administration of the Ministry.

ii. *Israel Land Authority*

The Israel Land Authority is in charge of all public land in Israel – rural and urban – and controls all the transactions involved. The authority was set up in 1959 by a merger between the Land Department of the Jewish National Fund and the Land Department of the Ministry of Agriculture into which various other bodies were also absorbed. Government land, abandoned property under state custody and the land holdings of the Jewish National Fund were thus placed under the control of one single authority.

The Land Authority is presided over by a Council headed by the Minister of Agriculture. This Council determines its land policy and its annual budget. In addition there is a Supreme Committee for the approval of rural leases, composed of senior staff members and representatives of the farming community.

The amount of lease payable is fixed by the Government. The present rates are two percent of the value of non-agricultural land or of the value of the farm unit and its projected income. The amount is thus not determined by the area leased, but by the income the farmer's family derives from the land of which it pays two percent.

The Israel Land Authority has two major divisions and three regional offices. The two divisions are the Land Title and Registration, and the Land Use Division. These administrative Divisions also comprise mapping, land registration and survey service and a map library. The Authority has its own legal service to deal with land claims.

The duties of the Land Authority are to hold and register the land under its control and to either sell or lease it. Every sale requires the prior approval of the Minister of Agriculture who is responsible to the Government for the operations of the Authority. Other lands are leased out. The allocation of the lease of agricultural land to the various farmers is determined by the Joint Planning Centre by way of the settlement plans it draws up.

iii. *Production and marketing boards*

Many branches of agriculture which at one time produced less than the market demand, eventually produced more than the market could absorb. It obviously became necessary to introduce adequate planning so that farmers should produce crops and other produce according to market requirements, taking into account the low elasticity of agricultural products. This was difficult because it often meant changing output quotas to which farmers and settlements had formerly been entitled – and safeguarding the rights of farmers in new settlements which had not yet reached their full potential. Further difficulties were caused by a considerable disparity between old and new farm types and the fact that at the time no suitable organizational means were available for comprehensive preliminary surveys and tests. Yearly ad hoc arrangements had to be made.

In 1959 the Minister of Agriculture started appointing production and marketing boards. The cabinet and sometimes also the Knesset vested authority in these boards so that their decisions constitute regulations binding upon the farmers. As a rule the boards are composed of representatives of the Ministry of Agriculture, the various settlement agencies and any other institution with which the council may be affiliated according to the branch in which it deals (large marketing firms, consumers' associations, etc.).

The boards supervise the implementation and registration of the quotas allocated. To enable the younger settlements to develop they are given an overall allocation from the Settlement Department which is distributed according to settlement plans. So far, production and marketing boards have been set up for poultry and eggs, milk and dairy products, vegetables, sugar beet, cotton, plantations, citrus fruit, vineyards, flowers, beef, tobacco and groundnuts.

The Ministry's directives are transmitted to the boards through the deputy director general for economic affairs, who also sits on all the boards, presiding over some of them. The board is informed of the amount that it is profitable for the branch to produce each year and shares this out among the producers. Its actual function is confined to maintaining existing quotas and to the allocation of supplementary quotas, if possible. This is in a way a substitute for current planning. However, as long as there is no planning of farm income, the main objective of agricultural planning is lost. There accordingly is an increasing tendency towards branch planning and the planning of entire farms. However, the greatest drawback of the board is its exclusionist trend. The boards represent the present producers and hinder the development of new settlements by barring their entry into the production cycle. The boards have thus become the scene of conflict between old and new settlements which have not yet reached their full potential.

Owing to these developments there is a growing awareness of the need for a different system by which output would not be determined by each board separately, but by a central planning authority. Thus it will be possible to ensure that the farmer has a fair income, which is in fact the principle condition for rural stability.

iv. *The Jewish National Fund – (Keren Kayemet LeIsrael)*

The Jewish National Fund (JNF) was established by the Zionist Organization several decades before the establishment of the state for the acquisition and amelioration of land to be placed at the disposal of agricultural settlers. After the establishment of the State, the Keren Kayemet took on additional functions, but the title of the land was transferred to the Israel Land Authority which took care of all transactions connected with it. The JNF, in addition to soil preparation, took charge of afforestation activities.

The JNF is run by an executive board which determines its policy.

The JNF comprises three divisions: the Land Development Administration, the Finance and Economics Division and the Organization and Information Division. Since the JNF finances its operations out of its own income, derived mainly from abroad, it gives the prominence to finance and public relations. A considerable portion of the funds consists of legacies which require administrative work, and the maintenance of relations with potential donors. The Land Development Administration of the JNF has a number of departments:

Afforestation Department, which plants trees on land which is not fit for agriculture. The Department owns the forests it plants and is responsible for their upkeep. The work teams and supervisory staff are attached to the Department's district offices.

Soil Preparation Department, which is responsible for land betterment and construction works which it carries out on its own behalf and on behalf of other departments of the JNF. It builds roads in inaccessible and sparsely populated districts, helps in building roads in agricultural settlements and provides dirt tracks through the forests.

Drainage and Water Department, which carries out large-scale operations such as swamp drainage. One of its major feats was the drainage of the Huleh. Now that little remains to be done in this sphere, this Department has been amalgamated with the Soil Preparation Department.

Survey Department, which serves the other departments as the need arises.

Since the JNF's operations are mobile by nature, it was impossible to have autonomous district offices for each department although each district has a JNF office.

The Land Development Administration is run by a board which draws up the budget for the approval of the JNF executive and controls current activities.

Since rural and agricultural development is a dynamic process, changes take place from time to time in the structure and functions of the agencies concerned. As we have seen, considerable changes have indeed taken place in the organizational structure of the Ministry of Agriculture and of the Settlement Department. As soon as more settlements become economically and socially consolidated further changes will occur.

Thus, for example, the Settlement Department took care of 480 agricultural settlements in 1965. By the end of 1966, 110 of them had left its auspices and during 1967 more are expected to follow. These settlements, which have sufficiently matured, no longer need the special care provided by the Settlement Department. This requires modification of the organizational structure of both the Settlement Department – whose charge they leave – and of the Ministry of Agriculture, whose responsibility they become. Other agencies concerned with rural and agricultural planning also undergo changes. The entire structure accordingly is in a constant state of flux.

The advance of recent settlements towards viability, the new settlements planned, the gradual shift to a regional rural structure and the growing integration of agriculture and other branches of the economy will undoubtedly require further changes to meet the new conditions that arise.

Achievement of the necessary structural adaptation of the various institutions dealing with the farm economy and the regulation of price policy, subsidies, investments, research and extension as will be outlined in the next sections of this chapter are a prerequisite for the development of agriculture.

B. MEANS OF PRODUCTION CONTROL

There are several means by which the development authority or the governmental framework responsible for the implementation of planning, can induce the farmer to farm according to a plan. They are economic – that is by pricing, by quota control, and by subsidy – and fiscal.

1. Economic means

i. *Price policy*

Farmers produce for a market in response to prices which are their guide to demand and to their expected profit. Experience has shown that the free play of this price mechanism, particularly in the short run, does not lead to the matching of supply with demand. Agricultural producers tend to react to price incentives as a body, with correspondingly drastic effect on markets, leading to production cycles. The reaction of farmers to low price for a product in one season, for example, is to produce less the next, leading to commodity shortages and a high price. The high price will induce many more farmers to enter production so that the market becomes depressed the next season. A resulting low price will again discourage future production so that the whole cycle begins once more.

The uncertainty for the farmer arising from these cycles is a main source of waste and loss in farming, leading to inefficient allocation of resources among farm enterprises. The small farmer, in particular, cannot plan a long-term production programme effectively, if he cannot assess the outcome of total profit and relative profit between branches. The farmer rations his capital. He is afraid to invest in the best and cheapest production methods because he may choose wrongly with respect to future markets. His loss may be heavy in terms of his investment; he will have lost profit from other enterprises in which he could and would have invested, had he known the relative market prices at the market time.

From the point of view of the consuming public, uncertainty at the farm level and hence an erratic flow of essential supplies is also a source of waste, leading to some goods being in short supply and high priced or the opposite.

The cost of losses in the farming sector and its correspondingly depressed level is also felt by those who supply agriculture with essential raw material and consumer goods. Fluctuations in the agricultural price level have a definite unsettling effect on the general price level and the cost of living. The general price level, reflecting the health and prosperity of the economy, is itself of vital importance to agriculture. Agricultural prices are among the first to fall if the general price level falls, since farmers, unlike factory owners, cannot stop or slow down production: the incidence of fixed costs is too high and maturing crops and livestock cannot long be held back from the market. Agriculture can only prosper if its buyers at home and abroad are prosperous and able to absorb its products. Full employment in a country as a whole is a necessary condition for an active market in agricultural goods. Losses in farming mean losses elsewhere and a reduction in secondary and tertiary employment generated by agriculture.

To keep an economy as a whole at full production must be a cornerstone of policy; it can go far in solving agriculture's problems. It is obviously desirable from all points of view and for all sectors of the economy, that the production effort of agriculture should be stable in its overall dimensions (which, in the main, it is) and that the supply of the various goods comprising this total effort should be in accordance with the needs of the economy. Cycles in production caused by price uncertainty detract considerably from this stability.

Besides the price mechanism, there is no way in which the farmer can be informed of the state of the market and hence plan his actions. If this information is to be of use to him initially, it must relate to the market price he will receive at the conclusion of production when ready to sell. This price in a free market is likely to be different from that ruling at the time he makes his production decisions.

Because government is able to examine future demands for various agricultural goods, it can try, through a controlled price structure, to pass this information on, so that the sum of the farmers' effort will lead to production of the desired quantities.

The limitations of price policy: It is not easy, however, to arrive at a controlled price structure which tends to match supply to demand. Both supply and demand are difficult to estimate. Price is only one factor facing farmers when they make their production decisions. Even if the price fixed for a particular crop leads to a planted acreage which should meet the projected demand according to yield norms, weather can lead to yield fluctuations which completely upset the supply projections.

Price supports can remove uncertainty as far as market fluctuations are concerned, but they cannot directly influence yield per unit which is just as important in determining income. They do have an important indirect effect on yields because the farmer can arrange his production to specialize in a few branches, become expert in them and make a maximum effort for high yields. Through investment in machinery, buildings, good seeds, fertilizers and other inputs, he can largely mitigate the negative effects of weather. It costs money to do this, but if there is a certain market for the final product, the farmer can realistically weigh up the benefit of investing in inputs to obtain high yields. When price uncertainty reigns, however, he is not sure whether high yields will mean low prices; he will likely play safe on inputs and have far less control over the success of the crops. Nor does he know which crops will bring high prices: he is forced to diversify his efforts at a consequent low efficiency in the hope that some products will return a good income.

Price guarantees therefore, directly increase the efficiency of the individual farmer and can therefore also be used to improve the overall efficiency of agriculture. The aim is to attain both production and economic efficiency. Planning seeks to attain both – that is, efficient farms working with economic use of resources, producing the flow of agricultural commodities needed by the nation. When supply and demand are in sound balance, the farmers should enjoy acceptable living standards and the consumer should be supplied with food and fibre at an acceptable price.

Price policy is an important aid to reaching this balance, but one which must be used carefully. Price supports quickly become vested interests of agriculture, and once set up are difficult to remove. Unless they are designed with inherent flexibility, they may lead to grossly inefficient allocation of agricultural and national resources. It is important to make a careful distinction between the basic issues at stake in formulating price policy and the aims it is intended to achieve in various situations.

Prices and the food problem: A plan designed to increase food production in a large and undersupplied market may include price incentives to encourage farmers to branch out into new crops or livestock production. However, an increase in the supply of cheap investment credit or working capital would probably in such a case be more useful as an incentive. High market prices on their own may have failed to increase output because the physical and capital structure of farming has been inadequate.*

* Expanding the physical and capital structure of farming is dealt with in detail in a later section.

At the same time as improving the capital structure, price guarantees on new production may be necessary. In a seller's market, it is not likely that the government giving the guarantees would be called upon to pay out heavily to back them up. There is a need to move resources into agriculture to step up its productivity: by underwriting its plans with firm price guarantees, the government can encourage farmers to invest. In Israel the guarantee for prices is based on the costs of production, including a profit margin. Such price guarantees are one means of raising supply to meet demand.

The effectiveness of a price increase in achieving more production depends on the elasticity of supply – the ratio between the percentage increase in supply, and the percentage increase in price which generated it. This ratio differs for different farm products; if large investments have to be made by farmers in order to increase supply, a large percentage increase in price may be necessary to encourage them to do this, and supply is inelastic. When the investment is made however, a comparatively small percentage increase in price may make a large input of variable factors to the investment worthwhile, generating a high percentage increase in supply. The change is to a high elasticity of supply.

Such may be the case in a country building up a dairy herd, for example. A change in the price of milk in the early stages will have to be large enough to encourage a significant change in the size of the national dairy herd, as education, building, stock, forage and feed production will all have to be increased. Once these facilities are established, a small increase in price is able to generate a large increase in supply as the farmer finds it worthwhile to aim for higher yields by increasing his variable costs – feed and labour input.

With field crops, a small percentage change in the price of one of them may so alter its profitability in relation to other field crops, that farmers as a whole will increase their acreage at the expense of the other field crops: it is this internal substitution possible in agriculture between one crop and another that leads to the great flexibility in the planning of individual farmers, and increases the difficulty of administering a price structure unless demand is greater than supply. A higher price may also lead to larger variable inputs of fertilizers and labour per land unit, the resulting increase in yields producing a larger crop increase than was anticipated by the government in fixing its guaranteed price to farmers.

Similarly, price guarantees may be used to encourage production in particular sectors, and re-allocating resources from lines of overproduction. By lowering guarantees on products in sufficient supply, and setting high guarantees on the product in short supply, farmers are encouraged to move over into new and unknown lines of production. By giving a guarantee, the government shoulders part of the risk which farmers may not be willing to undertake on their own.

An example of a special price structure, tied in with a particular credit scheme, was operated in the south of Israel to encourage farmers to start growing cotton. Without the incentive, they were unwilling to risk the crop. In 1958, the directed credit amounted to IL. 130 per dunam of cotton, as shown in Table 8-1 which compares credit advances for cotton with those given for other crops.[1A]

TABLE 8-1

Credit for selected crops

Crop	Size of advance IL./dunam	Terms of loan (months)
Potatoes	50	6
Vegetables (export and industry)	40	4
Groundnuts	65	6
Sugar beet	40	6
Cotton	130	7

The price of cotton was also subsidized on a cost-plus-profit basis. The Ministry of Agriculture together with farmers' representatives estimated the costs of production with a 'suitable profit'. The difference between this price and the lower price of imported cotton was paid by the government. Thus the spinning mills bought from the farmers at the price of imported cotton and the farmer was compensated by a government subsidy.

As the yields of cotton per dunam would rise and production costs per kilogram decline, the subsidy was to be reduced, for the desired goal of expanded cotton production would have been reached.

Prices and the farm problem: In a different situation, where the market is too small to absorb the products of agriculture, a farm problem results. Market prices and hence farm incomes are low. It is necessary to find new markets for agricultural produce, or encourage farmers to grow different crops for which a market exists. In this case, price policy works as previously described. It is not, however, likely to solve the problems entirely. There is a need to move resources out of agriculture, but social justice and the power of the farm vote do not always allow low prices in depressed rural areas to push people off the land indiscriminately. Pressure is exerted to support farm incomes so that the whole nation shares the cost of the re-adjustment until the number of people moving out of agriculture bring it into better balance with the rest of the country, or until the market expands sufficiently to absorb the products of agriculture. This latter alternative is the case in Israel. The increase in settlement has led to an agriculture with a production potential too large for the present market; thus production of certain commodities was restricted by imposing 'production quotas'. The urban population is, however, rapidly increasing due to immigration and natural increase, and will be in much better balance with agriculture in a few years time. It will take probably eight or ten years or more until the full economic potential of Israeli agriculture can be realized and its expensive irrigation land be used to produce high value fresh products for home and export markets. Until then, something of a farm problem will exist in Israel. As seen by the 1972/73 plan, expansion into new crops and for new markets will be undertaken. Pricing will be used to encourage this expansion and to support farm incomes as it has done in the past. There are three main ways in which prices are guaranteed for farm products in Israel: Prices fixed by law, minimum prices and two-price structure.

Prices may be fixed by law at a maximum designed to protect consumers in the interim if production is below the market needs, or to guarantee a minimum income to farmers if surpluses exist. The price to the consumer cannot be above the retail level fixed by the law, and the price to the farmer cannot be below the fixed wholesale price.

As a means of supporting farm incomes, price policy must be carefully controlled; guaranteed prices will otherwise lead to a flood of products for which there is no market and government will have to pay out the difference between the resultant low prices and its own guarantees. In the situation in Israel, the aim is not to increase supply to meet demand, but to guarantee farm incomes in the face of inadequate markets. The incentive which the guaranteed price gives to higher production must therefore be curbed by quota systems so that the government liability to meet its guarantees does not surpass the amount it budgeted for the purpose. These guaranteed or minimum prices act as a floor level, and are arrived at by discussion on the various products between government and farmers' representatives. The price is based on cost of production plus profit, and is fixed for one production period ahead. Should the market price be above this minimum, the government has no liability; if below the minimum, the government pays out the difference on the amount of production it has guaranteed.

In order to add to farmers' income above the minimum price level, a minimum price fund is in operation for certain products which show large seasonal variations. A levy is made on high prices and put in a fund to raise the price to the farmer in seasons when only the minimum price is reached on the market. The following example shows how the Vegetable Marketing Board of Israel has operated its minimum price fund in recent years.

The 'Vegetable Agreement' for minimum prices is concluded between the Vegetable Marketing Board and the Government, and renewed each fiscal year. The agreement provides for the disposal of surpluses which remain unsold or when the wholesale price drops below the agreed minimum. Most of the surpluses are sold to the canning industry; a certain quantity is given to welfare institutions and the rest is used as animal feed. Exceptionally large surpluses for which no use can be found are destroyed. For all these surpluses, farmers receive the agreed minimum price. If sold for canning or animal feed, the government pays the difference between the price received and the agreed minimum; if donated to institutions or destroyed, the government bears the entire loss. In recent years the agreement has included seven main vegetables: tomatoes, potatoes, cucumbers, carrots, green peppers, cauliflower and onions – together, some 75% of Israel's vegetable production. Not included are the vegetables grown on limited areas for which there is small likelihood of surpluses.

Regarding potatoes, for which there is no immediate question of surplus since unpurchased quantities are stored against the seasons of scarcity, the minimum price refers to the entire crop in order to guarantee orderly supply throughout the year: the minimum price is paid for each ton stored. In marginal seasons, in which production falls behind demand, sale is directly to the market without intervening storage.

Part of the agreement is a series of directives laying down national growing areas for each of the above seven varieties: the Vegetable Board fixes an area for each farmer for each of these varieties, in accordance with the national allocations. Surpluses are thus reduced to a minimum.

In 1961/2, outlay for the disposal of vegetable surpluses was IL. 2.37 million, and IL. 1.8 million through the guarantee of minimum potato prices – a total of IL. 4.17 million. In 1963/4 the total was IL. 11 million, and in 1964/5, IL. 10.3 million.[2]

Another means to limit government liability is to create different price structures for the same product artificially. Milk is the best example. It is sold at a high price on the liquid milk market, and surpluses are diverted at a much lower price for processing. The farmers' total income from the two-price structure is compared to the amount guaranteed by government, which pays any negative difference.

ii. *Controlling and directing production*

Before production: In many countries and for many products, control of production is not undertaken in the full sense. Price guarantees may be given for limited quantities of produce, but their use should leave the market as free as possible. Price policies which are not severely limited, to restrictive production quotas may leave government with large surpluses on its hands, and transfer the farm problem to the nation as a whole without leading to any solution. Price supports, once granted, become the vested interest of the farming population which presses to increase rather than reduce them. A government in tending to provide a price floor to guarantee farm income must therefore act with extreme caution from the outset. The price floor must be used:

a) to improve, not burden, the efficiency of agriculture as a first step to easing the shift of resources necessary to bring agriculture into balance;

b) to keep the liability of the taxpayer for farm subsidies within reasonable limits.

Some form of production control is necessary if the market situation is such that guaranteed prices alone are insufficient to effect the changes wanted. It inevitably implies a strong administrative apparatus and considerable bureaucracy. Applying and checking on controls is costly, tends to be clumsy and is often inaccurate. Care must be taken that the institutions, marketing boards or other bodies undertaking the administration are properly organized for the job on hand, have proper representation from the interested parties of government, growers and consumers, and the necessary authority to carry out their work. Such institutions in Israel are described later in this chapter.

Israel uses production control for three specific purposes:

a) to limit the government's subsidy liability;

b) to allocate the acreage of crops with guaranteed prices among farmers registered as the holders of farm types suited to the crops. In this way, each farmer can draw up a farm plan guaranteeing him an income. This, in effect, represents a sharing of the limited market among the over-numerous farmers, and undoubtedly leads to grave inefficiencies in production patterns. However, the social obligation to maintain farm income, particularly in the new settlements, is such that the practice must continue until the market becomes large enough to enable liberal production policies and more freedom at the farm level.

An example of this kind of production quota in Israel is dairy farming. Milk quotas are allocated only to farm types designated as 'dairy farms':

for each such unit, a certain production quota is fixed so that the country-wide sum of such quotas is in accordance with the national milk production policy determined by the government. Veteran settlements with large herds and an annual production of over 24000 litres had their production frozen at the stated output. For new farm units in young settlements, with limited production, the quota was fixed at a level above actual output, enabling them to increase output to the level of the annually established quota. Their production is usually frozen when they reach 24000 litres a year. Farms not defined as dairy types or which exceed their quota do not receive the subsidy on produced milk resulting in non-profitable production. As a result, not many farmers stray from their quotas.

Another example is poultry farming. To strengthen hill settlements they have in recent years been allocated large egg quotas, while egg production in farm units with production alternatives has accordingly been reduced. Those exceeding their quota do not receive the guaranteed egg price which includes the subsidy.

Quotas operated on a seasonal basis can be used to extend the time over which crops are produced in relation to market demands. Thus, if early production in southern Israel is tied in with late production in the north, the total volume is better geared to a stable market demand.

This use of quotas is very important in so far as it affords a means of using price policy to solve the farm problem where it is most difficult: on the smaller and weaker farms. If guaranteed prices are offered to the farming industry as a whole, the larger and more efficient farmers working at low cost expand production, clog the markets and qualify for the minimum prices which to them represent a large profit. The small farmers, which in many countries grow a large part of household food needs, only benefit on the amount of produce they sell. The bulk of the large subsidy payments therefore goes to farmers who do not need them, and the smaller farmers for whom they were designated do not benefit as much as they could. They also feel the competition of the large farmers and are not encouraged directly to expand their commercial farming.

By restricting the acreage which larger farmers can work in particular crops and allocating acreage to smaller farmers together with appropriate extension methods and encouragement, the small farmers can gain a larger share of the market. The quota system reaches to the heart of the problem – though nonetheless accompanied by difficulties in costly administration and marketing. Imbalance due to unequal distribution of resources among farmers will always lead to corresponding income differences whatever price policy is adopted. Israel has used the quota system extensively to guarantee farm income to all her farmers though it must be pointed out that the farms do not differ greatly in size: most are family farms (the kibbutzim, for this purpose, can also be regarded as composite family farms) and little can be said here on the applicability of the system to other countries.

c) The third use of production control is to direct the production of particular products to the areas most suited to them and away from others which could more profitably grow different crops or livestock. The survey of Israel's natural resources described earlier show the wide variation in climate, topography and soils in the country, Some areas are suited to one type of production, some to another. By price arrangements specific to certain areas, and by allocating larger quotas to favoured areas, production is concentrated in the areas most suited to it.

Through active planning, direction and encouragement, Israel is developing her agriculture on a regional basis. Part of the country's area was already well established and producing before the establishment of the State. The fertile plains with cheap water resources, a ready market in fast growing towns for milk and livestock products and a soil and climate suitable for citrus led by natural development to dairy, citrus and poultry farming. The new settlement after the establishment of the State was planned in relation to regional and market factors. Various farm types were developed to meet regional needs: fruit and tobacco cultivation in the hills; industrial crops in the south; high value crops for early season export where climate was suitable in the Jordan valley and southern Negev; beef and sheep grazing in extensive hill areas.

A characteristic example of such regional adaptation is the cultivation of deciduous fruits and vines. The natural area for these products is the hill area, where production alternatives are in any case lacking. In many low-lying areas, it is indeed agrotechnically feasible to grow these products, but the fruit is in some cases of low quality and there are in all cases production alternatives. It was therefore decided that the planting of vines or deciduous fruits would require a government licence. Such licences were given only to hill farms. Transgressors can be summoned to court and their groves uprooted. There are therefore very few cases of non-licenced planting.

There are a number of severe problems connected with quota systems operating in relation to guaranteed prices. The difficulty is in deciding on a basis for quota allocation. If an acreage basis is chosen for crops and price uncertainty thus removed, for example, farmers are free to intensify their operation by heavy application of inputs such as fertilizers. Given 'normal' weather conditions, it is likely that the average yield will be higher than the norm accepted at the time the price was fixed. The government is therefore faced with the payment of heavy subsidies. In the U.S.A. schemes to retire land from production and stem the flow of surplus products to the market failed because farmers intensified on the acreage left and average yields rose considerably.[3] They were helped in this by the money they had received for retiring part of their land from production. Because land is only one of a large number of variable factors important in production, and can be substituted to a large extent by capital, labour and enterprising management, it is unreliable as a basis for allocation schemes.

But no other basis readily presents itself for crop production. If the guarantee is made on a tonnage basis, what is the farmer to do with his surplus produce in years of high yields? If one product is produced in surplus and fetches a low price on the market, this also affects the demand and price of products for which it can substitute. Very low priced tomatoes will attract demand from higher priced cucumbers to an extent that may affect the cucumber marketing situation considerably. If surpluses are left outside the price guarantee system, they find their way to the markets by other ways and depress prices for the entire crop. The subsidy to be paid under the minimum price agreements will thus be raised.

For livestock, the problem is somewhat different, although probably as difficult in other ways. Farmers may be granted quotas adapted to the size of their enterprise and levels of input. But stock multiplies itself; stock farmers want to use this increase to raise the size of herds or flocks, and they therefore exert pressure on government to raise the quota allocation. Such pressure is difficult to resist in a democracy. Technological advances, better feeding methods and management continually tend to raise the output per unit: by adopting them, the farmer finds that he overshoots his quota. He is not likely to take kindly to an order to reduce the size of his herd, particularly if he has built it up because of the encouragement of price guarantees. Raising quotas over the market capacity forces government to search for other markets, often unprofitable, or to hold large stocks of the surplus products.

Control after Production: If the restrictive measures on production are unsuccessful and supply oversteps demand under the minimum price agreement, there are certain controls which can keep market prices up so that the difference between them and the minimum price is not so great that government liability becomes excessive. These measures are: (a) storage: (b) flexible grading; (c) dumping; (d) conserving.

Storage: The supply of many farm products is seasonal, but a demand for them exists all year round. Prices are low at harvest time, and high for the rest of the year. Storage absorbs produce in excess of market requirements at harvest and releases it throughout the year as demand necessitates. A firm price is thus maintained throughout the year, making allowance for the costs of storage, which may be considerable.

Two examples are potatoes and apples. At the height of the harvest, with relatively low prevailing prices, part of the crop is stored and later redirected to the market. With regard to apples, in Israel harvest occurs in the summer and early autumn; however, apples are now available all the year round because of storage facilities: otherwise, prices would drop below the profitability level during the harvest months. Table 8-2 is a breakdown of apples marketed in Israel during 1962/3 and 1963/4.

TABLE 8-2

Marketing of apples 1962/3 and 1963/4 [4]

Month	1962/3	1963/4
January	3386	3768
February	2762	3350
March	2529	3696
April	1913	2153
May	1184	1677
June	754	988
July	1495	2231
August	2980	4681
September	5463	7474
October	4573	3841
November	3901	3503
December	3208	3618
Total	34148	40980

Early varieties in Israel are harvested in July and the latest in October. From November until June, large quantities of apples are marketed which have been put away in cold storage.

The storage life of some products is long, given protection from damp and vermin. It is desirable to keep turning over stocks, however, as deterioration of quality may take the eventual price well below that of new produce. Apart from emergency stocks to guard against war or famine and surplus stocks, the storage period is usually from times of low price at one harvest to the next. Aside from price considerations, some seasonal farm products are essential all the year round and adequate reserves of such staples must be kept to feed the population or supply fibre factories. In times of undersupply or famine, storage schemes operate to ration the limited supplies throughout the year, and direct them to areas or sections of the population as required.

Flexible grading: Grading standards lend themselves to alteration and can be used to hold produce back from the market. This practice is widely used in vegetable marketing. In times of low price, grades are made stricter and only top quality produce is sold. The second grade produce is diverted elsewhere. When prices are high, this second grade produce is also sold as first grade. Such activities represent monopoly practice and can be broken unless all producers are included in an organized marketing scheme which handles all of the produce. Farmers with only low grade produce find it profitable to sell outside the organized scheme and therefore bring the market price down by their competition. If such schemes are to work, an element of compulsion, legal or through co-operative discipline (rarely successful) must be used. There is danger in such methods, both for sections of the farming community and for the consuming public, unless their interests are safeguarded by representation on the administration or marketing boards which carry out the grading and pricing.

Dumping: Produce held back from the market in order to keep prices up often has no alternative use and therefore has to be dumped. Dumping is wasteful, but may be unavoidable, if fair prices for farmers' produce are to be obtained. It should be regarded as a last resort and its necessity should stimulate efforts to bring supply more into line with demand. Dumping is a very poor solution for chronic oversupply, year after year, in which case, resources should be transferred to other production. However, it is sometimes the only way of avoiding ruinous prices in a year of especially good harvests.

Conserving: Perishable surplus produce is sometimes preserved by industrial processing and stored without a particular market in view, or sold at a considerable loss. Milk and eggs lend themselves to this expensive form of storage which is necessary to take surplus supplies off the market and protect prices. It is a better alternative than dumping, but, again not satisfactory in solving problems of chronic oversupply. The type of industry which processes surpluses for storage, differs from agricultural industry which operates on a commercial basis for a direct market, and which is in direct competition with the fresh market for prime produce.

iii. *Subsidies*

There are two kind of subsidies to farmers: direct and indirect.

Direct Subsidies: A direct subsidy is an amount paid out by government to farmers as a result of a price guarantee scheme: it is the difference between market returns and the sum guaranteed – in other words, the sum necessary to underwrite government encouragement of agriculture through which the cost of maladjustment between the supply and demand for agricultural products are, to the extent of the subsidy, transferred from the farmer to the taxpayer. The discussion on minimum prices and quota systems has included many points relative to this subject, but a few further points should be mentioned:

The difference paid out when market price falls short of the minimum price is usually called a *producer subsidy*. For some products in Israel, notably eggs, the wholesale price is not determined by the free market but is controlled. The government guarantees a minimum price to the farmers for eggs on a quota system and can therefore estimate its subsidy liabilities before the production period. Vegetables are left to find their own price on the free market, although grading is used to some extent to regulate the amounts reaching the market. Since vegetable quotas are allotted on an acreage basis, and since a minimum price is guaranteed, the government does not know its liability until after conclusion of the season. It may have no liabilities and pay no subsidies to producers.

In the early years of the State, the Israel government wanted to encourage milk production in the interests of good nutrition. However, the consumers were unwilling to pay the price at which the farmers were willing to produce. Therefore, the government offered a minimum price to farmers for milk and sold it, at a loss, to the consuming public. The difference was a *consumer subsidy* which was designed to gain an increase in consumption. The difference between producer and consumer subsidies is very fine, for underlying the idea of increasing consumption of farm products rather than reducing supply, is likewise the desire to support farm incomes.

If a government, in time of war or food shortage, encourages farmers by price guarantees and cheap long- or short-term credit, to invest in their farms and increase production, the farmers' response may be greater than predicted. This is especially true today, when technological advance is continually raising the output per unit of input. However, once the government is committed to a price which bears a fair relation to the farmers' costs at the time of the pricing, as well as to the market demand, it cannot cut prices to reduce over-production to an equilibrium level without causing loss to those who invested because of government encouragement. It must therefore bear the extra cost itself through subsidies, seek other outlets for the surplus production, or aim at raising consumption by lowering the market resale price. Adjustment downwards can only be achieved in the long term, during which time positive efforts to raise consumption may have equal success in solving the problem.

The effectiveness of a consumer subsidy depends on the elasticity of demand for a product – the ratio between the percentage increase in demand and the percentage change in price causing it. A small lowering of price will lead to an increase in total revenue, since more of the product will be sold if the demand is elastic, and a loss in revenue if it is not. If demand is elastic, a small percentage change in price will lead to a larger percentage change in demand, and total revenue, which is price multiplied by amount sold, will rise.

If demand is inelastic, i.e. not sensitive to price *change*, more revenue may be gained by monopoly practice through withholding produce from the market to gain a higher price per unit over less units. This is the justification for storage schemes.

Direct subsidies to agriculture in Israel in 1963/4 amounted to IL. 99.9 million[5]) or about 8 percent of the total value of agricultural production for that year. In 1964/5 direct subsidies amounted to IL. 92.8 million or 7 percent of the value of agricultural production which was IL. 1337 million in that year.

Table 8-3 is a breakdown of direct subsidies in 1963/4 and 1964/5.

Indirect Subsidies: Subsidies to agriculture in actual practice take several forms other than the price guarantees and consumer subsidies mentioned. Government may want to expand a certain branch of farming, or undertake

TABLE 8-3
Direct subsidies in 1963/4 and 1964/5 (million IL.)[5]

Branch	1963/4	1964/5
Eggs	26.97	24.94
Poultry meat	6.96	5.54
Milk	31.78	32.76
Cotton	10.74	10.26
Vegetables and potatoes	11.02	11.60
Beef	3.06	1.12
Carp fish	2.14	1.76
Other branches	7.23	4.83
Total direct subsidies	99.91	92.81

a new settlement. The first large long-term investments in new buildings, equipment and stock are beyond the financial capabilities of small and new farmers so that cheap long-term credit facilities are usually made available. Cheaper credit to the farming community is in itself a subsidy. The first production will most likely be of high cost due to inefficient methods and initial difficulties. A subsidy on production in this period is justified until the branch or new settlement reaches a viable stage and takes its place in the farm economy. The argument is similar to that advanced for 'infant industry'; while an industry is establishing itself it needs protection from competition until it reaches a scale at which its costs become no more than those of its competitors, and the subsidy can be withdrawn.

Subsidies are sometimes given to reduce the price of raw materials – feeds and fertilizers – in order to encourage their use. Such subsidies are often a much cheaper and more positive way of raising production than a guaranteed price. Their main use, however, is in encouraging farmers to take advantage of a favourable market by increasing production.

Inputs must be balanced with one another in order to obtain good yields without waste of expensive input components: a good cow cannot yield properly unless fed a balanced diet; the best seeds cannot take advantage of ideal irrigation methods if plant nutrients are missing in the soil; nitrogen application is wasted unless the phosphate supply is adequate. Subsidies on feed and fertilizers cheapen their price and make their use more attractive to farmers. They therefore have two positive effects – they make total inputs more economic, and increase production.

Indirect subsidies can lead also to poor allocation of resources if the full economic effect is not studied. There is much discussion in Israel about the price of water. Some maintain that since water is nationally owned, it should be distributed throughout the country at a uniform rate, with areas of cheap water supply subsidizing those of high cost. Even if the rate is not uniform, they say that some levelling out should take place to the advantage of the hills and south where water pumping is expensive because of heights or distance. However, such a policy would not always be in the best interests of the nation nor of agriculture: it might lead to waste of water in areas of scarcity, and militate against the planning of agriculture in them on economic lines. One national water price, or a high subsidy to high cost water areas, would mean that farmers in those areas would grow crops using large quantities of water without any economic justification, or would grow crops not suited to the area. A drastic lowering of water prices in the hills would lead to the expansion of economically unsuited farm branches and would de-emphasize the need to develop a paying localized agriculture. Moreover, 'levelling' water prices would be a very heavy burden on the government, or on farmers elsewhere. Raising the price of water in regions of cheap supply would adversely effect the economics of farming there, preventing its development on the most economic lines. A policy of water subsidization, if necessary at all, cannot therefore be based on one national price without imposing severe strains on agricultural development.

Israel adopted a policy of differential yet subsidized prices so that the price of water in the hill regions was higher than on the plain, yet below the real cost of production.

2. Fiscal Means

Investment and credit funds are important means by which government can aid the implementation of national development plans for agriculture. Investment, through the provision of long-term cheap credits or grants to farmers, industries and organizations associated with agriculture, is the means by which the physical framework suitable for carrying out the plan is built up or improved. Short-term credit or working capital is necessary to enable farmers to make use of the resources at their disposal for current production.

Investment and credit policy, and the organization necessary to obtain the efficient use of limited funds for agricultural improvement, present many problems to developing countries over and above the major problem of obtaining funds. Israel, in building up her agriculture and settlement, was more fortunate than many other countries in that certain funds were available. The many priority calls upon them left no room for waste in their application, and a large part of the country's success in its agricultural development has been due to clear policy and the organization set up to use finances effectively. No-one pretends that the methods used were ideal, or that no wastage was involved, but the work was efficiently carried out in an extremely short time. A description of the policy and organization used may be instructive to other workers faced with similar problems.

A central authority can apply an investment and credit policy to fulfill a plan, providing that

a) it has funds at its own disposal;

b) it can influence existing finance systems and current sources of credit in the economy to work for the realization of the plan.

i. *Investment*

Investment is used here in the sense of the increase of durable capital goods in agriculture, both on the farms and in the industries and organizations which serve them. The scope and part played by investment policy necessarily differs from country to country and situation to situation. It ranges from the creation of completely new farming regions to replacement of a machine by a better one in an already heavily capitalized and mechanized system. Its aim is to add to the stock of capital used for agricultural purposes in order to improve it and increase its capacity and efficiency. Investment policy is bound up with the same two issues discussed in connection with price policy – investment purely for agricultural production, and investment to support, improve and strengthen farmers and rural areas. The desired balance seeks to combine investment for optimum economic efficiency of agricultural resources with investment guided by social principles affecting rural population. If family farming is the structure upheld as the most desirable from a social viewpoint, the production advantages of large-scale farming are limited in their application. If, for various reasons, large-scale farming is made the cornerstone of agricultural policy, then the social cost of adapting traditional peasant or family farming systems to it must be borne by the nation as well as by the peasants. The injection of capital into agriculture by investment schemes must be attached to a plan which is drawn up with the eventual social structure of the nation in mind: investment leads to change, and the social consequences must be foreseen in the plan.

If instead of family farming Israel had decided in favour of large-scale farming under state supervision or though private enterprise, the investment policy, budgeting and implementation would have been organized accordingly. The resulting capital structure, its physical and architectural form and location would have been completely different from that which actually exists.

Investment policies and development of farm branches: The principle of diversified farming in Israel had had a marked influence on the methods and investment policies adopted for establishing and developing new settlements. In the past, settlement authorities followed a policy of simultaneously developing all farm branches included in the final plan. A miniature image of the final farm composition was created with the intention of gradually developing the farm to its final planned stage by additional investment. This could be likened to an expanding circle representing the developing structure of the farm with areas divided into radial sectors, the size of each representing the relative scale of the several enterprises. After the founding of the State, the policy had serious drawbacks.[6]

Firstly, the available initial investments, spread over all the planned farm branches, were insufficient to establish any one branch as a profitable enterprise, able to compete on a modern market. The farm could not provide both a living and a surplus for the investments needed to raise the various branches to a profitable scale of operation; no one branch was strong enough to support the others while all expanded together. Only when they had all reached a final stage of development did any one of them have sufficient resources for effective operation.

Secondly, investment must be in total production units; a cowshed must from the outset be built to a certain size and pattern; it is wasteful to build in small stages. Together with the building, cows must be provided, or the capital is invested at cost with no corresponding income.

Thirdly, if investment on an uneconomic scale is made at an early stage, the farmer must of necessity build up one or two branches to a profitable level while neglecting others. Investment in the neglected branches is wasted, yet it remains in short supply for the branches on which the farmer is concentrating.

It became clear that simultaneous development of all branches of a new farm was not sound policy. The alternative adopted since 1956 by the Settlement Authority was to expand investment in building up one branch to its optimum scale, and then concentrating on a second complementary branch. The release of investment was to be in accordance with the gradual planned development of farm types based on regional possibilities, and according to the progress of the settlers.

Farm investments, conditions of investment and return of capital: Most investments in agricultural development in Israel have been made from public sources. A number of factors have caused this predominance of public investment: In the first place, most new settlers have been immigrants who lacked all possibility of self-investment. The settlement institutions had to provide them not only with the necessary productive investments but also with all other investment requirements. Secondly, it was essential to achieve speedy agricultural development because of immigration to Israel and the rapidly growing population and because of the economic necessity to create exports and reduce imports. These factors made it imperative not only to expand and develop already existing settlements, but also to establish new settlements in areas for which no farming experience had been gathered. The new setlement demanded a high investment ratio which certainly was more than could be provided by the settlers themselves; initial investment therefore had to be aided by public funds.

During the ten-year period 1949–1958, gross investment in agriculture and irrigation (excluding farmers' housing) totalled IL. 960 million at current prices, or IL. 1430 million at 1958 prices.[7] Of this IL. 960 million, 84% (IL. 814 million) was from public funds:

	IL. millions
Investments of the Jewish Agency's Settlement Department	500
Direct investments from the Government's Development Budget	104
Investment in plant and institutions from the Government's Development Budget	210
Total	814

Table 8-4 details overall public and private capital investment in agriculture for the period 1954/5–1958/9.[8]

TABLE 8-4

Financing of investment in agriculture and afforestation by public and private sources 1954/55–1958/59* (IL. million)

	1954/55		1955/56		1956/57		1957/58		1958/59	
	mill.	%	mill.	%	mill.	%	mill.	%	mill.	%
Public sources	98	78.4	93	65.0	89	53.3	111	55.2	134	70.9
Private sources	27	21.6	50	35.0	78	46.7	90	44.8	55	29.1
Total	125	100.0	143	100.0	167	100.0	204	100.0	189	100.0

* Agricultural year

During the first years of the State, from 1949–1953, when 100 new settlements were established annually, public investments were relatively larger than in the period 1954–58, averaging over 80% of total investment. Investments for the ten years 1955–1964 totalled IL. 1692 million according to 1963 prices.

The division of investment into the main agricultural branches was as follows[9]:

	IL. millions
Fruit plantations	381
Land reclamation, irrigation, drainage, afforestation	509
Farm buildings	293
Machinery and equipment	403
Livestock	106
Total	1692

As noted above, the two principal public investors in agriculture have been the Settlement Department of the Jewish Agency and the Government through the Development Budget of the Ministry of Agriculture. The investments of the former are both quantitatively greater and differ in kind: while the Settlement Department finances the establishment of new settlements and supports them until full growth and consolidation have been achieved, Government investments are used to expand farming activities of existing settlements, or to finance regional projects (pumping stations, for example), etc.

The Settlement Department invests approximately $15000 in each farm unit. The investment programme is determined by the farm type decided upon for a specific village, and the budget is given to the settler in the form of equipment and stock suited to the selected farm type. These investments include producer and consumer investments (such as settler housing); they also include common village property such as public buildings (school, synagogue, clinic, store), public buildings connected with production (feed store, sorting shed), approach roads, etc. These investments are debited to the village, each settler being charged with the total divided by the number of settlers (i.e. heads of nuclear families). Of a total investment, consumer investments (housing, public buildings, approach roads, etc.) account for some 25%, and planning, extension and administration services total a further 15%. Investment in direct production means, therefore, totals only 60% of the investment in each farm unit. This budget is made available over a period of 10 years, providing production assets as the settler becomes ready to exploit them.

The legal ownership of the property remains vested in the Settlement Department until the capital invested is repaid. Repayment begins only when the total budget of $15000 has been advanced. The period of repayment is 30 years, and interest of 3.5% is charged over that period. No interest is charged until the whole investment has been made.

In addition to the direct investment of $15000 per farm unit, the Government (Ministries of Labour and Agriculture) and the Settlement Department together invest an additional indirect investment of about $7000 per farm unit in public works, such as regional water schemes, roads, forestry, etc. An additional sum, according to regional conditions, is invested by the Jewish National Fund, which bears responsibility for land reclamation and clearing. The sum is relatively low in the plains but high in the hills, where great efforts must be expended on preparatory works before the soil is fit for intensive cultivation. These sums are not repaid by the settler and may be regarded as government and public expenditure in infra-structure. On the other hand, community services and investments in community buildings in the villages or rural centres are debited to each farm unit, to be eventually paid for by the settler.

To utilize resources to the maximum, however, an additional investment of 30% of the amount advanced by the Settlement Department is required. This additional investment is provided by the settler himself, either through the investment of accumulated profit, or through a special development loan offered by the Ministry of Agriculture after the Settlement Department has completed its capital investment. This loan is available at an interest rate of 7% and is repayable over 10 years.

The financing of agriculture by means of the Government's Development Budget is carried out in two ways:

Funds for general development works, such as soil conservation, drainage, research, etc., is directly made over to the appropriate institutions and companies. On the other hand, loans for the development of various farm branches are made to the farmers themselves through the Israel Bank for Agriculture. A distinction must thus be drawn in the Development Budget between the usually short-term loans to farmers and the loans to institutions and development projects which are more in the nature of investments and are thus long-term.

The terms of loans to farmers vary according to the agricultural branches which are to be developed. The policy adopted by the Ministry of Agriculture has been to ease the terms of loans for branches which the government is interested in expanding, and to make them more difficult for those branches which it is less interested in developing. At the same time, the loan does not cover the full investment required. Until 1955, for example, loans for farm buildings did not cover more than 40 to 50% of the required investment. Since 1956, participation has been increased to 60 or 75%. In order to receive the loan, the farmer must prove that he has invested a certain sum, and the balance, to the percentage limit laid down, will be awarded as a loan. The settler may not reduce the total investment by erecting only part of the farm buildings he has presented in his loan proposal, enjoying full support for those which he does erect. By such a loan policy the farmer is encouraged to invest his own capital in agricultural development.

As pointed out, loans on easy terms are granted to certain branches in order to encourage their expansion. For example, since 1955/6, loans for banana planting have been given for a three-year period, with repayment commencing one year from the data of the contract signing. For citrus, on the other hand, the loans are granted for 10–12 years, repayment commencing only 5 years after the contract is signed. The differences in the loan terms are in line with a policy favouring citrus more than bananas.

The investment policy of both the government and the Settlement Department is thus designed to encourage the implementation of development programmes. The following example, again in the field of citrus, illustrates the effect of investment policy in achieving the expansion of a particular branch:

Citrus plantations for new settlements are financed by the Settlement Department through its regular settlement budget, according to the 'key' for each farm type. However, plans for agricultural expansion in Israel called as well for an increase in the area of plantations within existing settlements and the private sector. Priority was given to the development of citrus orchards in the Development Budget of the Government: special long-term loans were made available at easy rates of interest and terms of repayment.

The terms of the loans were as follows during the year 1955/56:

i) Costs of the saplings: 35% of the planting costs in the first year and 80–90% of the investment in irrigation network, were included in the loan.

ii) The loan was given at a 6% interest rate over (a) nine years for the irrigation equipment; (b) seven or eight years for the plantation loan.

iii) Repayment started three months after signing of the loan contract.

Some changes were made after 1956. The loan on irrigation equipment was reduced to only 60% of the total cost, but the time period was extended to fifteen years; the interest rate was set at $6\frac{1}{2}$; the repayment period for the loan for the orchard itself was extended to 10–12 years; repayment on the irrigation equipment loan started two years after the signing of the contract, and repayment on the orchard loan started four to five years after the signing. The easier conditions after 1956 reflect the aim to increase the plantation area as rapidly and effectively as possible.

Investment in new settlement: The bulk of public investment in agriculture in Israel went largely to new settlement since the existing settlements by and large financed their own expansion, although special credit terms and price incentives were given. With new settlement, the optimal structure close to specifications as laid down by the planners can be achieved since there is no need to first dismantle an existing system or superimpose new methods on an outdated base. Agrarian reform is the term applied to modernizing a traditional agricultural pattern, work in which Israel has had little experience except for her present efforts to improve the local Arab agriculture. The principles and practice of new agricultural settlement however do not greatly differ from those of agrarian reform, once land tenure and legal systems enable improvement. The need for community development, for building and advancing a commercial framework, for physical planning to create sound production units, for farm planning and for integrating the work of the new or improved farms into the national economy, are essentially the same for both. The agrarian reformer therefore has much to learn from the settlement planner and vice-versa. The planner should certainly avoid creating situations which the agrarian reformer is struggling to remedy.

Investment policies necessarily reflect the general agricultural policy of a nation. They are usually concerned with achieving the best results for minimum outlay, and in this respect particularly, new settlement and agrarian reform are very close. An understanding of Israel's investment policies, therefore, may be useful to those whose main concern is not in new settlement but in agrarian reform and improvement.

ii. *Short-term agricultural credit*

The need for short-term credit for agriculture depends on the relative weight of the various farm branches in total production. The credit needs of a particular branch vary according to the cost of production, the distribution of cost over the production period, and the length of this period (from the processing of the raw material to the sale of the produce). Dairy and poultry production have a short-term need for working capital, since the time from feeding to milk or egg sale is short. (The cost of the stock, buildings, etc., are regarded here as requiring long-term credit, i.e., investment funds.) The working capital turns over quickly and the same sum can be used several times per year. For vegetables the period in which working capital is tied up varies, but is usually not very long. Industrial crops, however, are at least half a year in the ground; the credit turnover is small. The basic amount of working capital needed for the operation of the 'national farm' depends therefore on what it produces, since this determines how many times one sum of money can be used in a year. For a national plan, it is important to assess how much working capital will be needed while the production structure is being erected and when it is working at full capacity. Shortage of credit means that resources cannot be used to the maximum, and high rates of interest make production expensive and distort the relative economic returns of farm branches. In planning agricultural development, the provision of working capital is as important as investment and price incentives.

Countries with an advanced commercial agriculture using traditional sources of short-term credit are usually concerned with means by which this supply can be augmented and made available to farmers at low rates of interest. The institution of chattel mortgage in England was designed to enable farmers to borrow on the security of crops and livestock. It was not successful because of the inherent conservatism and dislike of publicity which characterizes the British farmer. This emerged in an enquiry carried out in 1923 by the Ministry of Agriculture[10], and was embodied in the Agricultural Credits Act of 1928. The first part of the Act deals with long-term credit and the second with short-term credit, including chattel mortgage. The aim was to improve the position of farmers in borrowing from the banks and to help them break away from high cost, short-term merchant financing. The merchant provides seeds, fertilizers, store, cattle, etc. at the beginning of the production period on condition that the farmer markets through him. While this method has great advantages in that credit is available at the proper time, in the form required and for the period required, the charge is never precisely known. The farmer loses his freedom to take advantage of market prices and is in a very weak position regarding the merchant. Chattel mortgage did little to reduce the use of this type of convenient but expensive credit; it is mentioned here to show the concern of British agricultural policy with the needs of farmers for short-term credit even though channels such as merchants finance and bank loans also existed.

In a country striving to initiate a plan which will build up agriculture and expand its cash sales both absolutely and in relation to subsistence consumption, the vital part played by short-term credit is clear. Important are not only recognition of the need and budgeting the sums required, but also a smoothly working apparatus to provide the farmer with the money he needs when, how and for as long as he needs it.

The very nature of agricultural production makes it risky to finance its short-term credit needs. Money lent for seeds, fertilizers, feed or other current expenses can be all too easily swallowed up by natural disaster or ruinous market prices. Agricultural stock-in-process is generally regarded as poor security for bank loans; but for many small farmers, it represents all they can offer. Commerce and industry are usually in a better position to give collateral as security, and therefore farming is in a poor competitive position in attracting finance. The results of lack of credit are similar to those discussed under price uncertainty and reflect the same factors. The farmer cannot borrow sufficient funds to make a maximum effort for efficient production, or is afraid to borrow enough; he rations his capital among too many farm enterprises in the hope that those which fail will be more than covered by those which succeed. His fear of getting into debt may keep him from breaking out of subsistence agriculture into more profitable and productive commercial agriculture. As it is for the individual farmer, so for the national farm. Means have to be found of supplying agriculture with adequate short-term credit so that its resources will be economically and fully exploited in terms of national demand for agricultural products. Such means must be based on the acceptance of maturing crops and livestock as security for short-term loans. If this is done, the way is open for financing agriculture through normal banking channels at accepted rates of interest and return.

Many banks offer loans to farmers, formally or through an overdraft, on the basis of reputations built over many years. A farmer with a sound business and a good bank record has little difficulty in obtaining the credit he needs. But not all farmers have either a business which can be commercially called 'sound' or a good bank record: regrettably, the less money or reputation one has, the more difficult it is to borrow. How can poor farmers be made credit worthy as far as banks are concerned? Certain sums of money are usually available to farmers, even of very poor standing, but the usurious rates of interest usually put the farmer into the hands of the moneylender in one or two seasons.[11] The risk of dealing with many small farmers, the difficulty of assessing the credit worthiness of each, makes the operation too unprofitable for normal financing through the banks at reasonable rates of interest. The financial standing of each individual farmer is not sufficient to qualify him as a customer for bank loans.

In many countries, co-operative credit societies have been formed to pool resources of members and borrow from private or institutional sources, using the joint resources of the association as security. Such credit co-operatives which lend to members at comparatively low rates of interest, are in the best possible position to judge the risk of lending to a particular farmer, since the members are well known. The credit societies, built on the democratic lines of co-operatives, thus adequately supervise the credit. They provide agriculture with units large enough to draw funds from the wider capital market without individual farms having to take on a joint-stock structure. Their role in underdeveloped countries is important because of their power to help peasants move beyond the subsistence level once they have been organized on workable lines suited to local conditions.

In Israel, the older co-operative villages satisfy the credit needs of their members, their guarantee being the undertaking of each member to market solely through the co-operative. Thus, the individual member has no dealings with outside financial or marketing institutions; he runs an account with the co-operative which obtains the necessary loans for the village and markets all its products. Usually, the village is a member of a regional purchasing agency comprising several villages, in order to gain the advantages of bulk purchasing and therefore better business terms. The agency gives the village an open credit for the amount of its needs for working capital and is repaid as and when produce is marketed. Each member is allowed to draw a certain sum of credit from the co-operative, which he repays out of incoming returns from market sales. The system is highly efficient, cheap, and very well suited to the organized, stable co-operatives. The trust of the members in the co-operative is complete; many actually hold sizeable amounts of money in it. They use it as a bank and receive interest on their balance just as they pay interest on outstanding loans.

The purchasing agency, since it works for several villages, has a large turnover and therefore little difficulty in obtaining credit from agricultural suppliers, banks or other sources. The individual farmer is thus able to obtain his needs for working capital because he is a member of a co-operative integrated into a wider co-operative framework of sufficient size to obtain commercial credit facilities.

The new moshav settlements created during the first years of the State's existence were badly undercapitalized, both in respect to long- and short-term credit. They were forced to seek many and varied forms of credit from many different private and public institutions, often at high rates of interest and at the cost of unwieldy administration and an impossible financial policy. At the same time the balance sheets of many kibbutzim – both old and those settled since 1948 – showed numerous financial institutions and just as many private creditor or co-operative societies with which they did business. Difficult times hit all these under-capitalized settlements hard; they lived from hand-to-mouth, with consequent piling up of debts and payment of interest on loans taken to pay interest on other loans. Floundering in debt, they were poor risks for underwriting by sound financial institutions in order to get them on their feet. In 1956 the Government and the Jewish Agency stepping in with the 'guided kibbutz' policy, consolidated all the loans into one long-term loan, and allowed development on a sound financial and administrative footing.

iii. *Stages of credit administration in Israel*

Situations such as that of the floundering settlements soon made the Government and the Settlement Department realize that agricultural credit for new settlements presented unique and difficult problems which had to be attacked in a comprehensive and orderly fashion. Experience showed that new settlements had neither the financial strength nor the ability to run themselves properly in the first stages: thus, a system allowing the gradual development of credit facilities within the villages was evolved and put into operation.

Stage One: The settlement budget of each settler included a sum of IL. 2000 for short-term revolving credit. On this basis, a second or 'B-Budget' was opened by the Jewish Agency Settlement Department in addition to the 'A-Budget' through which long-term production resources are allocated. Each village formed a co-operative, thus undertaking to market all produce collectively. This was the first guarantee of loan repayment. A farm plan was drawn up for each farmer which served as a basis of short-term credit availability; the monthly credit needs of the plan were ascertained by drawing up norms for the monthly credit needs of each crop. The farmers were then entitled to draw on this credit throughout the growing season in the form of credit slips exchangeable in the village general agricultural store or against water bills. The slips, given in coupon books, each coupon with its value stamped on it, were not cashable in the food shop. The settlers signed for each instalment as they received it, and the account was kept in two copies, one with the Settlement Department, the other with the village book-keeper. As the co-operative marketed produce, the list was submitted to the Settlement Department which was entitled to deduct up to 50% of the sale proceeds, under its agreement with the farmer, against his outstanding loan.

Important features of this first stage were:

1) Funds were available as and when needed;

2) Credit allowances were calculated on the basis of the farm plan and were therefore sufficient to carry it out; a monthly crop report for each farmer was eventually instituted to enable planners and administrators to check that the money was actually used for the crops planned;

3) The slip system meant that the credit had to be used on the farm;

4) The system was simple in operation and left the new farmer free to attend to agricultural tasks;

5) Credit advances were automatically recovered through the tie-up of credit and marketing;

6) In poor seasons less than the possible 50% was deducted if the family would otherwise have been left without subsistence.

The system also had its dangers and drawbacks: mainly, the paternalism implicit in handing out credit when necessary and not giving the farmer personal responsibility for its return. This was however only a first stage and cleared the way to smooth farm development; it parallelled the first land allotment to the new settlers and their first steps as independent farmers. In this period they still had recourse to supplementary sources of income in forestry, public works projects, etc.

Stage Two: As the scope of the farm expanded through the allocation of additional resources, banks were used as the means of granting credit to certain villages. This method was not applied to all the villages, but only to those in which co-operation was not advancing as it should and in which the members refused, formally or in practice, to market through the village. The Settlement Department opened a fund to be administered by the banks on a cash basis with the farmers: the money was in fact from the same source as the credit slips, but the farmers did not know this. The bank offered them credit, on bank terms, at the recommendation of the Department and on the basis of the farm plan. The loans were made available as and when needed, and the farmers had to provide two guarantors, usually members of the same village, to sign bills of repayment with them for dates corresponding to the harvesting months. With the cash, they bought the raw materials, and they repaid in cash.

In some villages which had access to this arrangement, few of the members actually took up the loans. Enquiry revealed that they had turned to private credit sources, largely from relatives or friends within the village or outside it, since the necessity of providing guarantors was distasteful to them. Therefore family and personal finance replaced institutional finance in these cases, lessened the burden on the Settlement Department, and rendered the settlers more independent.

Stage Three: The third stage was the formation of purchasing agencies to take over the functions of credit administration from the Settlement Department. The Department backed the agency with its own funds until such time as the villages were established and the agencies could function as they did for the older settlements. The co-operative provision of credit was then at its full development. The villages and their members had become fully aware of the part credit plays in their operations and had been educated in its administration. Each member had an account with the village co-operative and was entitled to credit as he needed it. He marketed through the co-operative, and his debt was automatically cancelled by incoming returns.

iv. *The regional credit administration*

In the regions, the Settlement Department operates on a decentralized basis. Each regional office is responsible for financing and administering its credit schemes and co-ordinating them with the central finance body of the Department. It must be able to plan in advance both for credit needs and the return of money for relending: this is achieved by means of the regional farm planning teams. The planners receive the quotas of the different crops for their districts and allocate them among the villages in accordance with the need for an adequate farm plan for each settler. The particular preferences of each village are taken into account as far as the quotas and rotational principles allow. In winter a seasonal plan is drawn up for each of the districts and villages for the following spring and summer. The autumn–winter programme is drawn up in spring and summer. Each farm is planned in detail by the planner and agricultural instructor, but the administration works on a village basis in assessing overall credit needs. The monthly, two-monthly and quarterly needs are worked out from the crop pattern as it emerges for each village and each district according to the monthly expenditure norms for each crop. The regional administration then knows the amount of credit it has to make available via credit slips, banks or purchasing agencies, and it budgets accordingly. At the other end of the production process, the same plan indicates when crops will be marketed, and hence the monthly spread of incoming returns from loan repayment.

Certain crops and livestock have a long return period and, as with cotton, there may be a delay between harvest and payment. Marketing boards contract to buy up the crop and help in its financing by a series of advance payments to the grower. Such payments are usually administered by the marketing board, not the region. The system is a very useful way of providing short-term credit in certain cases and for certain crops.

C. RESEARCH, EXTENSION, SOCIOLOGY

The third set of activities through which a government can directly influence the implementation of its plans for agricultural development comprises research, experimentation and extension and application of sociological insights. All are concerned with the expansion and dissemination of knowledge which directly aids the community in carrying out its tasks, and with the farmer in the community context. They are particularly important in the context of development, which begins and continues by the addition of new activities to those already established. Planned development seeks to initiate and continue the process of change as it has been defined in the plan, by working out the methods of development and making these known to those whose task is to implement it. Through research and experiment it is possible to broaden the existing body of knowledge according to the needs of the plan, while through education and extension, the knowledge necessary for functioning of the development process is brought to the participants: these pragmatic and almost absurdly narrow definitions are used to co-ordinate these four activities with the special needs of the development machinery: research and education as aims in themselves, though vital, are not the subject of this discussion.

1. Spreading knowledge

Every plan throws up problems which have no answers: and answers must be found. The problems vary from the adaptation of imported seed varieties to local conditions to the best way of forming a co-operative from a traditionally individualistic social structure. Organizational and economic problems arise which must be solved if the plan is to be given substance. New knowledge must be sought, and this is done through research and experimentation; for agricultural development, this research cannot be confined to discovering new materials and better genetic material: social and organizational research under particular local conditions is probably more important. In this connection, W. Arthur Lewis says: "Economic growth depends both upon technical knowledge about things and living creatures, and also upon social knowledge about man and his relations with his fellow men. The former is often emphasized in this context, but the latter is just as important since growth depends as much upon such matters as learning how to administer large-scale organizations, or creating institutions which favour economizing efforts, as it does upon breeding new seeds or learning how to build bigger dams."[12]

In newly developing countries, research programmes should be intensely practical aimed directly at the problems in hand. Pure research is not an activity which these countries can profitably undertake; their efforts must be in applied research, seeking ways in which the enormous body of technical and social experience existing in the world today can be applied to their own conditions.

Schultz discusses the production of techniques in agriculture and comes to the conclusion that it lends itself to economic analysis[13]: "It is our contention that a new technique is a valuable (scarce) resource that has a 'price' and that this resource is not given to the community or to the producer as a free good; on the contrary, it entails costs some of which are borne by the community and some by producers, as a price that is paid to acquire and apply to resource. Therefore, a new technique is simply a particular kind of input and the economies underlying the supply and use are in principle the same as that of any other type of input. We do not wish to imply that every human activity entering into the development of new techniques can be explained wholly by considerations of cost and revenue; our belief simply is that a large part of the modern process of technological research from 'pure' science to successful practice can be explained by economic analysis. Resources of the community, consisting of both public and private funds, are 'invested' in research, motivated, it is true, to acquire new scientific insight to be used, however, among other things, mainly to increase the ratio of output from a given input in making something that society wants. Accordingly, the allocation of resources to such research and to the application of the research results is not unrelated to the prospective returns from such effort set against anticipated costs."

He goes on to examine the costs and returns to research in agriculture in the U.S., and concludes by reaffirming a belief that[14] "the returns from these inputs to society are large; the additions to the social product to be had from a further increase in expenditures for agricultural research are substantially larger than are the returns from (most) alternative uses. Vaguely and intuitively the public process realizes this to be true; the public wants these gains and this explains the motivation that supports these expenditures."

W. Arthur Lewis, in discussing knowledge as a factor in economic growth, attempts to assess the scale of minimum research expenditure in under-developed countries: "There is no doubt that one of the main deficiencies of underdeveloped countries is their failure to spend adequately upon research, and upon the development of new processes and materials appropriate to their circumstances. Part of the reason for this is institutional. In industrial countries private entrepreneurs spend great sums of industrial research, because they hope it will pay them to do so. The underdeveloped countries, on the other hand, are agricultural. Where their agriculture includes large commercial companies, these companies have invested in research (e.g. rubber, bananas, sugar) either individually or collectively, but, in all that part of their agriculture (the major part) which is not organized on this basis, there are no private interests financing research. It follows that almost the whole of the research expenditures needed in these countries (i.e. excluding mining and commercial agriculture) has to fall upon the public purse. Whereas in industrial countries research can be thought of primarily as a matter for private interests, with the government plugging gaps, in underdeveloped countries research is primarily a matter for governments, and ought to be one of their major fields of activity.

"How much ought they to spend? This is of course an unanswerable question. Current expenditure on industrial research and development in the United Kingdom is estimated at a little under one per cent of income generated in industry. In the United States industrial research is at a similar level, while agricultural research is a little less than one half of one per cent of the net value of agricultural output. On the same basis it would not be unreasonable if the underdeveloped countries were to spend on research of all sorts (technical, social, health, etc.) a sum equal to between $\frac{1}{2}$ and 1 per cent of their national income (not to be confused with government expenditure). There is no firm basis for such a suggestion. All the same, current expenditures, which do not reach a fraction of this level, are clearly too low."[15]

The technical revolution in agriculture was discussed in Chapter 4 and clearly shows the advantages which scientific method and experimentation have brought to agriculture in the past. These are proof enough of the economic value of research. Most of the research work has and is being done on an ever-increasing scale in the economically advanced nations of the world. The newly developing areas have the results at their disposal, but the minimum amounts suggested for research expenditure in these countries is not therefore less important. The results of the universities and experimental stations of one country furnish the raw material for research in other countries, research which must scientifically seek means of adapting useful data from abroad to particular local conditions.

A plan requires various types of decision from research personnel. Some items are immediate issues which have to be examined before the plan can get fully under way, e.g., the economic effects of a new crop and the measures needed to gain the most from it – optimal marketing arrangements, types of packaging materials, industrial capacity of processing. Other items are subject for long-term research which proceeds with implementation, measures its economic and social progress, and draws conclusions designed to improve farming systems, organization, or community development techniques. Another aspect of research is enquiry into the incomes of farmers and the factors which determine them, designed both to raise the income further by applying the research results and advising on social and investment outlets for the new income.

Research projects are long- or short-term; as far as development is concerned, they should be carried out to pose and answer particular questions in order to boost progress. This means that research must be combined with institutions which can request that certain problems be investigated and which are able to apply the results of the research directly. Without this connection, research becomes academic and separate from the real issues of development planning and implementation. The tie-up in Israel is related in the last section of this chapter, where institutions are discussed.

2. Agricultural education

Agricultural education in the rural sector is a prerequisite for the sound development of agriculture and rural life. There can be little progress towards better farming, land settlement, co-operative enterprise or any other aspect of life which the development plan seeks to improve unless a suitable educational process produces the men on whom these achievements can be based. When agriculture is the main source of livelihood for a population, its importance is obvious; when only a small percentage of workers are engaged in agriculture, they must be trained to work efficiently. An essential part of a development plan is therefore a clear educational policy, geared to meet development needs.

Trade and commerce offer the openings which are most attractive to those graduating from a general elementary school system. Youth with even a minimal education cannot see its place within a poor village and prefers to seek its fortune in the towns. This process clearly militates against the interests of a nation striving for basic development in the countryside and the necessary corps of leadership.

An agricultural plan, at the village and regional level, needs the following staff groups:

1) village instructors in agriculture, co-operation and home economics;

2) regional specialist instructor in the different branches of agriculture, co-operation, home economics, irrigation, community development, marketing and credit organization;

3) agronomists, engineers and administrators to undertake the planning, its implementation and administration.

Central and local government budgets for education are usually restricted and rarely sufficient to cover the need. Agricultural education competes with the need to train youth for other vocations essential to a developing economy. It must be given its due weight, in spite of the fact that agricultural extension necessarily demands considerable funds if the immense benefits accruing from it are to be obtained. The competition is at the secondary school level, but in rural areas and in towns serving them, agricultural secondary schools combining a general education must be given priority. Without them, school leaders will be oriented away from agriculture into jobs which use primarily the reading, writing, arithmetic and other educational attainments with which a general secondary school provides them.

Most of the graduates of an agricultural school will seek employment in agriculture, particularly if their number is tied in with the staff needs of the plan, so that jobs are waiting for them. It may be desirable to build trade schools which combine general secondary education with training for a variety of trades among which agriculture predominates. Secondary pupils are thus given a choice of trade and the 'overhead' of general courses is used to the full. Farms attached to the schools practically demonstrate the methods taught, and the curriculum should include a fair proportion of chemistry, soil science, biology, co-operative organization, and elementary civics aimed at a basic understanding of community development. Emphasis should be laid on the particular farming systems of the locality in question. These schools supply young people who can go to work directly as agricultural instructors. Their education should be continued by means of courses in the 'dead season', a system which maintains their professional interest and increases their capabilities.

The best of the pupils should be given further training in agricultural colleges or farm institutes for a period of one or two years, qualifying them as regional specialist instructors. Once again, their relative number is determined by the needs of the plan and the budget available. It is clear from the experience of many countries that investment in agricultural education is extremely profitable. It cannot be skimped because without it investment in 'real' production resources cannot be properly used. Skilled people are very often the limiting factor in development and only by relating training to the needs of the plan and its financial estimates can the best allocation of funds for agricultural education be made.

The training of agronomists depends on national plans for university study. It is often possible to obtain help from abroad under the various technical assistance schemes, thus filling important gaps for which locally trained personnel are not available. This can be regarded only as a temporary measure, for although technical assistance is used by very advanced countries in some fields, a country cannot rely on it for the main corps of its experts, which must eventually be formed by people trained at home who know local conditions and problems, who understand the culture and national life, and who are permanently settled. Visiting experts cannot form the backbone of a development programme, which needs people of high calibre, familiar with the country, who have primarily its interests at heart. In any university, an agricultural faculty must be awarded a priority concomitant with the urgency and size of the task of developing agriculture.

It is important that the particular needs of agricultural education be accorded proper attention at government level. The needs of the plan for personnel will otherwise not be met by the educational system. An inter-ministerial committee of the Ministry of Agriculture and the Ministry of Education furnishes the means by which the requirements of agriculture can be met through national education programmes. To formulate a clear policy regarding overseas training, a commission consisting of the Ministry of Agriculture, the Treasury and the Prime Minister's Office or the Foreign Ministry could be of very great use to ensure that opportunities for study abroad are taken up by the right people and in the right fields. It is also important that students going abroad should go with particular aims in mind and even with a particular post waiting for them on return. Otherwise they may not be willing to undertake the long years of study and preparation after which they would probably find that other contemporaries who have not studied, have meanwhile advanced far beyond them.

3. Extension methods

Extension methods and organization are factors of prime importance in advancing agriculture and farmers, especially in a country like Israel in which most farmers were neither previously engaged in agriculture nor were from rural backgrounds, or in developing countries where subsistence agriculture is practised, agricultural methods tending to be primitive.

In recent years Israel has acquired experience in both extension organiza-tion and methods, especially in adapting methods to the social and ethnic background of the settlers. Trials during the first years of Statehood in giving a uniform type of instruction to farmers without considering their social background led to failures and attempts were thus made to change extension techniques. As a result, the extension institutions were reorganized and methods adapted to varying conditions.

i. *Extension organization*

Extension work in Israel's settlements is the responsibility of the Extension Service, which was until 1965 a joint undertaking of the Ministry of Agri-culture and the Jewish Agency's Settlement Department. Since 1966 it has passed to the Ministry of Agriculture alone. Local agricultural instructors are located in the new villages during their establishment and for a few years thereafter; although not employees of the Extension Service, they nonetheless receive agricultural directives from the Service.

The Extension Service is decentralized, functioning through regional bureaus. Their number is determined by administrative considerations and natural conditions, such as soil, climate, etc. Each regional bureau deals on the average with some 80 settlements.

The Director of the regional extension bureau is directly subordinate to the Director of the Extension Service and maintains direct contact with the various branches of the Service. The local director draws up annual and current extension plans and is responsible for running the office to the satisfaction of the Service and local farmers.

To cover the area served by the bureau, it has at its disposal a staff of instructors for the various agricultural branches, according to the relative size of each branch. In the case of branches too small to warrant the employment of a special regional officer, the bureau receives aid from the central or another regional service. For those branches in which there are a number of instructors there is a professional group, e.g. an irrigated crops or a horticulture group, in which the branch instructors and those connected with the services (mechanization, plant protection, etc.) participate. One of the instructors co-ordinates the group activities, which provide an overall perspective of the professional problems of the branch in question and co-ordination with the service branches.

The local instructors participate in the central professional team of the Extension Service. In addition to the teams of instructors for each branch, an instructor in farm management is employed by certain bureaus: his task is to help the instructors achieve an overall view of their branch and aid veteran farmers in determining the most effective use of production means at their disposal. This economic instruction is important because of the need for wider perspectives in view of each instructor's tendency to overemphasize his own branch, ignoring its actual importance in the farm pattern as a whole.

Connected with each of the bureaus is an executive headed by the bureau director, meeting monthly to review current problems such as: the monthly programme, evaluation of activities carried out, the responsibilities of each of the instructors, etc. This review serves to keep the executive in close touch with the actual progress of work.

The bureau is linked with farmers through an advisory board on which farmers' representatives and the executive members serve. In addition, there are a number of professional agricultural committees for the various branches connected with each bureau, usually representing the more important regional farming activities. Some six to eight members serve on each committee, with either an instructor or a committee member co-ordinating its activities.

The bureau works according to an annual plan. At an early stage in its preparation each of the teams reviews the problem of its particular branch and determines its aims in the field of extension for the coming year; at this stage, the committees, Jewish Agency planners and the village instructors participate in the discussions. On the basis of this analysis, the entire team of instructors meets with the bureau director in order to determine the programme of activities for each branch. The final version of the plan takes into account the need to give appropriate degree of emphasis to each subject, in accordance with its importance to the farmers of the region. With regard to settlements which require it, the bureau draws up a village instruction programme, discussed jointly with the village institutions and other bodies connected with the village in question. The annual programme as a whole is ratified by the advisory board and the directorate of the Extension Service. The programme determines in which villages the bureau will conduct a full survey of professional and economic activities and also of observation and demonstration farms, in which the instructors closely follow the various stages in the development of a particular crop which requires further clarification in the light of regional conditions. The results of the observation are brought to the attention of the farmers and form the basis for professional directives.

In addition to the annual programme, a monthly programme is drawn up which lays down the precise order of the instructor's activities for the month in question, in addition to group activities for that month. The underlying assumption is that the annual programme must be re-examined each month to adapt it to changing conditions such as weather, problems which arise in the course of implementation, etc. The monthly programme emphasizes correct division of the instructor's work between advanced and struggling farms.

The bureau is aided by the various divisions of the Extension Service. The branch co-ordinators or their deputies and members of the Instruction Division carry out tours of inspection in terms of a monthly programme fixed in advance, during which they advise and direct the instructors with regard to special problems.

The Extension Service lays special emphasis on the training and advancement of its instructors. In accordance with the results of a half-yearly survey of instructors, they are sent to advanced intensive courses for further study, in terms of a programme worked out in advance. In addition there are team meetings, shorter study courses, etc., designed to raise the professional standards of instructors. All-in-all, the instructor spends one month a year at training courses, which include study of instruction methods as a professional discipline, in order to provide him with appropriate means of communication with farmers, and understanding of study motivations and the incentives to change in various groups, etc.

The Divisions of the Extension Service's central office are instruction, services and instruction services, departments of home economics and field service. The last two departments are separate units that do not form part of the overall structure.

The instruction division has responsibility for directing professional activities in all branches and implementing national programmes such as courses, demonstrations, study days, etc. The Division is divided into professional groupings which cover the various farm branches and crops and their relevant services branches, represented by the national instructors for irrigation, farm management, mechanization and plant protection. In addition, the group contains a co-ordinator of activities whose task is to plan and implement national activities in co-operation with all members of the group.

Each farm branch has an instruction co-ordinator whose task is to direct instruction activities including preparation of instruction objectives for the annual programme, annual planning of national and area activities and month-by-month programming of these activities. He must also plan for the preparation of audio-visual means and the publications necessary to carry out the instruction plan. Each branch has a professional advisory board made up of the branch instruction co-ordinator, a representative of the appropriate Department in the Ministry of Agriculture, a representative of the research services and of the farmers. In addition, a representative of each regional bureau participates in this team.

The Services and Instruction Services Division runs the following departments: manpower, budgets, publications, audio-visual, and secretariat.

ii. *Instruction methods*

Experience gained in instruction has shown that over and above the importance of the professional aspect of the work – the quality of instruction – instruction methods and communication techniques between the instructor and the farmer are of the greatest importance. In order to know how to approach the farmer, the instructor must understand his mentality, thinking processes and general intellectual level: on such a basis he can adopt the most effective methods of communicating with him.

Experience gained in Israel and elsewhere has demonstrated that the impact of oral instruction is less than that achieved by the use of visual methods – pictures, demonstration, etc. This accounts for the increasing emphasis placed on visual instruction methods, though it must be pointed out that oral instruction can be even more effective, provided that it is given in the form of a personal discussion between the instructor and the individual farmer. Since this is usually impossible because of the great disproportion between the number of farmers and of instructors, concrete demonstration sessions on a village basis are important. Demonstration plots are set up in the village for the purpose, usually situated on the farms of outstanding settlers who serve as examples of what can be achieved by correct cultivation methods. The plots are directly supervised by the instructor, with the settler carrying out his instructions as exactly as possible.

If it becomes necessary to demonstrate a particular method in all its stages or to become acquainted with new equipment (new cultivating or harvest machines, etc.) the instructor must bring the implements in question to the village and demonstrate their approved use under actual work conditions.

Another method is that of field trips which take settlers from one village to neighbouring settlements in order that they become acquainted at first hand with the successes and failures of others and the reasons for them.

Sometimes instruction is given to an entire area, including a large number of villages, at the same time by means of 'field days' in which demonstrations are given in a number of branches and with different implements. The field days usually include the demonstration of new implements and products. They also serve as occasions for social meetings between settlers from different villages, which is a further incentive for participation.

The usual method of instruction by means of lectures has been improved by the inclusion of film shows illustrating the methods discussed. The search for new and different instruction methods has been intensified in view of the varied social and ethnic make-up of Israel's farmers, different methods being appropriate for different groups. Lectures are suitable for settlers on a fairly high educational level; demonstration methods best for those with a lower cultural level. The ethnic origin of a settler is also a factor in determining the most appropriate instruction method.

Research undertaken by rural sociologists in Israel has shown that when more than one extended family group is present in a settlement, a demonstration poultry run should not be established on the farm of a settler of only one of these groups. Separate demonstration runs must be set up on farms of representatives of each of the various family groups. Instruction subjects must be selected with similar care: it has become clear that in order to gain the trust of the settlers in the instructors and in the methods they propose, instruction should commence with subjects and branches in which the settlers have special interest. Only afterwards can the instructor pass on to other subjects which for one reason or another appear to the settlers to be less significant. There is not always agreement between the various groups in the village on the importance of the subjects, and the instructor must take into account the various facets of social structure. Sociological survey shows that instruction has often failed to achieve its aim because the instructor regarded each village as a single social unit, without taking into account variations in the settlers' level of knowledge or their social affiliations. Instruction activities must therefore always be accompanied by continuous social survey, study of the character traits of the settlers, inter-group relations within the village and other factors which the sociologists – and the appropriately trained instructor – can identify.

Research of this kind is important not only for the advance of new settlers who have only recently begun their agricultural careers but also – and perhaps even more so – in order to effect changes in the habits and methods which have become ingrained in older settlements. The introduction of new methods, improved varieties, new cultivation techniques, etc. is often confronted by apathy or antagonism by the experienced farmer who has taken over his methods from previous generations. Instruction is thus of prime importance in the developing countries in the process of transit from subsistence to market agriculture. Social research in this field aims at selecting the most effective method of introducing production methods in a way that will be accepted by farmers willingly and with understanding.

4. The sociologist in the administrative framework

In Chapter 4 on the Human Resource, we discussed the necessity for sociological research and insight, both in dealing with individuals to be made into willing new farmers, and in modernizing farm communities of traditional farmers. We discussed some of the ways in which sociological work has been useful in Israeli experience. Here it remains to examine the operational framework of sociological work in Israeli settlement – both its organizational links and its functional links to the settlement administration.[16]

Sociologists or social-anthropologists were first used in the early stages of settling the Lakhish region in 1954 by the Settlement Department of the Jewish Agency. At first, one or two sociological workers with a special status answerable to the director of the department acted more or less as 'troubleshooter' in the settling of people from so many different backgrounds. They became a means of reconciling the demands of the settlement staff and the view points and customs of the settlers. When it was realized that the sociological viewpoint could help solve thorny problems which often meant the success or desertion of farms and even villages, it became routine for extension workers in the field to be given courses in sociological orientation.

In 1958 sociological work came into its own in a more comprehensive way, with the establishment of a Council for Social Affairs run jointly by the Department of Sociology of the Hebrew University and the Settlement Department. Its joint chairmen are the Head of the Sociology Department of the Hebrew University and the Head of the Settlement Department.

The Council has two wings – one for study and research, composed of a team from the University – and another for counselling and instruction made up of regional sociologists.

The following are some of the problems tackled by the study and research wing: Influence of the settler's demographic composition on their adaptation to the farm; the influence of various cultures on the economic and social development of the agricultural settlement[17]; the assimilation of agricultural know-how by settlers from different ethnic background; the cultural background of the settlers and their motivation for farming[18]; the relation between the ethnic origin of the settlers and their adaptation and leaning to certain agricultural branches.

At the present writing there are two comprehensive research projects underway which make use of the regional sociologists as well as those attached to the study wing. The first deals with the problem of second generation continuity in moshavim, and the second deals with regional co-operation and its focal points in the rural centres.[19] Conclusions from both studies should have an important effect on rural development policy. The projects are aimed at determining the methods by which the rural way of life may be preserved and continued, while at the same time ensuring the settlers an appropriate standard of living.

The functions of the regional sociologists were defined as follows:

a) Research – The regional sociologist, in consultation with the administration of the region, formulates a research subject, constructs its theoretical framework, and assembles data which he processes and sums up.

b) Counselling – The sociologist is at the service of the regional administration for counselling on current problems.

c) Policy Formulation – The sociologist participates in formulation of regional policy in meetings of the various institutions of the region.

d) Instruction – The regional sociologist delivers lectures on sociology to settlement workers in order to provide them with sociological perspectives.[20]

e) Contact with Institutions – In the course of his work the sociologist maintains contact with various institutions such as settlement movements, government ministries, etc., – whether for the purpose of assembling data, or as a representative of the region on various committees, or for other professional matters.

f) Professional Contact – The regional sociologist participates in the study programmes of the study wing of the Settlement Department – both in formulating its research programme and in enriching its knowledge in the field of rural sociology.

The Directorate of the settlement authority is linked with the professional sociologist at the executive level through study of the work done by the regional sociologist and drawing of conclusions from it. The second link is that with the research undertaken to clarify general problems not only for current needs, but also for future policy formulation.

There is no advance guarantee that the conclusions and recommendations of the sociologists will be accepted. Final decision must take into consideration a range of facts and conclusions reached by various professionals, each in his own field. The sociologist is only one among many professional advisors to the director. Though the value of his proposals cannot remain in doubt, the director must weigh them against the proposals of other advisors.

In order that the directors of development machinery may follow the progress of research projects, it is essential that there be direct and frequent contact between them and the sociologists. For the sake of the research itself, it is essential that the director participate in all phases of the programme. For this, he should have the necessary professional training, so that he can be at home in the entire complex of problems under discussion and assist the researcher in selecting the direction of his study.

Although the director can anticipate practical contribution to settlement work by the sociologists, it should not be a precondition for any particular project. Not every project leads to tangible results from the point of view of the director's objectives, though every project should be of use in the long run, if only in increasing the sum of available knowledge or contributing to the experience of the sociologist.

In conclusion, Israeli experience has proven that sociology is an important instrument in any organization dealing with development in general and with rural development in particular. The effective use of this discipline depends to a large extent on the guarantee of organizational conditions that ensure direct and constant contact between the sociologist and the director.

NOTES

1. 'The Settlement Department – Functions and Structure', The Jewish Agency Settlement Department (1962), (mimograph in Hebrew).

1A. 'The Position of Agriculture in Israel', Ministry of Agriculture and Bank of Israel, (Jerusalem, January 1960), p. 153, table 128 (in Hebrew).

2. 'Report on Agriculture submitted to the Knesset by the Minister of Agriculture' (February 1964), p. 17 (March 1966), p. 37 (in Hebrew).

3. See: JOHNSON, V. W., and BARLOWE, R.: *Land Problems and Policies* (McGraw-Hill Book Company Inc., New-York 1954), pp. 88–89.

4. 'Report on Agriculture submitted to the Knesset by the Minister of Agriculture' (February 1965), p. 104 (in Hebrew).

5. 'Report on Agriculture submitted to the Knesset by the Minister of Agriculture' (March 1966), p. 40 (in Hebrew).

6. See: WEITZ, R.: *Agriculture and Settlement*, op. cit., pp. 193–194 (in Hebrew).

7. 'The Position of Agriculture in Israel', op. cit., p. 119.

8. 'Report on Investments in Israel Agriculture, for the period 1954–1959', prepared for the Food and Agriculture Organisation of the U.N. Ministry of Agriculture (Tel-Aviv, June 1960), (mimograph).

9. 'Report on Agriculture, Presented to the Knesset by the Minister of Agriculture' (March 1966), p. 14 (in Hebrew).

10. Ministry of Agriculture and Fisheries, Economic Series No. 8 (London 1926).

11. See also: WEITZ, R. (Editor): *Rural Planning in Developing Countries* (Routledge and Kegan Paul, London 1965), p. 84.

12. LEWIS, W. A.: *The Theory of Economic Growth*, p. 164.

13. SCHULTZ, T. W.: *The Economic Organisation of Agriculture*, p. 110.

14. SCHULTZ, T. W.: Op. cit., p. 118.

15. LEWIS, W. A.: Op. cit., p. 175.

16. See also: WEITZ, R.: 'Sociologists and Policy Makers', Transactions of the Fifth World Congress of Sociology, Vol. 1. (September 1962), pp. 59–75.

17. WEINTRAUB, D. and LISSAK, M.: 'Social Integration and Change' in: Agricultural Planning and Village Community in Israel, UNESCO, Arid Zone Research, XXIII (Paris 1964).

18. WEINTRAUB, D. and LISSAK, M.: Op. cit.

19. 'Preliminary Comparative Analysis of Regional Cooperation Patterns in 12 Rural Regions', in: Regional Cooperation in Israel, Settlement Study Centre, Publications on Problems of Regional Development, Publication No. 1 (Rehovot 1966).

20. See: 'Sociological Comments and Recommendations on Extension Work', Settlement Department (1960), (mimograph in Hebrew).

INDEX OF SUBJECTS